Statistics in Finance

Arnold Applications of Statistics Series

Series Editor: **BRIAN EVERITT**
Department of Biostatistics and Computing, Institute of Psychiatry, London, UK

This series offers titles which cover the statistical methodology most relevant to particular subject matters. Readers will be assumed to have a basic grasp of the topics covered in most general introductory statistics courses and texts, thus enabling the authors of the books in the series to concentrate on those techniques of most importance in the discipline under discussion. Although not introductory, most publications in the series are applied rather than highly technical, and all contain many detailed examples.

Other titles in the series:

Statistics in Education Ian Plewis

Statistics in Civil Engineering Andrew Metcalfe

Statistics in Human Genetics Pak Sham

Statistics in Finance

Edited by

David J. Hand
Open University, UK

and

Saul D. Jacka
Warwick University, UK

A member of the Hodder Headline Group
LONDON • SYDNEY • AUCKLAND

Copublished in North, Central and South America
by John Wiley & Sons Inc.
New York • Toronto

First published in Great Britain 1998 by
Arnold, a member of the Hodder Headline Group,
338 Euston Road, London NW1 3BH
http://www.arnoldpublishers.com

Copublished in North, Central and South America by
John Wiley & Sons Inc., 605 Third Avenue,
New York, NY 10158

British Library Cataloguing in Publication Data
A catalogue record for this book is available from the British Library

Library of Congress Cataloging-in-Publication Data
A catalog record for this book is available from the Library of Congress

Publisher: Nicki Dennis
Production Editor: Liz Gooster
Production Controller: Rose James
Cover Design: M2

ISBN 0 340 67719 8
ISBN 0 471 24634 4 in North, Central and South America only

Typeset in 10/11pt Times by AFS Image Setters Ltd, Glasgow
Printed and bound in Great Britain by J W Arrowsmith Ltd, Bristol

Contents

List of contributors

Philip Booth
Department of Actuarial Science and Statistics, The City of University, Northampton Square, London, UK

Dr Jonathan Crook
Credit Research Centre, Department of Business Studies, The University of Edinburgh, Edinburgh, UK

Professor Eric Ghysels
Department of Economics, Penn State University, Pennsylvania, USA and CIRANO, Montreal, Canada

Christian Gouriéroux
CREST and CEPREMAP

Professor Steven Haberman
Department of Actuarial Science and Statistics, The City University, Northampton Square, London, UK

Professor David Hand
Department of Statistics, The Open University, Milton Keynes, UK

Dr David G. Hobson
Department of Mathematical Sciences, University of Bath, Bath, UK

Professor Stewart G. Hodges
Financial Options Research Centre (FORT), University of Warwick, Coventry, UK

Dr Saul D. Jacka
Department of Statistics, University of Warwick, Coventry, UK

Professor Joanna Jasiak
York University, York, Ontario, Canada

Dr Paul King
Institute of Actuaries, Oxford, UK

Professor Damien Lamberton
Department de Mathematiques, Université de Marne-la-Vallée, Noisy le Grand Cedex, France

Professor Kevin Leonard
Department of Health Administration, University of Toronto, Toronto, Ontario, Canada

Professor Dilip Madan
Department of Finance, College of Business and Management, University of Maryland, College Park, Maryland, USA

Valentin Patilea
Institut de Statistique, UCL, Belgium

Professor Éric Renault
INSEE and CREST, Paris, France

Dr Arthur E. Renshaw
Department of Actuarial Science and Statistics, The City University, London, UK

Professor Lyn Thomas
Department of Business Studies, The University of Edinburgh, Edinburgh, UK

Olivier Torrès
GREMARS, Université de Lille 3, France

Preface

This is not a textbook. There are no exercises at the ends of the chapters. Rather, it is a collection of chapters describing various aspects of the application of statistical methods in finance. Its aim is to interest and attract statisticians to this area, to illustrate just some of the many ways that statistical tools are used in financial applications, and to give some indication of problems that are still outstanding. The hope is that statisticians will be stimulated to learn more about the kinds of models and techniques outlined in what follows – both the domain of finance and the science of statistics will benefit from increased awareness by statisticians of the problems, models and techniques applied in financial applications. For this reason, extensive references are given.

The level of technical detail varies between the chapters. Some present broad non-technical overviews of an area, while others describe the mathematical niceties. We hope this will illustrate the range of possibilities available in the area for statisticians, while simultaneously giving a flavour of the different kinds of mathematical and statistical skills required. Whether you favour data analysis or mathematical manipulation, if you are a statistician there are problems in finance which are appropriate to your skills.

Because of this variety in content and level, readers are encouraged to dip into the different chapters. It is not necessary to read the book from the beginning to the end in order to gain benefit from it – indeed it would probably be a mistake to set out to do so.

David J. Hand
Saul D. Jacka

1

Introduction

David J. Hand and Saul D. Jacka

The development of statistical theory and methods has always benefited from the areas in which the discipline of statistics is applied. In particular, the different problems arising in the different application areas have frequently led to the development of new statistical methods. So, for example, agricultural applications led to early developments in experimental design, medical applications led to survival analysis and cross-over trials, social science applications led to structural equation models, psychological applications led to factor analysis, and industrial applications led to quality control techniques. Of course, once a method or class of methods has been developed, it rapidly spreads out and is applied in domains quite different from those that provoked its genesis, so stimulating further development and application. This ubiquity of potential application is one of the strengths of statistics, something which makes it such a powerful tool. Nevertheless, the different application domains retain their own unique flavours. The statistical tools they use as a matter of course vary markedly, as do the ways in which they use familiar and common tools.

The domain of finance has not been previously delineated as an application area for statistics. Of course, statistical methods have been applied in this area for a long time: actuarial work goes back for decades, formal stochastic models for financial markets have attracted considerable media interest over the last ten or twenty years, and work on credit scoring has gone on behind the scenes in banks, certainly for the last twenty years. So the area has been active even if, until recently, it was not accorded the same status as medical statistics or psychological statistics or industrial statistics. Perhaps we should comment here that the *mathematics* of finance has long been recognised as a distinct area and, inevitably, its coverage overlaps considerably that of the *statistics* of finance. Nonetheless, we think it useful to identify financial statistics as a unique sub-domain of statistics in its own right. Like any other statistical domain, it will eclectically adopt techniques from throughout the body of statistics. However, the peculiarity of its applications, the unique questions it has to answer, the distinct problems of finance which do not occur or have not yet been recognised to occur in other application domains, will require the development of new classes of statistical tools, methods, and models. Our aim in putting together this collection is to draw to the attention of statisticians some of the different problems, the different types of models, and the flavour of the issues of financial statistics. We hope that the collection will stimulate statisticians, perhaps working in other areas, or perhaps seeking applications to which to apply their unique skills, to examine the area of financial statistics. Both the domain and the discipline of statistics will benefit from such attention.

As noted above, this book is a collection of chapters illustrating some of the problems and issues that arise in financial statistics. In particular, it is not a textbook. It is intended to be a taster, showing the range and scope of financial statistics. In pursuance of this aim of illustrating the breadth of the field, some of the chapters are technically quite demanding, while others are less so. This serves to demonstrate the scope for different kinds of statistical applications. The statistical tools applied range from advanced stochastic models, for example those using Itô calculus, through models which apply formal statistical model building techniques such as generalised linear models, supervised classification techniques, and kernel smoothing methods, to less formal data summary and description methods. In some cases the models are quite abstract, while in others the constraints of the real world – perhaps relating to data quality, legal restrictions, or doubts about the models – must be taken into account.

The book is divided into three parts. Part I contains two chapters describing what are perhaps the oldest applications of statistical methods in finance, namely actuarial applications. The first of these chapters, by Philip Booth and Paul King, shows how actuarial work has developed from its unique role in investment management, performance measurement, and investment index construction to include work in areas such as asset-liability modelling, investment manager selection, risk control, and derivatives management. This chapter also contrasts actuarial work with modern financial economics, and describes some portfolio selection models which have been developed in the finance literature, noting that important insights may be obtained but also cautioning that some of the more abstract developments suffer from over-formalism and lack of realism. The authors develop further the ideas they believe to be most important and go on to look at the theory of immunisation and asset-liability modelling.

Although actuarial applications are amongst the oldest applications of statistical methods in finance, the methods used are often amongst the most modern. Thus, Chapter 3, by Steven Haberman and Arthur Renshaw, shows how generalised linear models are applied in actuarial work. They define actuarial science as being concerned with the financial management of financial security systems – systems which reduce the effect of random adverse events. Actuaries are thus concerned with managing uncertainty. The chapter outlines the statistical basics of generalised linear models and describes their application in graduation, the process by which observed probabilities are adjusted to permit accurate inferences, multiple state models, risk classification, premium rating, and claims reserving in non-life insurance.

Part II of the book is concerned with credit, especially consumer credit, an application that has been described as 'the most successful application of statistics in business in the last twenty years'. Once again, statistical methods have long been applied in this area but, until recently, most applications have used fairly basic methods. However, in response to dramatic changes in the market (more varied requirements from consumers), the competitive environment (such as supermarkets entering the consumer banking fray), and the range of financial products available (for example, mortgage loans which may also function as current accounts), vastly more sophisticated statistical models are needed for accurate prediction. These new developments have been made feasible by the availability of powerful computers and associated software. The huge datasets available in this domain make statistical methods essential.

Chapter 4, by David Hand, describes this background to consumer credit models, outlining the different problems which must be tackled, the sort of data with which statisticians have to deal, special problems arising from distorted sampling and changes due to population drift, and how advantage can be taken of sequential data acquisition. This chapter also touches on the legal issues, which provide constraints on the structure of the statistical models.

Chapter 5, by Lyn Thomas, gets down to the statistical technicalities of systems which aid the credit granting decision. The basic decision-theoretic structure is outlined, and then a range of different methods which have been used is described. These methods include classical linear discriminant analysis, linear and logistic regression, classification trees, nearest neighbour methods, mathematical programming techniques, neural networks, expert systems, and genetic algorithms.

Chapter 6, by Kevin Leonard, addresses performance in consumer credit scoring systems. After setting the context, he describes such issues as what measures of performance to take and feed back to management and the impact of exceptions, overrides, policies, and procedures on scorecard performance.

Chapter 7, the final chapter in this part of the book, by Jonathan Crook, introduces an economic flavour, relating issues of consumer credit to business and economic cycles. A huge amount of research has been carried out in an effort to understand economic cycles. By virtue of this, this chapter is necessarily rather technical, in that it first introduces the theoretical economic models and then describes the stochastic aspects. Various economic models are outlined, and the statistical evidence is examined.

Part III of the book is concerned with an area of statistical applications in finance that has attracted much media interest in recent years: financial markets. It opens with a technical chapter by Saul Jacka, which presents a brief historical overview of the use of probability in measuring uncertainty in a market of traded assets, reminds the reader of the basics of stochastic calculus, introduces and generalises the basic Black–Scholes formula, describes changes of measure, and introduces American options.

Chapter 9, by Stewart Hodges, introduces the ideas of financial economics in a relatively informal way, thus outlining the basis for understanding the role and valuation of derivatives. The chapter focuses on the roles capital markets play in the economy, the state-preference framework for representing securities, and the paradigm of expected utility maximisation.

Chapter 10, by Damien Lamberton, describes American options – options that may be exercised at any time before their expiry date. An obvious question here is, what is the optimal stopping time? This chapter introduces the theory of optimal stopping, in both discrete and continuous time, and describes the basic properties of American option prices.

Chapter 11, by Saul Jacka, summarises the theory of term structure models, surveys some recent advances in the modelling of forward rates, and discusses the pricing of derivatives. A generalisation of one-factor affine models is presented, and a sketch of an approximation scheme for pricing derivatives in this context is given.

Dilip Madan, in Chapter 12, discusses statistical models for default risk: the risk assumed by one party in a contract when it fulfils its part of the contract but agrees to wait for the complementary delivery. Such arrangements are common, but the chapter focuses on financial contracts, where the exchanges in each direction are sums of money.

Chapter 13, by Eric Ghysels, Éric Renault, Olivier Torres and Valentin Patilea, surveys some recent non-parametric methods for pricing derivative contracts. The authors point out that explicit analytic formulae for pricing do not always exist, so that numerical methods must be deployed. Moreover, even if they do exist, the formulae may be too complex for numerical calculation. There is often also a large body of data. Each of these factors favours the adoption of non-parametric and semi-parametric methods, such as kernel smoothing approaches. They suggest that 'non-parametric statistical techniques provide a way of filling the gap between [the] Black–Scholes [model] and the real world'.

Chapter 14, by David Hobson, discusses stochastic volatility. Volatility is a measure of how rapidly the price of an asset fluctuates – the variance, per unit time, of the logarithm of the price. The basic Black–Scholes model assumes constant volatility, an assumption contradicted by most option prices. This chapter explores the implications for option pricing when the assumption breaks down.

It can be argued that models of price movements should be based on stochastic processes using a deformed time scale. For example, the number of markets which are active around the world changes over time, as does the volume of trading, so perhaps the nominal time scale should be replaced by one that reflects this changing level of activity. The final chapter, by Eric Ghysels, Christian Gouriéroux and Joanna Jasiak, describes stochastic process theory and statistical inference for stochastic processes based on such deformed time scales – subordinated stochastic processes, as they are called.

Part I

ACTUARIAL MATHEMATICS

2

The Relationship Between Finance and Actuarial Science

Philip Booth and Paul King

2.1 Actuaries and investment

The actuarial profession has been involved with institutional investment for as long as it has existed (Bailey, 1862). Although the profession is more commonly associated with the liability side of institutional investors' balance sheets it has always been clear to actuaries that ensuring that financial institutions can meet their liabilities as they fall due requires the consideration of the nature of both assets and liabilities together. Such considerations have led actuaries to have a significant influence over the investment policies of life assurance companies throughout the history of the industry (Pegler, 1948). It was also an actuary, turned fund manager, who was widely credited with leading the shift in investment policies of pension funds away from bonds to equities.[1] The rationale for this shift was the changing nature of pension fund liabilities in an inflationary environment.

Actuaries have made major contributions in the mathematical treatment of fixed interest markets (Clarkson, 1978) and modelling the term structure of interest rates remains an important area of research (Chaplin, 1996), not least in the context of producing bond yield indices.

For many years the actuarial profession had a near monopoly of high-level mathematical skills in the world of investment management, and the qualification as a Fellow of the Faculty or Institute of Actuaries was one of the few formal qualifications that demonstrated its holder had been trained and examined in investment techniques. That situation has now changed. Fund managers now have their own professional organisations and qualifications and many others, such as academic mathematicians and financial economists, have applied their skills to investment problems.

Actuaries do still work as fund managers, mainly for life assurance companies, but the profession now concentrates increasingly on those areas

---

[1] George Ross Goobey is commonly acknowledged as having had a major influence on the shift of pension fund investment policies in the 1950s and 1960s.

where it feels that specifically actuarial skills can be used to 'add value'. Such areas include 'traditional' actuarial ones of performance measurement and investment index construction; newer areas, which are natural extensions of the working environment of actuaries, such as asset-liability modelling (Kemp, 1996a) and investment manager selection (St John Hall *et al.*, 1996); and areas where a combination of mathematical skills and pragmatic business sense enable actuaries to make a valuable and possibly unique contribution. This latter category includes such topics as risk control for fund managers (Rains and Gardner, 1995), derivatives management (Kemp, 1996b) and quantitative techniques (Griffiths *et al.*, 1996). Smith (1996) explains how the skills embraced by actuaries working in these areas are also being applied to more traditional actuarial areas.

2.2   Introduction

Whilst actuaries have long been involved in practical investment issues, not all professional or academic actuaries have welcomed, in its entirety, modern financial economics developed by academics working in the finance field. In particular, Clarkson (1996) criticises many of the assumptions and models. There are a number of reasons for this. It is possible that actuaries have found modern financial economic theory wanting at an academic level. Some of the over-formalisation of systems that are essentially economic, into complex mathematical systems, which can never fully represent the complexity of the financial sector of the economy, have not always been regarded as helpful. On one hand, the understanding of the fundamental economics can be lost in the complex mathematical representation. On the other hand, the complex mathematical representation can never come close to representing the financial system.

At an academic level, modern financial economics could also be criticised for concentrating on equilibrium phenomena. The interesting problems in financial systems result from the way prices and values chase constantly changing equilibrium positions and also from the way new information is reflected in prices often quickly, but not instantaneously. This is how investment managers can 'make money' even if, in a settled state, the market would reach the kind of equilibrium predicted by modern financial theory.

These academic criticisms are similar to the criticisms that the Austrian school of economists might level at some branches of economics, in particular at macro-economics and econometrics. If these criticisms are correct, and modern financial economics is flawed in theory, it would explain why many actuaries do not find modern financial theory useful in practice.

A different form of criticism of modern financial economics has been made at a more practical level. It is suggested that it has not addressed the problems faced by institutional investors. For example, market models have generally defined a risk free security as one which provides a guaranteed nominal return over a relatively short period. Institutional investors may be more concerned about real returns (relative to inflation or salaries) and returns over a long period. They would also regard an investment as risk free not if it provided a guaranteed rate of return but if it provided a return that would guarantee that the institution could meet its liabilities. Investment actuaries normally work for intermediating financial institutions which hold assets to meet liabilities. Thus, portfolio selection models, if they are to be useful, need to be able to give guidance on how to choose a portfolio of assets to meet a given set of liabilities.

In this chapter, we describe some of the portfolio selection models that have been developed in the finance literature. A critique is then given of the models, and their insights are also discussed. It is quite consistent to conclude that there are a number of important insights from portfolio selection models whilst, at the same time, concluding that some of the more abstract developments suffer from the over-formalism and lack of realism referred to above. We then take those ideas from portfolio selection models which we believe can be most usefully applied to actuarial investment problems and develop them further. Some of these developments have been carried out by actuaries such as Wilkie (1985) and Sherris (1992); while some of them have been carried out by financial economists such as Sharpe and Tinte (1990). In the penultimate section of this chapter, we look at the theory of immunisation, which can be regarded as a portfolio selection model that has the limited aim of minimising risk and is applied in a set of conditions that can be modelled relatively realistically in a mathematical form. We then look at asset/liability modelling, which is carried out widely by actuaries. Asset/liability modelling is an attempt to model by simulating conditions that are difficult to encapsulate in formal mathematical models.

2.3 Mean–variance models in finance

2.3.1 Outline of portfolio theory

The fundamental assumption of mean–variance models for portfolio selection is that investors can rationally select portfolios by considering only two properties: the expected return over a single period and the variance of that return. Thus, all possible portfolios available to an investor can be represented as a set of points in a two-dimensional mean–variance space (often more conveniently represented in terms of the mean and standard deviation rather than the variance).

The theory, often called Modern Portfolio Theory (MPT) was first published by Markowitz (1952) and has proved remarkably fruitful in providing a framework within which the portfolio selection problem can be discussed in an intuitive way by identifying the variance of return with the 'riskiness' of the investment. Its mathematical simplicity has meant that it can be easily applied and developed by finance practitioners as well as academic theorists. MPT is treated in many standard finance textbooks, for example Elton and Gruber (1995) and Sharpe (1978). Markowitz's original paper has spawned a large amount of subsequent research and extensions of the basic theory to more realistic situations. In the actuarial field, these developments have included the extension of the theory to encompass investors' liabilities as well as their assets (Wise, 1984a, 1984b, 1987; Wilkie, 1985; Sherris, 1992), see Section 2.8.

It has to be said that the simplifying assumptions as to the behaviour of investors in making investment decisions has led many to question the usefulness of MPT in real-world portfolio selection problems (Clarkson, 1989, 1996). Some of the points raised were discussed in Section 2 and will be considered again in Sections 6–9 but we begin by outlining the basic theory of mean–variance financial models.

It is helpful to consider the portfolio selection process in two stages. The first is to specify all the available portfolios in terms that are relevant to the investor's decision. This set of available portfolios is known as the opportunity set. The second stage is to describe how the investor can select a portfolio from the opportunity set that is optimal in some way. A point to note is that the description of the opportunity set is not unique and, in general, will depend on the investor's own circumstances and assumptions about the properties of the available investments.

We suppose that there are N available risky investments (a risky investment being one with non-zero variance of return). The data required are the vector of expected returns, E, and the covariance matrix C. The elements of E are E_i ($i = 1 \ldots N$), the expected return on the ith investment. C has elements $\sigma_{i,j}$ ($i = 1 \ldots N, j = 1 \ldots N$) where $\sigma_{i,j} = \sigma_{j,i}$ and $\sigma_{i,i} = V_i$, the variance of the return on the ith investment. The expected returns should be expressed net of tax and expenses and should be total returns, including both increases in capital value and interest or other income received. Ways of dealing with intra-period cash flows are discussed in Wilkie (1985).

We now consider the portfolio created when the investor invests a proportion x_i of his or her wealth in the ith investment.

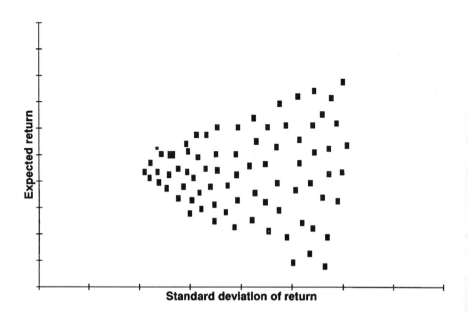

Figure 2.1 Graph showing the opportunity set of possible portfolios

The expected return on this portfolio is

$$E_P = EX^T \tag{2.1}$$

where X^T is the transpose of the vector of the portfolio proportions x_i.
The variance is given by

$$V_P = XCX^T \tag{2.2}$$

The opportunity set can be represented diagrammatically by plotting standard deviation versus expected return for all possible portfolios as shown in Fig. 2.1.

Having specified the opportunity set, the investor's task is now to select the most desirable of the available portfolios. Before considering the preferences of the individual investor selecting a portfolio we make two assumptions common to all investors that considerably narrow down the choice. These assumptions seem intuitively reasonable and are:

(1) investors prefer more wealth to less. That is, given the choice between two portfolios of equal riskiness, they will always choose the one with the higher expected return;

(2) investors are risk averse. That is, given the choice between two portfolios with equal expected return, they will always choose the one with the lower risk.

Both of these assumptions appear uncontroversial, given our fundamental assumption that investors are concerned only with expected return and a single risk parameter.

The consequence of these two assumptions is that we can now ignore all

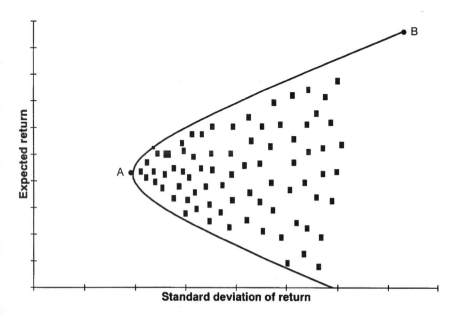

Figure 2.2 Graph showing the efficient frontier of portfolios

those portfolios in Fig. 2.1 which are not on the minimum standard deviation boundary of the opportunity set and do not have an expected return greater than the portfolio with the global minimum standard deviation. This leaves us with the portfolios lying on what is known as the *efficient frontier*, which is shown as the portion A–B of the curve in Fig. 2.2.

The concept of an efficient portfolio is an important one in finance and can be applied when we use definitions of risk that are different from those we are using here. A portfolio is efficient if it is not possible to find an alternative portfolio with the same (or higher) expected return but lower risk, or an alternative with the same (or lower) risk but a higher expected return.

In practical terms, the efficient frontier is calculated by minimising the variance for each value of expected return. The minimisation procedure can easily include constraints such as a restriction on short selling ($x_i \geqslant 0$ for all i) or limits on the concentration of investment ($x_i \leqslant a$ for all i and some specified value of a).

We have now completed the first stage of our task – the specification of the opportunity set (although we have glossed over the vital problem of how to obtain the required inputs of the expected returns on the individual investments and the covariance matrix). The next stage is the selection of a single optimum portfolio.

In theory this can be done by considering the individual investor's set of indifference curves in mean–standard deviation space. An indifference curve is the locus of all points such that the combination of expected return and standard deviation is equally valued by the investor. A representative set of curves is shown in Fig. 2.3. Clearly the investor gains more satisfaction from

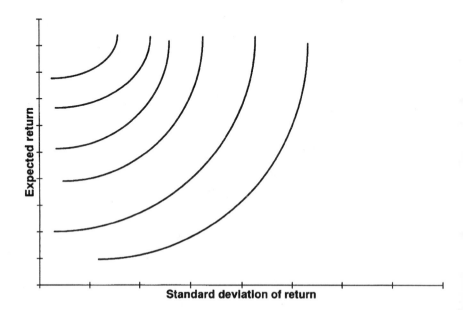

Figure 2.3   A representative set of indifference curves

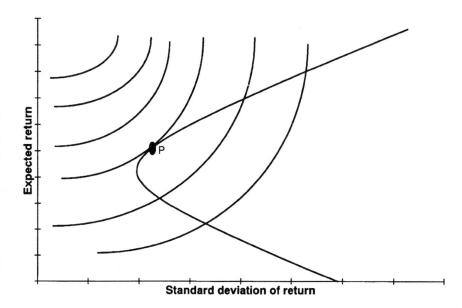

Figure 2.4   Graph showing the position of the optimum portfolio (*P*) for an investor who defines his preferences in terms of expected return and variance

points in the upper left corner of the diagram than from those in the lower right and so will choose the portfolio that lies on the indifference curve as far as possible to the upper left. In Fig. 2.4, this is the portfolio marked P. Portfolio P is the optimum portfolio for an investor who defines his or her preferences in terms of expected return and variance. The procedure will always produce such an optimum portfolio provided the investor is, even slightly, risk averse.

In practice, it is usually impossible to define a set of indifference curves for an investor and a portfolio is usually selected after examining the properties of several different efficient portfolios. For example, the investor might consider the expected return and standard deviation of a number of portfolios and choose the one that produces the combination that appears most desirable, without being able to express this preference in a mathematical way.

Our discussion so far has not allowed for the existence of a risk free invest-ment. However, for many practical purposes (especially in the short term) we can consider that investors can lend and borrow and know with certainty the amount that will be received at the end of the term or the amount that will have to be repaid. An example is the purchase or sale of short-term government securities (Treasury Bills). For simplicity, we will consider the case where borrowing and lending can take place at the same interest rate and in unlimited amounts.

In mean–standard deviation space the risk free asset is represented as a point on the vertical axis with zero standard deviation and an expected return of E_F (point F in Fig. 2.5). A linear combination of F and any risky asset will be represented as a straight line passing through the two points. A few

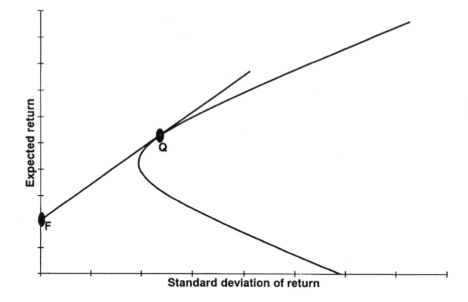

Figure 2.5   The efficient frontier with risk free bending and borrowing

moments' thought shows that the efficient frontier now becomes the straight line passing through F that is tangential to the original (risky asset) efficient frontier. The investor will now hold a combination of the risky portfolio Q and the risk free asset F. Combinations between Q and F represent a division of the investor's wealth between risky assets and a risk free 'deposit' while combinations beyond Q represent borrowing to purchase additional units of Q (gearing up).

2.3.2   The beta-line

A powerful result that can be derived from the portfolio selection model described above is that a linear relationship exists between the expected return on individual securities and their so-called 'beta factors'. This result is usually derived as a consequence of the Capital Asset Pricing Model (see Section 2.4) but Lewin *et al.* (1995) showed that it can be derived directly from the above model with suitable assumptions and definitions.

The result is:

$$E_i - r = \beta_{i,P}(E_P - r) \qquad (2.3)$$

where E_i is the expected return on security i;
  r is the return on the risk-free asset;
  E_P is the expected return on any portfolio on the efficient frontier;
  $\beta_{i,P}$ is a beta factor defined as $\text{Covar}[R_i, R_P]/V_P$, with R_i the return on security i and R_P the return on portfolio P.

As the beta of a portfolio is the weighted average of the betas of its constituent securities, a similar result holds for any portfolio Q, composed of the S_i

$$E_Q - r = \beta_{Q,P}(E_P - r) \tag{2.4}$$

where E_Q is the expected return on portfolio Q;
  r is the return on the risk-free asset;
  E_P is the expected return on any portfolio on the efficient frontier;
  $\beta_{Q,P}$ is a beta factor defined as $\mathrm{Covar}[R_Q, R_P]/V_P$.

One way of deriving this result is to consider combinations of portfolio P and security i. These are represented by the curved line P–i in Fig. 2.6. If the proportion of security i is y_i, the mean return and variance of the combination are given by

$$E = E[R] = y_i E_i + (1 - y_i) E_P$$

and

$$s = \{\mathrm{Var}[R]\}^{1/2} = [y_i^2 V_i + 2y_i(1 - y_i) C_{i,P} + (1 - y_i)^2 V_P]^{1/2}$$

Differentiating with respect to y_i and using the fact that

$$\frac{dE}{ds} = \frac{\dfrac{dE}{dy_i}}{\dfrac{ds}{dy_i}}$$

we have

$$\frac{dE}{ds} = \frac{(E_i - E_P)[y_i^2 V_i + 2y_i(1 - y_i) C_{i,P} + (1 - y_i)^2 V_P]^{1/2}}{y_i V_i - V_P + y_i V_P + C_{i,P} - 2y_i C_{i,P}}$$

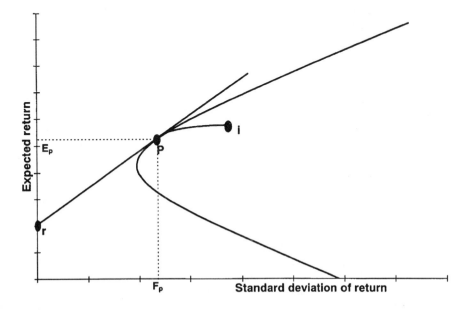

Figure 2.6   Combinations of portfolio *P* and security *I*

In particular, at the point $y_i = 0$, the gradient of the line P–i in mean–standard deviation space must be equal to the gradient of the efficient frontier (provided that E_P is not a corner portfolio – that is, a point at which a particular security just leaves or joins the efficient frontier). If the two gradients were not equal, P–i would cross the efficient frontier, which is not allowed.

Substituting $y_i = 0$ in the above gives

$$\left.\frac{dE}{ds}\right|_{y_i=0} = \frac{(E_i - E_P)\sigma_P}{C_{i,P} - V_P}$$

The gradient of the tangent to the efficient portfolio at this point is $(E_P - r)/s_P$, as is clear from Fig. 2.6. Equating the two expressions for the gradient at $y_i = 0$, and noting that $C_{i,P}/V_P = b_{i,P}$ and $V_P = s_P^2$, gives the desired result

$$E_i - r = \beta_{i,P}(E_p - r)$$

2.3.3   The inputs to the portfolio selection model

We mentioned in Section 2.3.1 that we had glossed over the problems of providing the inputs to the portfolio selection process. These problems are considerable in practice. Suppose for example we are considering a portfolio constructed from the ordinary shares of 100 companies. We require 100 different estimates of expected returns plus all the elements of a 100×100 covariance matrix. Since $s_{i,j} = s_{j,i}$, this is equal to 5050 different elements. In general, for N securities, $N(N + 3)/2$ different data estimates are needed. It should be noted also that these are not in a form corresponding to the way investment analysts traditionally work. Individual analysts usually concentrate on a particular sector, such as utilities. The analyst may be able to provide estimates of future return on the shares of one of the companies he or she follows, and may even be able to give a rough estimate of the expected performance of that share relative to the rest of the sector or even the market as a whole. However, analysts would not usually be able to give a reliable opinion on performance relative to individual shares in a sector other than their own.

One approach to overcoming this problem is to attempt to use the historical covariance matrix to predict the future covariance structure of the market. The covariance matrix based on historical data may then be combined with analysts' estimates of expected returns. This approach has the disadvantage of requiring the assumption that the covariance structure is independent of the returns. There is considerable debate as to whether expected returns, variances and covariances are independent of each other and statistically stable. If they are not, this is a serious practical impediment to efficient frontier analysis. Certainly if the opportunity set includes assets which do not have long quotation histories, reliability of data is a serious problem.

If asset classes, rather than individual securities, are being considered as the elements of the portfolio it may be desirable to generate the required data by stochastic simulation using interrelated time-series models of the individual asset classes. In this case the covariance matrix is implicit in the model used but it may not be possible to extract it analytically. The simulation approach also has the advantage that it will incorporate any time varying relationship between the returns and the covariance structure captured by the model. Such a stochastic modelling approach is widely used in actuarial work and stochastic asset models are the subject of considerable research work world-wide, see Section 2.11.

A final approach is to seek to reduce the problem by adopting a model for security returns that reduces and simplifies the data required to specify the covariance structure. One of the earliest such models was the single-index model discussed in the next section.

2.3.4   The single-index model

The single-index model models the return on a security as

$$R_i = a_i + b_i R_M + e_i \tag{2.5}$$

where R_i is the return on security i;
    a_i and b_i are constants;
    R_M is the return on the market;
    e_i is a random variable representing the random component of R_i not related to the market.

Under the model, e_i is uncorrelated with R_M and e_i is independent of e_j for all $i \neq j$. For any particular security a and b can be estimated by time series regression analysis.

The expected return and variance of return on security i and the covariance of the returns on securities i and j are given by

$$E_i = a_i + b_i E_M \tag{2.6}$$
$$V_i = b_i^2 V_M + V_{e_i} \tag{2.7}$$

and

$$C_{i,j} = b_i b_j V_M \tag{2.8}$$

where V_{e_i} is the variance of e_i.

The first of these three equations will be seen to be identical in form to the main result of the Capital Asset Pricing Model (CAPM) discussed below. However, it should be emphasised that the single-index model is purely empirical and is not based on any theoretical relationships between b_i and the other variables.

Equation (2.7) models the variance of the return on security i as the sum of a term related to the variance of the return on the market and a term specific to security i. These two terms are usually called systematic and specific risk respectively. Systematic risk can be regarded as relating to the market as a whole, while specific risk depends on factors peculiar to the individual security. It can be shown that in a diversified portfolio consisting of a large number of securities the contribution of the specific risk on each security to the total risk of the portfolio becomes very small and the contribution of each security to the portfolio's total risk is only the systematic risk of that security. Thus, it is only the systematic risk, measured by b_i of a security, that should be expected to be rewarded by increased return. Investors can diversify away specific risk and do not therefore require compensation for accepting it.

Equation (2.8) shows that in this particular model any correlation between the returns on two securities comes only from their joint correlation with the market as a whole. In other words, the only reason that securities move together is a common response to market movements. There are no other possible common factors.

Although many studies have found that incorporating more factors into the model (for example industry indices) leads to a better explanation of the historical data (see, for example Connor, 1995), correlation with the market is the largest factor in explaining security price variation. Furthermore, there is little evidence that multi-factor models are significantly better at forecasting the future correlation structure. See Elton and Gruber (1995) for a discussion of the empirical evidence.

The use of the single-index model dramatically reduces the amount of data required as an input to the portfolio selection process. For N securities, the number of data items needed has been reduced from $N(N + 3)/2$ to $3N + 1$. Furthermore, the nature of the estimates required from security analysts conforms much more closely to the way in which they traditionally work. In addition, considerably simplified methods for calculating the efficient frontier have been developed under the single-index model although, with increasing computer power, this is of less importance than it was at the time the model was first published.

2.4   The capital asset pricing model

The assumptions used so far in deriving the results of portfolio theory are as follows.

- Portfolios are evaluated by looking at the expected returns and standard deviations of return on the portfolio over a one-period horizon.
- Investors always prefer a portfolio with a higher expected return to one with a lower expected return, other things being equal.
- Investors are risk averse. A portfolio with a lower standard deviation of return is always preferred to one with a higher standard deviation, other things being equal.
- Individual assets are infinitely divisible, so can be held in any proportion of a total portfolio.
- Taxes and transaction costs are irrelevant.

In this section we describe the capital asset pricing model (CAPM) in the standard form developed independently by Sharpe (1964) and Lintner (1965).

CAPM requires some extra assumptions.

- All investors have the same one-period horizon.
- All investors can borrow or lend unlimited amounts at the same risk-free rate.
- The markets for risky assets are perfect. Information is freely and instantly available to all investors and no investor believes that they can affect the price of a security by their own actions.
- Investors have the same estimates of the expected returns, standard deviations and covariances of securities over the one-period horizon.

● Unlimited short sales are allowed.

Some of these assumptions are clearly unrealistic in practice.

If investors have homogeneous expectations, then they are all faced with the same efficient frontier of risky securities. If, in addition, they are all subject to the same risk-free rate of interest, the efficient frontier collapses to the straight line in mean–standard deviation space which passes through the risk-free rate of return on the expected return axis and is tangential to the efficient frontier for risky securities. This is shown in Fig. 2.7.

All rational investors will hold a combination of the risk-free asset and the portfolio labelled *M* in Fig. 2.7. Because *M* is the portfolio held in different quantities by all investors it must consist of all risky assets in proportion to their market capitalisation. It is commonly called the 'market portfolio'. The proportion of a particular investor's portfolio consisting of the market portfolio will be determined by their risk-return preference.

The fact that the optimal combination of risky assets for an investor can be determined without any knowledge of their preferences towards risk and return (or their liabilities) is often known as the *separation theorem*.

The line denoting the efficient frontier is called the Capital Market Line and its equation is

$$E_P - r = (E_M - r)s_P/s_M \tag{2.9}$$

where E_P is the expected return of any portfolio on the efficient frontier;
   s_p is the standard deviation of the return on portfolio P;
   E_M is the expected return on the market portfolio;

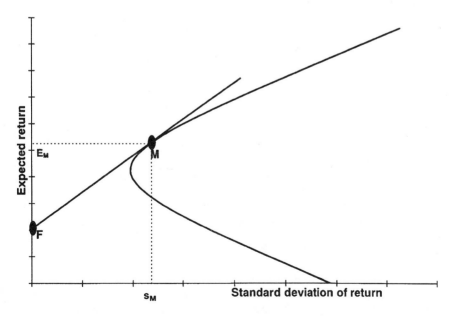

Figure 2.7   Position of market portfolio

s_M is the standard deviation of the return on the market portfolio;
r is the risk-free rate of return.

Thus, the expected return on any efficient portfolio is a linear function of its standard deviation. The factor $(E_M - r)/s_M$ is often called the market price of risk.

If we repeat the beta-line analysis of Section 2.3, substituting the market portfolio M for portfolio P and defining the beta of security i relative to the market portfolio, we obtain

$$E_i - r = \beta_i(E_M - r) \tag{2.10}$$

where E_i is the expected return on security i;
r is the return on the risk-free asset;
E_M is the expected return on the market portfolio;
β_i is the beta factor of security i defined as $\mathrm{Covar}[R_i, R_M]/V_M$.

This is the equation of a straight line in expected return – beta space, called the Security Market Line. It shows that the expected return on any security can be expressed as a linear function of the security's covariance with the market as a whole. Since the beta of a portfolio is the weighted sum of the betas of its constituent securities, the Security Market Line equation applies to portfolios as well as individual securities.

2.5   Risk adjusted discount rates

2.5.1   Capital project appraisal

Making an investment may mean purchasing a government bond or it may mean renting factory space and purchasing plant and machinery. In both cases the investment is made in anticipation of future cash flows and, in comparing different possible investments, we need to be able to place a value on those cash flows.

In the case of a freely traded security we are given a value by the market. However, many other assets or potential investment projects are not traded and there is no market value. This does not mean that the portfolio selection model outlined in Section 2.3 cannot be used by an investor seeking to construct an optimum portfolio. Given estimates of the expected returns, variances and correlations on all possible investments, the investor can still use portfolio theory to construct an efficient portfolio of any type of asset. Of course, in practice, investors do not reconstruct their entire portfolio every time a new opportunity becomes available and they need to assess the potential new investment for the effect it would have on their existing portfolio.

A common way of placing a value on potential capital project investments by firms is to calculate the net present value (NPV) of the cash flows. We can allow for the riskiness of the cash flows by using a discount rate that is risk adjusted.

A discussion of how to determine the correct discount rate in the evaluation of a capital project can be found in the paper by Lewin *et al.* (1995). One suggestion in that paper is to use the beta value of the project to determine the discount rate to be applied. In this case the beta is not determined relative to the market portfolio but relative to the portfolio of investment projects already undertaken by the firm. The discount rate to be used is the expected return on the project given by E_i, where

$$E_i - r = \beta_{i,P}(E_p - r) \qquad (2.11)$$

where E_i is the expected return on project i;

r is the return on the risk-free asset;

E_P is the expected return on any portfolio of projects on the efficient frontier;

$\beta_{i,P}$ is a beta factor defined as $\text{Covar}[R_i, R_P]/V_P$.

Here, it is assumed that the existing portfolio of investment projects is optimal and lies on the firm's efficient frontier. From the expected return and variance of the existing portfolio the equation of the company's beta line can be estimated. If the covariance of the return on the proposed project with that of the return on the existing portfolio can be estimated then the required discount rate can be calculated.

Of course there are a large number of practical problems in implementing this proposal. Amongst these are the assumptions that:

- the projects are divisible so can be undertaken on any scale;
- all projects can be assessed over the same timescale;
- the company can estimate the expected returns on each project and the covariance of the returns with the return on the total portfolio.

The method proposed also does not help to determine an appropriate discount rate for a totally new project requiring the setting up of a new company.

2.5.2 Valuing financial services businesses

One type of investment that may be considered by an institution is the purchase of a block of financial services business (such as a block of non-profit policies) or a whole financial services company. Mehta (1992) has proposed that such businesses should be valued by applying a separate discount rate to the cash inflows generated by each type of business, together with the outgoings. The appropriate discount rate is determined by comparing the nature of the cash flows with those generated by tradable asset classes and using the market return on the most similar asset class as a basis for setting the discount rate.

2.6 Insights and limitations of portfolio selection models

2.6.1 Introduction

In this section, we look at some of the insights and limitations of the models that have been discussed in the earlier sections. In later sections we look at how some of the limitations can be overcome. In many senses, Markowitz's work formalised what was generally accepted common sense and came to conclusions with which most statisticians would be familiar. That work was then developed to try to draw stronger conclusions about investment markets and pricing. It is these developments about which many of the criticisms could be made.

2.6.2   Insights of portfolio selection models

The first useful insight is that investors can look at the risk and return characteristics of portfolios and trade risk and expected return. Thus, a portfolio which has a higher risk would only be acceptable if it had a higher expected return. If all investors behaved in a similar way, investments should be priced in such a way that higher risk investments should always have a higher expected return.

The second insight is that, just as it is possible to develop a numerical measure of return, it is also possible to develop a numerical measure of risk. Standard deviation of returns is one possible measure. If standard deviation is the appropriate measure of risk, following on from the above, investments which have a higher standard deviation of return should have a higher expected return.

Thirdly, it is useful for investments to have a low correlation of returns with other investments, as this helps to reduce the risk of a portfolio of investments. Coming from this is the, perhaps more profound, observation that it is not actually the risk of an investment that will determine how it is valued by an investor but the contribution that an investment makes to the risk of the investor's portfolio. Thus, if we consider two investments, one of which has a high variance of returns but a low correlation of returns with other investments (for example an overseas equity investment), this may be less risky, from a portfolio perspective, than an investment which has a lower variance but a higher correlation with the other investments in a portfolio.

2.6.3   Limitations of mean–variance models

There are also a number of limitations of the models that we have discussed, particularly of those which try to determine how investments will be priced one against another. The first problem lies in the definition of risk. In most of Markowitz's work, risk was defined as the standard deviation of the return from an investment. However, there is evidence that investors are more concerned with what might be called 'downside risk'. Downside risk is the risk of underperforming a particular benchmark or target return. One such measure of downside risk is semi-variance, defined as:

$$\int_{-\infty}^{L} (L - r)^2 p(r) \, dr$$

where r is the level of return; L is a limit above which no adverse consequences are assumed to arise; and $p(r)$ is the probability density function of r. It might be thought that a downside definition of risk would best represent the risk to an investor. However, the use of a downside measure can be somewhat restrictive and, in many situations, it is found that, if an investor orders investments according to a downside measure of risk, such as semi-variance, he or she will get the same results in practice as if the investments are ordered according to the standard deviation of return. This is demonstrated, for example, in Levy and Markowitz (1979) and Booth (1995).

A realistic measure of risk may best be obtained by reference to utility theory: see for example, Booth (1997). Utility theory tries to put a numerical measure on different levels of wealth that could be held by an investor. Normally, the level of utility is expressed using a mathematical function called a utility function. If an investor has a logarithmic utility function, for example, it would mean that he or she regarded a given proportionate increment in wealth as being of the same value, whatever the starting value of wealth. It is possible to derive measures of investment risk which are compatible with different utility functions. Theoretical work, for example by Pratt (1964), confirms that standard deviation of return is not an appropriate risk measure for most utility functions that are regarded as reasonable in representing an investor's wealth preferences.

In fact, Markowitz's portfolio selection models do not depend, in theory, on the use of a particular measure of investment risk. In his work, he discussed a number of alternatives (for example, semi-variance). However, to obtain practical results and to develop market pricing models it is very difficult to use measures of risk that do not have the neat mathematical properties of standard deviation. For example, it is easy to calculate the standard deviation of returns from a portfolio of investments if we have information about the standard deviation of returns of, and covariance between, the elements of the portfolio. It is not possible to calculate other measures of risk for a portfolio so easily. The properties of standard deviation enable an investor to use a portfolio selection model easily and enable the mathematics of the capital asset pricing model to be developed easily. However, the reader should be aware that, for a particular investor, using the standard deviation of returns as a measure of risk may be inappropriate when selecting a portfolio. If investors, in general, do not regard standard deviation as a reasonable measure of risk, pricing models may not in fact have predictive power.

The second objection to the assumptions behind the models that we have discussed is that institutional investors do not necessarily view risk in terms of the performance of the investments themselves. Instead, they may be more concerned about the risk of not meeting liabilities. If this is the case, portfolio selection models can be adapted to be more suitable for use by investors (see Sections 2.8 and 2.9). However, market pricing models are unlikely to have predictive power as investors will have different liabilities and view risk differently.

The use of historical data to predict the risk and return from different investments is also problematic. Anderson and Breedon (1996) show that the volatility of investment markets has varied significantly in the last 50 years and in a way that is inherently unpredictable. This makes the estimation of the standard deviation of a portfolio of investments difficult. In addition to the standard deviation of returns from different investments, to calculate the standard deviation of a portfolio it is necessary to have information about the covariances between different investments. These are likely to be more unstable than the standard deviations. Furthermore, for portfolio selection models to be applied on a global basis, reliable information is required on the standard deviation of returns and covariance between different investment markets. Even if the parameters to be estimated were stable, there is very little historical information on which to base estimates for many markets.

All of the above problems pertain to the use of both portfolio selection and asset pricing models. Some of them can be overcome if the models are adapted (as we shall see below). Some of them, such as the lack of stability of the parameters to be estimated, cannot be overcome easily. However, there are further problems with the use of the capital asset pricing model. First, all investors do not have a common time horizon for decision making. Some investors will have a short time horizon (such as banks or non-life insurance companies); other investors will have a longer time horizon (such as pension funds). Partly because of this and partly because different investors have different liabilities, all investors will not have the same risk free security. A short-term investor may regard treasury bills as a risk free investment, whereas a long-term investor with real liabilities may regard index-linked gilts as a risk free investment.

The other assumptions underlying the capital asset pricing model, which may be unrealistic, are also commonly made in economic modelling. These are that all investors may borrow or lend unlimited amounts at the risk free rate of interest, that all securities are completely divisible, that there are no differential taxes and no restrictions on short selling, and that information is freely available to all investors.

Taken together, these criticisms of some aspects of modern financial economics explain why the techniques have not always found favour for use in financial modelling by actuaries. This does not mean that the models are not useful nor that they cannot be adapted for use by institutional investors. In the next section, we look at the primary considerations which drive actuaries when taking investment decisions. In Sections 2.8–2.11, we then develop statistical and mathematical techniques, using some aspects of the models discussed in Sections 2.3 to 2.5, to develop financial models which can be used by actuaries.

2.7   Principles of asset allocation

The factors an actuary takes into account when setting investment policy can be described more easily than they can be quantified. For most institutions, investment risk is defined in terms of the risk of not meeting liabilities. These liabilities may be well defined, such as the liabilities from a non-profit life insurance contract. In other situations the liabilities may be less well defined. For example, the liabilities of a pension fund depend on unknown future salary levels. A low risk investment strategy would try to minimise the variability of the contribution rate necessary to meet the liabilities.

Thus, sometimes liabilities will be fixed and known with reasonable certainty. In addition, sometimes future liabilities will depend on a number of factors such as: future salary levels, future inflation levels or, in the case of unitised funds, future investment market levels. The actuary cannot possibly predict the development of future financial conditions. Instead, the process of managing risk requires that the actuary manages the possible adverse effects of future changes in financial conditions. The best way to do this is by asset/

liability matching. This involves ensuring that the financial characteristics of the liabilities are similar to those of the assets. If there are unforeseen changes in financial conditions, the risk to the investor will then be limited. A life insurance company may therefore wish to hold fixed interest bonds to ensure that fixed non-profit liabilities are met. A pension fund would wish to hold real investments (which tend to rise in value when the price level rises) to protect itself from an increase in its liabilities (prospective pensions linked to salaries at retirement) when the price level rises. In the case of other liability types, the situation is a little more difficult because the nature of the liabilities is not always well defined. With with-profit life insurance liabilities, there is some investment discretion because the benefits will depend on the investment performance of the insurance company (under-pinned by guarantees). The insurance company therefore invests so that it can meet those guarantees but also so that it can have the type of risk profile that is required by its policyholders. A non-life insurance company, writing policies such as motor and household policies will have generally short-term liabilities, although some can have a long run-off period. The liabilities are likely to increase with inflation: however, they are likely to be due over a short period and inflation is relatively predictable over such time horizons. Mathematical and statistical models that can allow for the link between asset and liability characteristics will be discussed in Sections 2.8 to 2.11.

An institutional investor will also want to match the term of the assets and liabilities to ensure that it is protected from changes in interest rates. If an institution invests in short-term assets and has long-term obligations to meet, a fall in interest rates could lead to insolvency. Similarly, if the investments are long term and the liabilities short term, there is a risk that the capital values of investments may fall. The theory of immunisation (see Section 2.10) was developed to manage this problem using mathematical techniques.

Institutional investors will also wish to avoid investments which tend to have volatile return profiles if their liabilities are short term. To do otherwise would leave the investor open to the risk that asset values could fall when investments had to be realised. Long-term investors will be less concerned about short-term capital value volatility.

Thus, in an institution's attitude to risk, we see some differences from the assumptions made in some of the traditional finance models. Risk must be seen in relation to an investor's liabilities. Therefore, different investors can see risk very differently. Some investment types may be risky for some investors and not risky for others. Treasury bills, often assumed to be the risk free investment when finance models are applied, may be very risky for long-term investors such as life insurance companies. Furthermore, the investor may be much more worried about adverse outcomes which lead to insolvency (or an inadequate contribution rate to a pension scheme) than about the whole distribution of investment returns, which is used to calculate the standard deviation of returns.

After having satisfied itself that investment risk is controlled, the institution will wish to maximise expected return. This helps to lower the contribution rate in the case of a pension scheme. In the case of an insurance company, higher investment returns will help provide higher returns to shareholders or higher payouts to policyholders The application of the efficient portfolio concept to institutional investors would lead the investor to try to maximise the return for a given risk or minimise risk for a given return. This broadly reflects practice. However, two qualifications are worth making. Firstly, not all institutions view risk in the same way. It is therefore possible that, because of the behaviour of investors and their influence on market pricing, some investments will provide both higher risk and lower return or lower risk and higher return. An example of this is that many pension funds would regard equities as less risky than conventional gilts because of the inflation hedging characteristics of equities However, pension funds would normally assume that equities would provide a higher expected return than gilts. Other investors may regard conventional gilts as less risky than equities: thus the market becomes segmented with different investors being attracted to different categories of investment. Actuaries, in practical work, also often assume that markets are not efficient. Thus, for example, when equity markets fall and a valuation is carried out, the actuary may assume that higher expected returns will be achieved from the assets in the future. Much of modern portfolio theory can accommodate this, although some of the more abstract ideas do not.

Finally, actuaries have to take into account a number of practical matters when taking investment decisions. In particular, not all investments are perfectly liquid and marketable. This affects different investors in different ways. A short-term investor, such as a non-life insurance company would be very concerned to ensure that all but a very small proportion of its investments were in liquid and marketable investments. [Note: for the purposes of this chapter, liquidity will be defined as the ability to sell an investment quickly (actuaries do not find the definition of liquidity which is often used in economics and which relates to how close an investment is to cash, very helpful). Marketability will be defined as the ability to buy or sell a significant amount of the investment without affecting the market price.] A longer term investor would be less concerned with liquidity.

Taxation will also affect investment decisions. This can be difficult to allow for in equilibrium market pricing models. Markets can become segmented so that particular investors will be attracted to particular investment types because of their tax position.

Thus, in conclusion, it can be said that actuaries believe that there are weaknesses in some of the models of modern financial economics which limit their applications and explanatory power. Different investors see risk differently; they have different time horizons; most investors see risk in terms of the risk of not meeting liabilities; and there are institutional considerations and frictions in investment markets. In the following sections, we show the analytical techniques which can be used by institutions taking an actuarial approach to portfolio selection problems.

2.8　Developing mean–variance models: the inclusion of liabilities

Wise (1984a, 1984b) and Wilkie (1985) looked at the portfolio selection problem from the perspective of an investor who had unmarketable actuarial liabilities. Wilkie (1985) placed the problem firmly in the context of a traditional mean–variance portfolio selection model but with two additional elements, which aimed to give the model greater relevance. The first additional element was that the model considered the mean and variance of the ultimate surplus of a pension fund after having met liabilities. This meets one of the fundamental objections to traditional portfolio selection models. The second additional element was that the price of the portfolio was regarded as one of the decision variables. It was shown in Booth (1997) that the price of the portfolio was directly related to the initial surplus of the pension fund where this was measured as the market value of assets less the actuarial value of the liabilities, valued on a given valuation basis.

This latter element adds an extra dimension (literally) as we now have three decision variables and need to draw three-dimensional efficient frontiers. The decision variables become the expected value of the ultimate surplus; the variance of the ultimate surplus and the initial surplus. The most obvious way in which the third dimension comes into play is if we look at the trade-off between expected ultimate surplus and initial surplus. The pension fund has a choice. It can invest more assets in the pension fund initially (increasing the price of the portfolio) leading to a higher expected ultimate surplus for a given variance of surplus. Alternatively, it can invest less initially (reducing the price of the portfolio) leading to a lower expected ultimate surplus for a given variance of surplus. The trade-off between expected ultimate surplus and variance of ultimate surplus is determined by the risk preferences of the investor. The trade-off between the price of the portfolio and the expected ultimate surplus is determined by the time preferences of the investor.

2.9 Generalising mean–variance models

Sherris argued that the Wilkie framework could be simplified and considered as a traditional utility maximisation problem. The idea of utility theory has already been mentioned in Section 2.6. Sherris' contention was that an insurer or pension fund sponsor could be regarded as having a particular utility of wealth function, where wealth related to the level of assets after all liabilities had been paid off. The objective of the investing institution is therefore to choose the portfolio to maximise expected utility. However, only a limited range of utility functions can be used if analytical techniques are to be used to find investment portfolios which maximise expected utility. Otherwise numerical techniques have to be used.

In this mathematical analysis, it will be assumed that the investor has an exponential utility function of the form

$$-\exp(-S/r)$$

where S is the ultimate surplus of the investor after all liabilities have been met and r is a parameter known as the risk tolerance parameter. r will be different for different investors, producing a utility function of similar functional form for all investors who have an exponential utility function but with different aversion to risk.

The aim of the investor is therefore to choose the proportions to be invested in different categories of investments, in order to maximise

$$E[-\exp(-S/r)]$$

This is equivalent to minimising

$$E[\exp(St)] \tag{2.12}$$

where $t = -1/r$

This is therefore equivalent to minimising the moment generating function of the random variable S. For simple assumed probability distributions for the random variable S, it is possible to develop expressions that can be minimised analytically, to find the proportion of assets which should be invested in different asset categories. In most realistic cases, it would be necessary to use numerical methods. One case where an analytical approach would be used is where the surplus follows a normal distribution.

If the expected value of the ultimate surplus is E and the variance of ultimate surplus is V, the asset proportions should be chosen to minimise

$$\exp\{Et + Vt/2\}$$

or

$$\min\{\exp[-E/r + V/2r]\} \tag{2.13}$$

that is, minimise the moment generating function of the normal distribution. Clearly, the precise form of the function to be minimised depends on the risk tolerance parameter, r.

We will consider two asset classes, 1 and 2. The amount invested in the ith asset class will be denoted by w_i (it is not difficult to extend the idea to more asset classes). The expected accumulation factor (i.e. 1 plus the rate of interest) from asset class i will be denoted by E_i. The variance of return from asset i will be denoted by V_i and the covariance between asset i and asset j will be denoted by C_{ij}. The covariance between the liability and asset i will be denoted by C_{il}. E_l is the expected value of the liability and V_l is the variance of the liability value. The initial level of assets is A

$$E = A \times w_1 + A \times w_2 - E_l \tag{2.14}$$

that is, the expected surplus depends on the initial level of assets, the proportion invested in each asset and the expected return from each asset less the expected liability.

$$V = A^2 w_1^2 V_1 + A^2 w_2^2 V_2 + V_l + 2A^2 w_1 w_2 C_{12} - 2A w_1 C_{1l} - 2A w_2 C_{2l} \tag{2.15}$$

from standard statistical results from linear combinations of random variables.

We have to choose w_1 and w_2 to

$$\min[-E/r + V/2r^2] \tag{2.16}$$

where

$$w_1 = 1 - w_2$$

Differentiating the function with respect to w_1 and setting to zero, we obtain

$$w_1 = \frac{r/A(E_1 - E_2)V_2 - C_{12} + (C_{1l} - C_{2l})/A}{V_1 + V_2 - 2C_{12}} \tag{2.17}$$

Table 2.1   Sensitivity of proportions invested in asset 1 (w_1) to changes in the relevant variables

Variable change	Consequent change to w_1
Expected return from asset one increases	w_1 increases
Expected return from asset two decreases	w_1 increases
Variance of return from asset two decreases	w_1 decreases
Variance of return from asset one increases	w_1 decreases
Covariance of return between assets increases	indeterminate
Covariance of return between asset one and liability increases	w_1 increases
Covariance of return between asset two and liability increases	w_1 decreases

This approach has set up a model based on reasonable assumptions about the financial circumstances facing an institution. A utility function has been postulated; liabilities have been taken into consideration; the institution has to choose between investment types; and the investments have an expected return, variance of return and covariance. As far as the distinctions between this model and the financial economics models discussed earlier are concerned, the most important difference is that the covariance between the liabilities and the assets is taken into account. If the assumptions underlying the model are intuitively reasonable one would expect the results to be intuitively reasonable. We will now discuss the results of this model, in terms of the conclusions that can be drawn regarding the sensitivity of the proportions of assets invested in investment types 1 and 2 when the parameters of the model change.

It can be found from Equation (2.17) that the sensitivities are as shown in Table 2.1.

These sensitivities can be derived from differentiating Equation (2.17) with respect to the relevant parameters, and therefore show the sensitivity of w_1 to the relevant parameter if everything else remains constant. All the sensitivities are intuitive. The first two sensitivities show that if the relative expected return from asset one increases, the proportion invested in asset one increases. The next two show that if the relative risk of asset one increases (as measured by the variance of returns) the proportion invested in asset one decreases. This is similar to the results from the Markowitz model. In a two-asset model, there is no reason to expect the proportion invested in either asset to increase or decrease *a priori* if the covariance between the assets increases: this is the result found above. However, there is additional power in this model, compared with the traditional theory of finance models, as is shown by the last two results. If the covariance between either asset and the liability increases, more will be invested in that asset. This is a mathematical expression of the principle of matching discussed in Section 2.7. The more similar are the financial charac-teristics of an institution's assets and the liabilities, the higher their covariance will be and the greater will be the proportion invested in that asset.

The investment decision also depends on the risk tolerance of the investor. The investor is maximising expected utility and the actual proportions invested in the different assets will depend on the risk tolerance of the investor. Again, we find here an intuitive result. If the risk tolerance parameter of the investor (r) increases, then w_1 (the proportion invested in asset one) increases as long as: $E_1 > E_2$. Thus, the investor will invest more in the high return asset, if risk tolerance increases.

In practical terms, this model is a mean–variance model but with the incorporation of liabilities. However, in theory, the model can take a much broader view of risk. It was not initially assumed that the investor's decision would depend on the mean and variance of ultimate surplus. However, the assumption of the normal distribution (which is a two-parameter distribution defined only by its mean and variance) of surplus leads to the investment decision depending only on mean and variance. There are still a number of weaknesses of this model, despite its advantages over some of the models of conventional modern financial economics.

As has already been indicated, there are a limited number of utility functions and probability distributions of investment returns which can be used to find analytical solutions to the asset allocation problem. Numerical methods will therefore often have to be used. Estimating the parameters of the model can be very difficult because the statistical inputs may be statistically unstable over time. The model proposed has a very simple liability structure and may not capture fully the correlation structure between asset classes and between assets, liabilities and other economic variables such as inflation.

A major weakness of principle in the approach taken so far is that we have looked at a single time period. For example, the liability structure that has been assumed has a single liability to be met at a given time. In reality, there will be a complex series of liabilities, due over a number of time periods. Also, the investor can re-allocate assets over time. For example, as the liabilities become closer, a greater proportion of assets may need to be invested in cash. It is possible to develop the Wilkie/Sherris approach into a multi-period model. However, the assumptions needed to obtain analytical results become more and more restrictive. In practice, deeper investigations can be carried out using stochastic investment modelling which will be described in Section 2.11.

2.10   Immunisation

The classical theory of immunisation was set out at around the same time that Markowitz was publishing his seminal paper on portfolio theory (Redington, 1952; Markowitz, 1952). Redington considered the situation of an insurance company that was contracted to make a series of fixed monetary payments at specified times in the future and also expected to receive a fixed stream of premium income.

We assume that all payments are made and received at discrete intervals and let L_t represent the liability outgoing at time t ($t = t_1, t_2, \ldots t_n$). The liability outgoing is defined as the net amount of payments plus expenses minus premiums received. The company will also receive income from the assets held. We define the asset proceeds (interest or redemption payments) at time t as A_t.

The term structure of interest rates is characterised by the yields on the set of n zero coupon bonds redeemable at times $t_1, t_2, \ldots t_n$ having yields $i_{0.1}, i_{0.2}, \ldots i_{0,n}$.

The present values of the liability–outgoing and asset–proceeds are therefore

$$V_L = \sum_{j=1}^{n} v_{t_j}^{t_j} L_{t_j} \tag{2.18}$$

and

$$V_A = \sum_{j=1}^{n} v_{t_j}^{t_j} A_{t_j} \tag{2.19}$$

respectively, where

$$v_{t_j} = \frac{1}{(1 + i_{0,j})} \tag{2.20}$$

In general, the payments received will not match the contracted outgoing either in amount or timing. However, if the present value of the income stream is equal to that of the outgoing stream, the company would expect to be able to meet all payments as they fall due by investing excess income in liquid fixed income investments and reselling the bonds as required to make payments. Every so often, an insurance company will perform a valuation of its assets and liabilities at the interest rate assumed to be earned by the assets over the term of the liabilities.

So we assume that

$$V_A = V_L \tag{2.21}$$

The company will want to arrange its investments at the initial level of interest rates so that it does not suffer if rates change. It is subject to interest rate risk because, as the underlying term structure of interest rates changes, the present values of income and outgoing streams will be affected in different ways.

The classical theory of immunisation shows that, under certain assumptions, the assets can be invested to ensure that any small, parallel shift in the yield curve will result in a profit rather than a loss for the company. That is

$$V_A(t^+) > V_L(t^+) \tag{2.22}$$

where $V_A(t^+)$ and $V_L(t^+)$ are the values of the asset–proceeds and liability–outgoing immediately following a small parallel shift in the yield curve at time t.

The problem can be formulated as an optimisation problem by considering the objective function

$$f(v) = V_A(t) - V_L(t) \tag{2.23}$$

In the general case, we might want to minimise the variance of the objective function around an expected value of zero but in the special case, considered by Redington, of a flat yield curve, it is easy to show that a small shift in interest rates will always result in a profit if the following conditions hold

$$\frac{\sum_{j=1}^{n} v_{t_j}^{t_j} t_j A_{t_j}}{\sum_{j=1}^{n} v_{t_j}^{t_j} A_{t_j}} = \frac{\sum_{j=1}^{n} v_{t_j}^{t_j} t_j L_{t_j}}{\sum_{j=1}^{n} v_{t_j}^{t_j} L_{t_j}} = T \qquad (2.24)$$

$$\sum_{j=1}^{n} v_{t_j}^{t_j} (t_j - T)^2 A_{t_j} > \sum_{j=1}^{n} v_{t_j}^{t_j} (t_j - T)^2 L_{t_j} \qquad (2.25)$$

Equation (2.24) states that the duration (T) of the asset–proceeds must equal that of the liability–outgoing while Equation (2.25) states that the spread of the terms of the asset–proceeds must exceed that of the liability–outgoing. These results follow straightforwardly from considering the Taylor expansion of f around the current rate of interest.

There are problems in applying the Redington theory in practice (see Adams *et al.*, 1993) but, even in theory, there is a flaw in one of the underlying assumptions. The idea that things can be arranged to produce a profit in all conceivable circumstances sounds too good to be true and indeed the assumption of parallel shifts in the yield curve can be shown to be inconsistent with the principle of no arbitrage. Nevertheless, the fundamental concepts of duration matching and a greater spread of assets than liabilities have been an important guide to practical asset-liability management strategies for many years.

The extension of immunisation theory to other types of yield curve shifts and to stochastic interest rate models is discussed in De Felice (1995). Reviews of interest rate risk management from an actuarial perspective have been given by Tilley (1992) and Ang and Sherris (1995). Principles similar to immunisation are used in the hedging of portfolios with derivatives. Some of these ideas are discussed in Hull (1997).

2.11   Stochastic asset-liability modelling

2.11.1   What is asset-liability modelling (ALM)?

In Sections 2.8 and 2.9 we showed how traditional portfolio selection models could be extended to include liabilities. Most of these mathematical models are difficult to apply over multiple time periods. In asset-liability modelling (ALM) the period under consideration is usually divided into a number of sub-intervals. In long-term actuarial models these are usually years. The cash flows into and out of a fund are then considered explicitly for each year of the modelling period. It is these cash flows, both liability–outgoing and income from asset–proceeds, which are the fundamental output of the model, although other accounting quantities or valuation results can be calculated at each interval if required. ALM thus differs from many traditional actuarial techniques, where future cash flows are combined into a single present value.

The two essential components of an asset-liability model are:

- a time series investment model to generate the cash flows arising from the assets;
- a liability cash flow projection model.

It is possible to model the cash flows deterministically, considering only the expected value at each moment in time. Such an approach, however, tells us nothing about the possible variations in the outcome, and asset-liability models now invariably use stochastic cash flow models.

Stochastic asset-liability models can be constructed to consider problems for which analytical solutions cannot be found. Instead, numerical methods are used. In the simulation process, multiple scenarios are calculated using computer generated random numbers with the required distribution to provide the stochastic variation. The results of, say 10 000, simulations can be used to calculate the distribution of the desired output parameters.

An asset-liability model is used to calculate the net cash flow, X_t, into a pension fund during the period $t - 1$ to t, where

$$X_t = C_t + I_t - B_t \qquad (2.26)$$

with C_t, I_t and B_t being the contributions received, income generated by the assets held and benefit outgoing respectively in the period $t - 1$ to t.

The value of the assets at time t is A_t, given by

$$A_t = A_{t-1}(1 + i_t) + C_t - B_t \qquad (2.27)$$

where i_t is the rate of total return on the assets in the period $t - 1$ to t.

In the two equations above C, I, B and i are all random variables which are simulated in the asset-liability model.

At time t we may also wish to calculate the surplus of assets over liabilities

$$S_t = A'_t - V_t \qquad (2.28)$$

where V_t is the value of the liabilities and the prime on A_t indicates that it may be necessary to adjust the market value of the assets to be consistent with the liability valuation.

2.11.2   Advantages over traditional actuarial methods

A deterministic model can be used to examine different scenarios, typically including best and worse case combinations of events, and such scenarios can be used to try and quantify risk. However, such scenario testing gives no information about the probability of the different outcomes and the scenarios modelled are often chosen rather arbitrarily. It can also be difficult to choose the most appropriate combinations of events unless links between the economic and demographic variables are built into the model.

In contrast, stochastic asset-liability models include explicit links between economic and demographic variables and the output is a sample from the full distribution of possible outcomes, allowing statements to be made about the probabilities associated with different outcomes. However, asset-liability modelling is not without its practical and theoretical difficulties. We will look at some of these in Section 2.11.6.

2.11.3   The ALM Process

The goal of the ALM process is to define the distribution of possible outcomes for each possible set of the input parameters that can be controlled. An example of an outcome might be the statistical distribution of the funding level after a certain number of years. The input parameters that can be controlled, subject to certain constraints, include the contribution rate, the benefit structure and the investment policy. The set of attainable outcomes is known as the opportunity set. Some members of the opportunity set can usually be rejected immediately as not fulfilling certain pre-defined criteria. Those that remain are known as 'efficient'.

The output of a pensions asset-liability model could be plotted on a similar diagram to Fig. 2.1. For a given initial set of liabilities and a given value of assets, the stochastic model is used to calculate the expected value of the surplus and its standard deviation after (say) three years. The process is repeated a sufficient number of times to map out the efficient frontier.

In practice, of course, it is unusual to have such a simple optimisation problem and it is not necessary to restrict the calculations to two dimensions. The problem, for example, might be to minimise the initial value of the assets required at the same time as considering the size of the surplus. The solution to this problem is an 'efficient surface' as demonstrated by Wilkie (1985).

A schematic outline of the asset-liability modelling process is therefore:

(i)   specify the objectives and time horizon;
(ii)   choose the input parameters to be optimised;
(iii)   for a particular set of input parameters, run a number of simulations;
(iv)   record the distribution of the outputs;
(v)   vary the input parameter set and repeat the above two stages;
(vi)   repeat (iii), (iv) and (v) until the (efficient) opportunity set has been mapped out;
(vii)   discuss the opportunity set with the fund managers and decide on the most desirable combination of input parameters and outcome.

A more sophisticated model could incorporate decision rules to alter the strategy at defined intervals depending on the outcome so far. For example, if the funding level has fallen below a certain value at a particular point the investment strategy might be altered. This is known as dynamic or stochastic programming and has been shown to lead to an improvement in the strategies identified, at least in theory (Smith, 1996). A theoretical analysis of the benefits of portfolio rebalancing has been given by Wise (1996). However, the dynamic optimisation problem is vastly more complicated than the static one and much more work needs to be done in applying stochastic programming to actuarial problems.

2.11.4   Choosing the stochastic investment model

As has been mentioned, one of the components of an asset-liability model is a stochastic asset model, which can generate projections of the cash flows generated by the assets held. Stochastic investment modelling is an enormous subject in its own right and several different models have appeared in the actuarial literature. Smith (1996) lists six different types.

(i)   Random walk models.
(ii)   Chaotic models.
(iii)   Fractal models based on stable distributions for asset returns.
(iv)   Autoregressive models such as the Wilkie model (Wilkie, 1986, 1995).
(v)   Cointegrated models such as that published by Dyson and Exley (1995).
(vi)   Jump diffusion models such as that of Smith himself.

Although most of the models in use are proprietary and details are not usually published it seems that much current actuarial work in the UK is based on the Wilkie model. This model is based on the assumption that economic time series, such as the rate of inflation, security yields and total returns, are stationary over reasonably long periods of time. In other words they have a long-term mean value. Care and judgement must be used in assessing whether the long-term historical mean is an appropriate estimate of future behaviour and also in deciding whether current conditions form a suitable starting point for projections.

The model chosen must be appropriate to the use to which it will be put. For example, a model which exhibits appropriate statistical properties over the short term may not be suitable for use over the long term. For most actuarial applications it can be argued that it is more important that the model gives a realistic indication of the long-term statistical behaviour of investment returns, than that it attempts to give accurate estimates of the short term, cross-sectional distributions of the variables modelled.

2.11.5   Liability cash flow models

An asset-liability modelling exercise also needs to project future liability benefit outgoings, including that arising from future benefit accrual and new entrants to the scheme. Thus, it is necessary to keep track not only of the net cash flow in each year but also of the accumulating liabilities. The growth of the liabilities will depend on economic factors, which will require a link to the investment model and to demographic factors which may also be linked to economic factors.

The method chosen for the valuation of the liabilities will, of course, depend on the purpose of the modelling and should reflect the method and assumptions that would be used in practice. Whatever the model used, it is vital that it is consistent with, and linked to, the asset model.

The correlations between assets and liabilities are one of the most important factors in an ALM exercise. The strongest link between asset and liability models is likely to be via the rate of inflation This will affect index-linked benefits directly and also indirectly affect salary growth, which affects future pension liabilities. If the asset model does not generate an inflation rate explicitly it will be necessary to add an inflation series that is consistent with the returns on the various asset classes.

There may also be a link between economic factors and demographic factors such as the rate of withdrawal. Here, a balance has to be struck between the realism and the complexity of the model. The likely magnitude of the financial effects of stochastic variations in demographic factors and the strength of the evidence for the nature of the various links has to be considered before deciding on the benefits of the extra complexity. An overly complicated model will not only take longer to build and run than a more simplistic one but may also produce results that are more difficult to interpret.

2.11.6 Practical considerations

We have already touched on a number of practical points that have to be considered. The number of simulations is a trade-off between run time and statistical significance of the results. The complexity of the model is a factor here, giving a three-way trade-off. A highly complex model will take longer to run, allowing for fewer simulations. In turn, this will lead to results, the interpretation of which is less certain. On the other hand, an overly simplistic model may give beautifully clear results which fail to capture some significant feature.

Where assumptions have been made it is useful to have some feel for their significance. This is likely to be much more difficult for a complex stochastic model than for traditional actuarial models and is perhaps one of the main reasons why such techniques have not been adopted more quickly. It is important to bear in mind two types of risk that arise from the use of any model: parameter risk and model risk.

Parameter risk, the risk that the parameters such as expected future return are incorrectly specified, is widely recognised and can be quantified to an extent by running the model under several different assumptions. Model risk, the risk that the model itself is wrong, is harder to deal with and is possibly more dangerous. One defence of model risk might be greater publication of the various models being used, so that they can be subject to criticism by others. The major difficulties of stochastic asset models are discussed in Huber (1997). It is not simply the case that sufficiently sophisticated models have yet to be developed. A more realistic view is that real-world financial systems have behavioural features that cannot be captured by any statistical model. Professional judgement by model users will therefore always be necessary in choosing models and interpreting their results.

2.12 Conclusions

A number of texts deal with the development and application of financial economics. However, its practical application is limited because the economic assumptions on which mathematical and statistical models are based are often contravened in the real world. Actuaries have a long history of tackling such real world problems. They find many of the models of financial economics useful. However, they often need to be adapted to be applied to the kind of problems which financial institutions face in real life. Institutional investors face long-term liabilities and have complex risk profiles. Sometimes investment strategies can therefore only be investigated using simulation and numerical methods. Furthermore, the results obtained will only help the actuaries' understanding of the problem at hand. There will be many sources of potential error, some of which can be quantified but some of which cannot.

References

Adams, A. T., Bloomfield, D. S. F., Booth, P. M. and England, P. D. (1993) *Investment Mathematics and Statistics.* Graham and Trotman, London.

Anderson, N. and Breedon, F. (1996) *U.K. asset price volatility over the last 50 years.* Bank of England Working Paper, No. 57.

Ang, A. and Sherris, M. (1995) Interest rate risk management: developments in interest rate term structure modeling for risk management and valuation of interest rate dependent cash flows. *North American Actuarial Journal,* **1**, 1–26.

Bailey, A. H. (1862) On the principles on which the funds of life assurance societies should be invested. *Journal of the Institute of Actuaries,* **10**, 142–147.

Booth, P. M. (1995) The management of investment risk for defined contribution pension schemes. *Transactions of the XXV International Congress of Actuaries,* Vol. 3.

Booth, P. M. (1997) *The analysis of actuarial investment risk.* Actuarial Research Paper No. 93, City University, London, UK.

Brearley, R. A. and Myers, S. C. (1991) *Principles of Corporate Finance (4th Edn).* McGraw-Hill, New York.

Chaplin, G. B. (1996) *A review of term structure models and their applications.* The Faculty & Institute of Actuaries .

Clarkson, R. S. (1978) A mathematical model for the gilt-edged market. *Transactions of the Faculty of Actuaries,* **36**, 85–160.

Clarkson, R. S. (1989) The measurement of investment risk. *Journal of the Institute of Actuaries,* **116**, 127–178.

Clarkson, R. S. (1996) Financial economics: an investment actuary's viewpoint. *British Actuarial Journal,* **2**, 809–947.

Connor, G. (1995) The three types of factor models: a comparison of their explanatory power. *Financial Analysts Journal,* **51**, 42–46.

De Felice, M. (1995) Immunisation theory: an actuarial perspective on asset-liability management. In *Financial Risk in Insurance,* G. Ottaviani (Ed), Springer-Verlag, Berlin, pp. 63–85.

Dyson, A. C. L. and Exley, C. J. (1995) Pension fund asset valuation and investment. *British Actuarial Journal,* **1**, 471–557.

Elton, E. J. and Gruber, M. J. (1995) *Modern Portfolio Theory and Investment Analysis (5th Edn).* Wiley, New York.

Griffiths, J. P., Imam-Sadeque, I., Ong, A. S. K, Smith, A. D. and Wilkie, A. D. W. (1996) *Multi-factor techniques in active quant models. Proceedings of the 1996 Investment Conference.* The Faculty & Institute of Actuaries.

Huber, P. P. (1997) A study of the fundamentals of actuarial models. PhD thesis, City University, UK.

Hull, J. C. (1997) *Options, Futures, and other Derivatives (3rd Edn).* Prentice Hall International, Englewood Cliffs, New Jersey, USA.

Kemp, M. H. D. (1996a) *Asset/liability modelling for pension funds.* Presented to the Staple Inn Actuarial Society, 15 October 1996.

Kemp, M. H. D. (1996b) Actuaries and derivatives. *British Actuarial Journal,* **3**, 51–180.

Levy, H. and Markowitz, H. M. (1979) Approximating expected utility by a function of mean and variance. *American Economic Review,* **69**(2), 308–317.

Lewin, C. G., Carne, S. A., De Rivaz, N. F. C., Hall, R. E., McKelvie, K. J. G. and Wilkie, A. D. (1995) Capital projects. *British Actuarial Journal,* **1**, 155–249.

Lintner, J. (1965) Security prices, risk and maximal gains from diversification. *Journal of Finance,* **20**, 587–615.

Markowitz, H. M. (1952) Portfolio selection. *Journal of Finance,* **VII**, 77–91.

Mehta, S. J. B. (1992) Allowing for asset, liability and business risk in the valuation of a life office. *Journal of the Institute of Actuaries,* **119**, 385–440.

Pegler, J. B. H. (1948) The actuarial principles of investment. *Journal of the Institute of Actuaries,* **74**, 179–211.

Pratt, J. W. (1964) Risk aversion in the small and in the large. *Econometrica,* **32**(1–2), 122–136.

Rains, P. F. and Gardner, D. (1995) *Fund management risk control.* Presented to the Staple Inn Actuarial Society, 21 February 1995.

Redington, F. M. (1952) Review of the principles of life-office valuation. *Journal of the Institute of Actuaries,* **78**, 286–340.

St. John Hall, B., Black, P. F., Charters, G., Gardner, D. P., Griffiths, J. P., Hitchen, K. M., Jecks, K. M., Jung, G. and O'Brien, M. J. (1996) *Investment manager selection & style analysis. Proceedings of the 1996 Investment Conference.* The Faculty & Institute of Actuaries.

Sharpe, W. F. (1964) Capital asset prices: a theory of market equilibrium under conditions of risk. *Journal of Finance,* **19**, 425–442.

Sharpe, W. F. (1978) *Investments.* Prentice Hall, Englewood Cliffs, New Jersey.

Sharpe, W. F. and Tinte, L. G. (1990) Liabilities: a new approach. *The Journal of Portfolio Management,* Winter, 5–10.

Sherris, M. (1992) Portfolio selection and matching: a synthesis. *Journal of the Institute of Actuaries,* **119**, 87–105.

Smith, A. D. (1996) *How actuaries can use financial economics.* Presented to the Institute of Actuaries, 25 March, *British Actuarial Journal,* **2**, 1057–1193.

Tilley, J. A. (1992) An actuarial layman's guide to building stochastic interest rate generators. *Transactions of the Society of Actuaries,* **VLIV**, 509–538.

Wilkie, A. D. (1985) Portfolio selection in the presence of fixed liabilities: a comment on 'The matching of assets to liabilities'. *Journal of the Institute of Actuaries,* **112**, 229–277.

Wilkie, A. D. (1986) A stochastic model for actuarial use. *Transactions of the Faculty of Actuaries,* **39**, 341–403.

Wilkie, A. D. (1995) More on a stochastic model for actuarial use. *British Actuarial Journal,* **1**, 777–964.

Wise, A. J. (1984a) A theoretical analysis of the matching of assets to liabilities. *Journal of the Institute of Actuaries,* **111**, 375–402.

Wise, A. J. (1984b) The matching of assets to liabilities. *Journal of the Institute of Actuaries*, **111**, 445–485.

Wise, A. J. (1987) Matching and portfolio selection, Parts 1 and 2. *Journal of the Institute of Actuaries*, **114**, 113–133 and 551–568.

Wise, A. J. (1996) The investment return from a portfolio with a dynamic rebalancing policy. *British Actuarial Journal*, **2**, 975–1001.

3

Actuarial Applications of Generalised Linear Models

Steven Haberman and Arthur E. Renshaw

3.1 Introduction

In this chapter, we review the applications of generalised linear models to actuarial problems and we begin with a brief consideration of the nature of actuarial science. Actuarial science is concerned with the financial management of financial security systems, defined as 'mechanisms for reducing the adverse financial impact of random events that prevent the fulfilment of reasonable expectations' (Bowers *et al.*, 1986).

When considering such systems, we should be aware of their limitations. First, they deal with the consequences of random events that create losses which are measurable in monetary terms. Secondly, such systems do not directly reduce the probability of a loss occurring.

Examples of situations where random events may cause losses that can be measured in monetary terms would include the destruction of property by fire or natural catastrophe; a damage award imposed by a court as a result of a negligent event; prolonged illness or disability; death of a young adult; and survival to an advanced age.

One of the important tasks for an actuary employed by (or advising) a financial security system is the management of uncertainty. This management process can be broken down into a number of distinct stages; for example, one classification (based on the so-called 'actuarial control cycle') would be: identification of information sources, collection of data, analysis, model construction, sensitivity testing, prediction, monitoring the model assumptions in the light of emerging experience, updating the model. Statistics plays an important role in these stages and we will consider some examples of the successful applications of generalised linear models to highlight this point.

3.2 Introduction to generalised linear models

Generalised linear models are a natural generalisation of the classical linear model. The use of linear models in actuarial work is not new. For example, they have been an established part of the modelling of claim frequency rates and average claim costs in motor insurance – readers are referred to the early papers of Johnson and Hey (1971), Grimes (1971), Bennett (1978), Baxter *et al.* (1980) and Coutts (1984).

The use of generalised linear models in actuarial work can be traced back to the comprehensive monograph of McCullagh and Nelder (1983) who give a number of examples of the fitting of generalised linear models to different types of data, including average claim costs data from a motor insurance portfolio (originally modelled by Baxter *et al.* (1980) using a weighted least squares approach) and claim frequency data for marine insurance.

In this chapter, our objective is to demonstrate that generalised linear models have a wide area of application in actuarial work and are not confined, for example, to these early examples. In order to achieve this objective we review a number of distinct practical applications in actuarial work:

(1)   survival modelling in life insurance;
(2)   multiple state models in health insurance;
(3)   risk classification in life insurance;
(4)   non-life insurance premium rating;
(5)   claims reserving in non-life insurance.

Before describing these applications, we shall use the next section to describe briefly the nature and structure of generalised linear models.

3.3   Generalised linear models

The purpose of this section is to provide a brief introduction to generalised linear models (GLMs). A complete treatment of the theory and application can be found in McCullagh and Nelder (1989).

The basis of GLMs is motivated by the assumption that the data are sampled from a one-parameter exponential family of distributions. We first describe some of their fundamental properties in terms of a single observation y.

A one-parameter exponential family of distributions has a log-likelihood of the form

$$1 = \frac{y\theta - b(\theta)}{\phi} + c(y, \phi)$$

where θ is the canonical parameter and ϕ is the dispersion parameter, assumed known. It is then straightforward to demonstrate that

$$m = E(Y) = \frac{\mathrm{d}}{\mathrm{d}\theta} b(\theta) \quad \text{and} \quad \mathrm{Var}(Y) = \phi \frac{\mathrm{d}^z}{\mathrm{d}\theta^2} b(\theta) = \phi b''(\theta)$$

We note that $\mathrm{Var}(Y)$ is the product of two quantities. The quantity $b''(\theta)$ is called the variance function and depends on the canonical parameter and hence on the mean. This relationship is conveniently written as $V(m)$.

By way of illustration, some common distributions, which are members of this family, are presented below in terms of θ, $b(\theta)$, $V(m)$ and ϕ.

Normal: $\qquad\qquad \theta = m, \quad b(\theta) = \dfrac{\theta^2}{2}, \quad V(m) = 1, \quad \phi = \sigma^2$

Poisson: $\qquad\quad \theta = \log m, \quad b(\theta) = \exp\theta, \quad V(m) = m, \quad \phi = 1$

Binomial: Suppose $D \sim \text{binomial}(N, m)$. Define $Y = D/N$. Then

$$\theta = \log\frac{m}{1-m}, \quad b(\theta) = \log(1 + \exp\theta), \quad V(m) = m(1-m), \quad \phi = N^{-1}$$

Gamma: Suppose $Y \sim$ gamma with mean m and variance m^2/v. Then

$$\theta = -\frac{1}{m}, \quad b(\theta) = -\log(-\theta), \quad V(m) = m^2, \quad \phi = v^{-1}$$

More generally, a GLM is characterised by independent response variables $\{Y_u; u = 1, 2, \ldots, n\}$ for which:

$$E(Y_u) = m_u, \quad \text{Var}(Y_u) = \frac{\phi V(m_u)}{\omega_u}$$

comprising a variance function V, a scale parameter $\phi \, (>0)$ and prior weights ω_u. Covariates enter via a linear predictor

$$\eta_u = \sum_{j=1}^{p} x_{uj}\beta_j$$

with specified structure (x_{uj}) and unknown parameters β_j linked to the mean response through a known, differentiable, monotonic link function g with

$$g(m_u) = \eta_u$$

The special choice of link function $g = \theta$, so that $\theta(m) = \eta$, is called the canonical link function. Examples are

Normal: identity,	**Poisson:** log,
Binomial: log-odds or logit,	**Gamma:** reciprocal.

The suffixes or units, u, have a structure which is either intrinsic or imposed. The data comprise realisations $\{y_u\}$ of the independent response variables, matched to the structure of the units. Generally, in any one study, the detail of the distribution and link are fixed, while the predictor structure may be varied.

Model fitting is by maximising the quasi log-likelihood:

$$q = q(\underline{y}; \underline{m}) = \sum_{u=1}^{n} q_u = \sum_{u=1}^{n} \omega_u \int_{y_u}^{m_u} \frac{y_u - s}{\phi V(s)} \, ds$$

leading to the system of linear equations:

$$\sum_{u=1}^{n} \omega_u \frac{y_u - m_u}{\phi V(m_u)} \frac{\partial m_u}{\partial \beta_j} = 0 \quad \forall j$$

in the unknown β_js. These are solved numerically, e.g. Francis *et al.* (1993), McCullagh and Nelder (1989). Detail of the construction of standard errors for the parameter estimators, based on standard statistical theory, is also to be found in these references. We denote the resulting values of the parameter estimators, linear predictor and fitted values, for the current model c, by $\hat{\beta}_j$, $\hat{\eta}_u$ and $\hat{m}_u$ respectively, where

$$\hat{m}_u = g^{-1}(\hat{\eta}) = \sum_{j=1}^{p} x_{uj}\hat{\beta}_j$$

For members of the exponential family of distributions, the quasi log-likelihood equates to the log-likelihood. The maximal model structure possible has the property that the fitted values are equal to the observed responses, that is $\hat{m}_u = y_u$ for all u, and this is called the full or saturated model f.

The (unscaled) deviance of the current model c is:

$$D(c,f) = d(\underline{y}; \hat{\underline{m}}) = \sum_{u=1}^{n} d_u = \sum_{u=1}^{n} 2\omega_u \int_{\hat{m}_u}^{y_u} \frac{y_u - s}{V(s)} ds = -2\phi q(\underline{y}; \hat{\underline{m}})$$

in which the fitted values under the current and saturated models influence the formula through the lower and upper limits of the integral, respectively. The corresponding scaled deviance is:

$$S(c,f) = d^*(\underline{y}; \underline{m}) = \frac{d(\underline{y}; \hat{\underline{m}})}{\phi} = \sum_{u=1}^{n} 2\omega_u \int_{\hat{m}_u}^{y_u} \frac{y_u - s}{\phi V(s)} ds = -2q(\underline{y}; \hat{\underline{m}})$$

For fixed distribution, fixed link and hierarchical model structures c_1 and c_2, with c_2 nested in c_1, the difference in scaled deviance:

$$S(c_2,f) - S(c_1,f)$$

may be referred, generally as an approximation, to the chi-square distribution with $v_2 - v_1$ degrees-of-freedom, where v_1 and v_2 denote the respective degrees-of-freedom.

Two types of residuals (which are identical only in the case of the Gaussian distribution, for which $V(s) = 1$) are of interest, the Pearson residuals

$$\frac{y_u - \hat{m}_u}{\left(\dfrac{V(\hat{m}_u)}{\omega_u}\right)^{1/2}}$$

or the deviance residuals:

$$\text{sign}\,(y_u - \hat{m}_u)\sqrt{d_u}$$

where d_u is the uth component of the (unscaled) deviance above.

3.4   Survival modelling and graduation

3.4.1   What is graduation?

Graduation may be regarded as the principles and methods by which a set of observed (or crude) probabilities is adjusted in order to provide a suitable basis for inferences to be made and further practical calculations to be made.

One of the principal applications of graduation is the construction of a survival model, usually presented in the form of a life table.

We consider for the moment a set of age-specific crude probabilities of death $\mathring{q}_x$, or forces of mortality (i.e. hazard rates) $\mathring{\mu}_x$, which have been calculated from a set of observations. These values can each be regarded as a sample from a larger population and thus contain some random fluctuations. If we believe that the true q_x (or μ_x) were independent, then the crude values would be our final estimates of the true, underlying rates. However, a common, prior opinion about the form of these true rates is that each is closely related to its neighbours. This relationship is expressed by the belief that the true rates progress smoothly from one age to the next. So the next step is to adjust or 'graduate' the crude rates in order to produce smooth estimates $\hat{q}_x$ (or $\hat{\mu}_x$), of the true probabilities or rates.

The graduation methods suggested in the literature and used in practice tend to fall in the categories (a) parametric and (b) non-parametric. Parametric methods involve the fitting of a mathematical function to $\mathring{q}_x$ or $\mathring{\mu}_x$, with parameters being determined by a formal procedure like maximum likelihood. Although in the context of the assumed function, such methods are efficient, they are always liable to some degree of bias since no preassigned function will represent exactly the true (and unknown) values of q_x or μ_x. In contrast, non-parametric methods aim to give more stable estimates than the crude values by combining data at different values of x, but without presupposing any particular mathematical form for q_x or μ_x. Like parametric methods, they are liable to give biased estimates, but in such a way that it is possible to balance explicitly an increase in bias with a decrease in sampling variation. With non-parametric methods like kernel methods, the amount of smoothing of the crude data can be varied over a continuous range (for example by the choice of bandwidth). In contrast, the smoothness of parametric methods can only be regulated in discrete steps, for example by increasing the degree of polynomial or by increasing the number of knots in a cubic spline. In this process, the properties of such curves will also tend to change abruptly. However, parametric methods are able to achieve a higher degree of smoothness than non-parametric methods through their use of explicit mathematical formulae and may be more useful for extrapolation beyond the data range available.

We shall describe a flexible approach to (a), based on generalised linear models, which has been successfully implemented in a wide variety of contexts. For a discussion of (b), the interested reader is referred to Copas and Haberman (1983), London (1985) and Gavin *et al.* (1993, 1994, 1995).

The graduation process is an essential step in the construction of a survival model and ensures that the model displays the required degree of smoothness. Then, the functions of practical importance calculated from the model (and leading to life insurance premiums, reserves, surrender values and so on) share this important property of smoothness.

3.4.2 Graduation with respect to age

The Institute and Faculty of Actuaries' Continuous Mortality Investigation (CMI) Bureau has a collective responsibility for the construction of standard life tables for use by the UK actuarial profession. Its origin dates back to 1924 when the continuous collection of mortality data began. Forfar *et al.* (1988) give a comprehensive description of the graduation methodology currently used by the CMI Bureau to produce such tables. It is possible to reformulate and extend this methodology using the generalised linear and non-linear modelling frameworks.

The raw data comprise the number of recorded deaths a_x, accruing from matching exposures (or person years exposed to risk) r_x, over a range of ages x, in a specific calendar period, typically of 4 years duration. The target is either the force of mortality μ_x using *central exposures* or the probability of death q_x using *initial exposures*. The central exposure (or exposed to risk) is defined to be the total time that the contributing individuals are under observation in the cell defined by age x and the calendar period (and policy duration d if appropriate). To obtain the corresponding value for targeting the probability of death for the year of age $(x, x + 1)$, the central exposure is increased by a factor equal to the time $1 - s$, for each observed death, where $x + s$ is the exact age at death and $0 < s < 1$: this gives rise to the initial exposure (or exposed to risk). Practical issues involving the composition of such exposures are well documented in the actuarial literature (e.g. Chadburn *et al.*, 1994). The crude rates a_x/r_x estimate the respective targets, while the primary aim of a graduation is to provide smooth estimates for the target consistent with any underlying patterns in the raw data.

We model the actual numbers of deaths A_x as Poisson random variables when targeting μ_x and as binomial random variables when targeting q_x. Thus, for μ_x-graduations with responses $\{A_x\}$

$$m_x = E(A_x) = r_x\mu_{x+1/2}, \quad V(m_x) = m_x \quad \omega_x = 1, \quad \phi = 1$$

or equivalently, with responses $\{A_x/r_x\}$

$$m_x = E(A_x/r_x) = \mu_{x+1/2}, \quad V(m_x) = m_x, \quad \omega_x = r_x, \quad \phi = 1$$

and for q_x-graduations with responses $\{A_x\}$

$$m_x = r_x q_x, \quad V(m_x) = m_x\left(1 - \frac{m_x}{r_x}\right), \quad \omega_x = 1, \quad \phi = 1$$

The formulae underpinning a graduation are presented as predictor–link relationships with age x as the sole covariate. Thus, for μ_x-graduations we consider either the log link with $\{A_x\}$ as responses, so that

$$\log m_x = \eta_x = \log r_x + \log \mu_{x+1/2}$$

with inverse

$$\mu_{x+1/2} = \exp(\eta_x - \log r_x)$$

and offsets $\log r_x$ or the parameterised power link with $\{A_x/r_x\}$ as responses, so that

$$\mu_{x+1/2}^\gamma = \eta_x \quad \text{with inverse} \quad \mu_{x+1/2} = \eta_x^{1/\gamma}$$

For q_x-graduations with $\{A_x\}$ as responses, we consider either the log-odds link, so that

$$\log\left(\frac{m_x}{r_x - m_x}\right) = \log\left(\frac{q_x}{1 - q_x}\right) = \eta_x \quad \text{with inverse} \quad q_x = \frac{\exp\eta_x}{1 + \exp n_x}$$

or the complementary log-log link, so that

$$\log\left(-\log\left(1 - \frac{m_x}{r_x}\right)\right) = \log(-\log(1 - q_x)) = \eta_x$$

$$\text{with inverse} \quad q_x = 1 - \exp(-\exp\eta_x)$$

Also, for q_x-graduations, the parameterised family of link functions are available

$$\eta_x = \log\left\{\frac{(1 - q_x)^{-\gamma} - 1}{\gamma}\right\} \quad \text{with inverse} \quad q_x = 1 - (1 + \gamma\exp\eta_x)^{-1/\gamma} \quad (3.1)$$

reducing to the log-odds link when $\gamma = 1$ and the complementary log-log link as $\gamma \to 0$.

Orthogonal polynomial predictors

$$\eta_x = \sum_{j=0}^{r} \beta_j h_j\left(\frac{x - a}{b}\right)$$

with a suitable choice of a and b and an orthogonal basis $\{h_j\}$ play a central role, for reasons of convenience of computing and interpretation. Suitable choices for $h_j(x)$ would be $C_j(x)$, the Chebycheff polynomials of the first type generated by

$$C_0(x) = 1, \quad C_1(x) = x, \quad C_{n+1}(x) = 2xC_n(x) - C_{n-1}(x) \quad \text{for} \quad n \geqslant 1$$

or $L_j(x)$, the Legendre polynomials generated by

$$L_0(x) = 1, \quad L_1(x) = x, \quad (n + 1)L_{n+1}(x) = (2n + 1)xL_n(x) - nL_{n-1}(x) \quad \text{for} \quad n \geqslant 1$$

For μ_x-graduations, Forfar *et al.* (1988) focus on the formula

$$\mu_{x+1/2} = GM_x(r, s) = \sum_{i=0}^{r-1} \alpha_i h_i\left(\frac{x - a}{b}\right) + \exp\left\{\sum_{j=0}^{s-1} \beta_j h_j\left(\frac{x - a}{b}\right)\right\} \quad r, s \geqslant 0$$

comprising the identity link $(\gamma = 1)$ and non-linear Gompertz–Makeham predictor $GM_x(r, s)$, subject to the convention that $r = 0$ implies the absence of the first group of terms, and $s = 0$ implies the absence of the second exponentiated group of terms. When $r = 0$, the formula reduces to the log link in combination with a polynomial predictor. The cases $GM_x(0, 2)$ and $GM_x(1, 2)$ correspond respectively to the historically important Gompertz and Makeham formulae, usually parameterised in the form $\mu_x = \exp(\beta_0 + \beta_1 x)$ and $\mu_x = \alpha_0 + \exp(\beta_0 + \beta_1 x)$ respectively. Although non-linear, the case $GM_x(r, 2)$ $(r \neq 0)$ may be fitted iteratively as a GLM, e.g. Renshaw (1991).

For a given predictor structure in combination with the power link, it is possible to search for the optimum value of γ by refitting the same predictor structure for different values of γ, chosen systematically to construct a deviance profile.

For q_x-graduations, use of the log-odds link in combination with polynomial predictors, namely

$$\log\left(\frac{q_x}{1-q_x}\right) = \sum_{j=0}^{T} \beta_j h_j\left(\frac{x-a}{b}\right)$$

was used by the CMI Bureau to construct all the q_x-graduations published in CMI Committee (1976). Forfar *et al.* (1988), however, focus on the graduation formula

$$q_x = \text{LGM}_x(r, s) = \frac{\text{GM}_x(r, s)}{1 + \text{GM}_x(r, s)}$$

comprising the odds link in combination with the Gompertz–Makeham predictor $\eta_x = \text{GM}_x(r, s)$, which reduces to the log-odds link in combination with the polynomial predictor for the case $\text{GM}_x(0, s)$. When $r > 0$ and $s = 0$ or 1, the predictor is linear in combination with the odds link, otherwise when $r > 0$ and $s > 1$ the predictor is non-linear. Renshaw (1991) has suggested the graduation formula

$$q_x = 1 - \exp\left\{-\exp\sum_{j-0}^{T} \beta_j h_j\left(\frac{x-a}{b}\right)\right\}$$

comprising the complementary log-log link in combination with a polynomial predictor and which includes the Gompertz formula as a special when $r = 1$. The family of parameterised link functions defined by Equation (3.1) is also available in this case.

For q_x-graduations based on central exposures, the μ_x-graduation methodology may be applied on replacing $\mu_{x+1/2}$ by the approximation $-\log(1 - q_x)$ based on the identity

$$1 - q_x = \exp\left(-\int_0^1 \mu_{x+1}\,dt\right)$$

Thus, for the log link

$$\log m_x = \eta_x = \log r_x + \log\{-\log(1 - q_x)\}$$

with implied graduation formula

$$q_x = 1 - \exp\{-\exp(\eta_x - \log r_x)\}$$

and offsets $\log r_x$

It is common actuarial practice to tabulate the statistics

$$e_x = \hat{m}_x, \quad \text{dev}_x = a_x - e_x, \quad \sqrt{V_x} = \left(\frac{V(\hat{m}_x)}{\omega_x}\right)^{1/2}, \quad z_x = \frac{\text{dev}_x}{\sqrt{V_x}}, \quad 100\frac{a_x}{e_x}$$

as part of the diagnostics applied to the finally adopted graduation, where z_x are the Pearson residuals. The diagnostics may be augmented by a variety of residual plots, including the normal and half-normal plots, and a battery of tests (including the standardised deviations test, cumulative deviations test, serial correlations test, signs test, changes of sign test and grouping of signs test: see Chadburn *et al.* (1994) for a full discussion).

The methods outlined above may be extended in a number of ways, including the use of different predictor types, such as splines, smoothing splines or fractional polynomials (along the lines of Royston and Altman, 1994).

3.4.3 Graduation in presence of duplicates

Since the data used in the construction of actuarial life tables are generally based on policy rather than head counts, the effect of duplicate policies on the graduation process has long since been of interest to actuaries. For a review of the issues involved, readers should consult Forfar *et al.* (1988) and Renshaw (1992), who has reinterpreted the model in a GLM setting. The effect of duplicate policies is to induce over-dispersion in the GLM. This may be formulated by setting prior weights ω_x and interpreting their reciprocals as non-constant scale parameters $\phi_x (> 1)$. The reasons for this are based on the extensive studies of the properties of the so-called empirical variance ratios, defined by

$$vr_x = \frac{\sum_i i^2 f_x^{(i)}}{\sum_i i f_x^{(i)}}$$

which are possible estimates of the ϕ_xs. Here $f_x^{(i)}$ denotes the proportion of policyholders at age x who have i policies, such that

$$f_x^{(i)} \geqslant 0, \quad \sum_i f_x^{(i)} = 1 \Rightarrow vr_x \geqslant 1$$

There are a number of possibilities to allow for the effects of duplicate policies in the graduation process. Forfar *et al.* (1988), working outside the GLM framework, elect to transform the data by dividing both the actual numbers of deaths A_x, and matching exposures to the risk of death r_x, by the empirical variance ratios vr_x prior to graduation. Alternatively, within the GLM framework, the untransformed data could be modelled using (i) either the reciprocals of the variance ratios as weights when these are available (Currie and Waters, 1991), or (ii) using the joint modelling technique to generate the non-constant scale parameters ϕ_x for the more common case when the variance ratios are not available (Renshaw, 1992).

The latter approach, favoured by Renshaw, is to use a two-stage model, as follows.

(a)   Model the A_xs as independent response variables with

$$E(A_x) = m_x \qquad \mathrm{Var}\,(A_x) = \phi_x V_A(m_x)$$

and predictor link $g(m_x) = \sum_j u_{xj}\beta_j$.

(b)   Model the unknown parameters ϕ_x using an appropriate dispersion statistic with

$$E(d_x) = \phi_x \qquad \mathrm{Var}\,(d_x) = \tau V_D(\phi_x)$$

and predictor link $h(\phi_x) = \sum_j v_{xj}\gamma_j$.

Possible choices for d_x include the Pearson squared residual and the deviance squared residual. The u_{xj} and v_{xj} are known covariates while it is necessary to estimate the β_j and γ_j in order to fit the model. As a first step, an optimisation function is needed to play the role of the quasi log likelihood: further details of possible choices and the iterative nature of the fitting process are provided by Renshaw (1992).

3.4.4   Graduation with respect to age and time

Given deaths a_{xt} and exposures r_{xt} by individual year of age x and by individual calendar year t, the methods described above can readily be applied to models which incorporate both age variation in mortality and any underlying time trend in the mortality rates.

Renshaw *et al.* (1996) undertake a trend analysis for UK assured lives by targeting the force of mortality μ_{xt}, for the calendar period 1958 to 1990 and ages 22 to 89 years, both inclusive. While downward trends in mortality are well established in the general population and the insured subsection of the population, there is evidence that such trends may not be uniform across the age range. (For example, the advent of AIDS has meant a slowing down of the decreasing trend for young, adult males in the general population.) To allow for such an eventuality, bivariate polynomial predictors in age and period effects, used in combination with the log link, give rise to

$$\mu_{xt} = \exp\left\{\beta_0 + \sum_{j=1}^{s}\beta_j L_j(x')\right\}\exp\left[\sum_{i=1}^{r}\left\{\alpha_i + \sum_{j=1}^{s}\gamma_{ij}L_j(x')\right\}t'^i\right]$$

with

$$x' = \frac{x - c_x}{\omega_x}, \qquad t' = \frac{t - c_t}{\omega_t}$$

where c_x, c_t denote mid-points and ω_x, ω_t, denote semi-ranges. Such formulae may be interpreted as Gompertz type graduation formulae with multiplicative, age dependent, trend adjustment factors.

Similarly, Renshaw and Hatzopoulos (1996) have modelled recent trends in mortality for UK pensioners using complementary log-log link based formulae in combination with bivariate polynomials to target the probability of death q_{xt}.

An alternative to using age x and period t as covariates would be to take a diagonal perspective of the rectangular data grid and to define cohorts z by their year of birth, with $z = t - x$ and to use cohort z and period t, say, as covariates (or use z and x). This has been explored to a limited extent by Renshaw *et al.* (1996) with respect to the UK assured lives data set (described above) and this approach clearly impinges on the extensive demographic and epidemiological literature on age–period–cohort models: see, for example, Hobcraft *et al.* (1982), Osmond and Gardner (1982).

3.4.5   Graduation with respect to age and policy duration

The underwriting mechanism by which proposals for life insurance policies are assessed generally results in lower levels of mortality for the insured lives in the initial stages following policy inception age for age, than would otherwise be the case: this process is called temporary initial selection. As a result, it is normal to analyse mortality rates of insured lives specific for both age and policy duration.

Denote the resulting deaths and exposures by (a_x^d, r_x^d) at age x for policy duration $d = 0, 1, \ldots, d_+$. Typically, the 'ultimate' duration of interest is, at most, $d_+ = 5_+$ years in UK studies and is often set at $d_+ = 2_+$ years in practice, although in some North American investigations $d_+ = 15_+$ is common.

It is widely recognised that the separate graduation by duration of such a collective mortality experience, practised extensively by the CMI Bureau in the UK, can lead to anomalies whereby the graduated curves cross over, with heavier mortality being recorded at shorter durations compared with relatively longer durations, at specific attained ages. To deal with such an eventuality, the CMI Bureau makes *ad hoc* adjustments where necessary to ensure that the resulting graduations lead to rates that are ordered by duration, with age fixed.

To avoid such possible anomalies, Currie and Waters (1991) suggest fitting general structures

$$\log \mu_{x+1/2}^d = f_1^*(x) + f_2^*(d) + f_3^*(x, d)$$

with parameterised functions f_1^*, f_2^* and f_3^*, in a single stage, having first estimated the age specific 'centre of gravity' of the banded duration category d_+. This then gives them the flexibility to model the effects of duration d as a continuous variate (as well as a categorical factor) when specifying the nature of the parameterised functions f_2^* and f_3^*. This would appear to be an important option when it is found necessary to impose a monotonic parametric structure in duration effects.

Renshaw and Haberman (1996) have suggested a two-stage fitting process as an alternative way of avoiding these potential anomalies. Stage 1 involves the graduation of the data for the ultimate duration d_+, by any method thought suitable, resulting in the graduated values $\hat{\mu}_{x+1/2}^d$. Stage 2 involves the modelling of responses

$$Z_x^d = \log \left\{ \frac{\text{crude mortality rate at age } x, \text{ for select duration } d}{\text{crude mortality rate at age } x, \text{ for ultimate duration } d_+} \right\}$$

for which

$$E(Z_x^d) \approx \log \left\{ \frac{\mu_{x+1/2}^d}{\mu_{x+1/2}^{d_+}} \right\} = \eta_x^d, \quad V(\eta_x^d) = 1, \quad \omega_x^d = \frac{a_x^d a_x^{d_+}}{a_x^d + a_x^{d_+}}$$

and scale parameter ϕ, to target the predictor η_x^d. Denoting the resulting fitted values by $\hat{\eta}_x^d$, the graduations for the individual select durations $d = 0, 1, \ldots, < d_+$ then follow as

$$\hat{\mu}_{x+1/2}^d = \hat{\mu}_{x+1/2}^{d_+} \exp(\hat{\eta}_x^d)$$

By way of illustration, the empirical values z_x^d, for the UK male assured lives' experience based on the four-year observation period 1975–78, are plotted

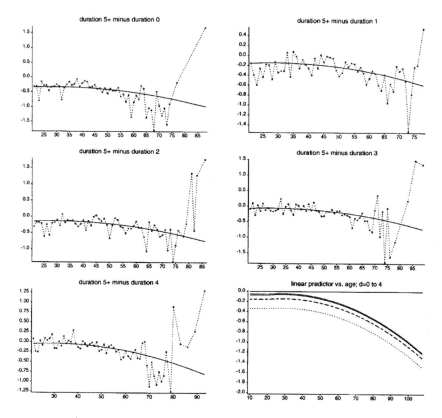

Figure 3.1 z_x^d response plots for policy durations 0,1,2,3,4 relative to policy duration 5+; male assured lives 1975–78 (UK)

against x for each $d = 0, 1, 2, 3, 4$ in Fig. 3.1. We have also superimposed on these plots the fitted values associated with the parameterised linear predictor structure

$$\eta_x^d = \alpha + \alpha_d + \beta(x - 20)_+ + \gamma(x - 20)_+^2$$

where

$$(x - 20)_+ = \begin{cases} 0, & x \leqslant 20 \\ x - 20, & x > 20 \end{cases}$$

while quoting two alternative versions for the parameter estimates and their standard errors in Table 3.1. We note in particular the ordering of these fitted curves with respect to duration d.

3.4.6 Graduation of 'amounts'-based data

The pensioners' mortality experience under UK life office pension schemes gives rise to two types of data sets, both of which are graduated as separate

Table 3.1 UK Male Assured Lives, 1975–78 study

Either	Or
$\alpha = -0.3404\,(0.04887)$	$\alpha = 0$
$\alpha_1 = 0$	$\alpha_1 = -0.3404\,(0.04887)$
$\alpha_2 = 0.1818\,(0.03773)$	$\alpha_2 = -0.1586\,(0.04816)$
$\alpha_3 = 0.1882\,(0.03726)$	$\alpha_3 = -0.1522\,(0.04792)$
$\alpha_4 = 0.2677\,(0.03627)$	$\alpha_4 = -0.07276\,(0.04763)$
$\alpha_5 = 0.2882\,(0.03575)$	$\alpha_5 = -0.05217\,(0.04761)$

with

$\beta = 0.003210\,(0.003384)$ $\qquad \gamma = -0.0001857\,(0.00006272)$

Deviance $= 352.55$ on 291 degrees-of-freedom

exercises by the CMI Bureau. These comprise the number of pension policies ceasing through death together with the associated exposures, the so-called 'lives' data, and the total amounts of pension ceasing through death together with the associated exposures, the so-called 'amounts' data.

Denoting the respective targets by μ_x^* or q_x^* for 'amounts' data, let

N_x = number of pension policies ceasing through deaths;
r_x = exposure to risk based on 'lives';
A_x = the amounts of pension arising from deaths;
e_x = exposure to risk based on 'amounts'.

Renshaw and Hatzopolous (1996) describe a multi-stage method for graduating 'amounts' data utilising GLMs based on the following properties of the first two moments

$$m_x = E(A_x) = e_x\mu_x^*, \quad \text{Var}(A_x) = \phi_x m_x, \quad \phi_x = (\psi + \tau)\rho_x$$

to target μ_x^*, and

$$m_x = E(A_x) = e_x q_x^*, \quad \text{Var}(A_x) = (\psi + \tau)\rho_x m_x - \frac{\tau}{r_x} m_x^2$$

to target q_x^*, subsequent to the graduation of the matching 'lives' data with scale parameter τ, and the modelling of the average claim amounts $\overline{X}_x$ as independent gamma response variables of a GLM with

$$E(\overline{X}_x) = \rho_x, \quad \text{Var}(\overline{X}_x) = \frac{\psi \rho_x^2}{n_x}$$

with weights n_x, scale parameter ψ and variance function $V(\rho_x) = \rho_x^2$ (where n_x is the observed value of N_x).

3.5 Multiple state models

In many countries, individual insurance policies providing an income to the policyholder while he/she is sick or disabled are becoming common: in the UK, these are called 'permanent health insurance' (PHI). A three-state Markov process comprising two transitive states H (healthy) and S (sick), and an absorbing state D (dead) has been proposed as the basis for modelling

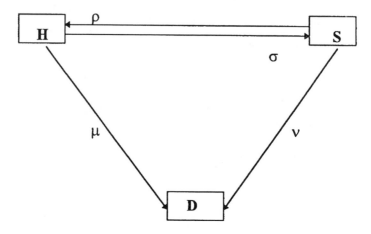

H - healthy, S - sick, D - dead

Figure 3.2   Possible transitions for multi-state model

the underlying transitions for PHI business, e.g. CMI Committee (1991). The process is represented in Fig. 3.2 in which a detailed knowledge of the four transition intensities $\sigma: H \to S$, $\rho: S \to H$, $v: S \to D$, $\mu: H \to D$ is needed for the calculation of premiums, reserves and annuity values. Under the usual conditions of PHI contracts, benefits become payable if state S continues to be occupied at the end of an agreed deferred period d, which typically is either 1, 4, 13, 26 or 52 weeks, and which commences immediately the policyholder enters state S (from state H). As a first step, the modelling of both transition intensities out of state S, ρ and v, as functions of age at sickness onset x and duration of sickness z, and the modelling of the sickness inception intensity σ as a function of age x, is possible using data relating to PHI policies made available by UK insurance companies. Renshaw and Haberman (1995) have formulated this modelling approach within the GLM framework.

For the graduation of ρ_{xz}, we denote the reported number of transitions out of state S by I_{xz} with matching central exposures e_{xz}, and then formulate

$$m_{xz} = E(I_{xz}) = e_{xz}\rho_{xz}, \quad V(m_{xz}) = m_{xz}, \quad \omega_{xz} = 1$$

with an equivalent assumption for v_{xz}. For ρ_{xz}, relationships

$$\log \rho_{xz} = \beta_0 + \beta_1 x + \beta_2 z + \beta_3 \sqrt{z} + \beta_4 xz + \beta_5 x\sqrt{z}$$

comprising the log-link in combination with a polynomial in the variates x and $\sqrt{z}$ are found to be particularly effective in the case of a deferred period of one week. Relationships

Table 3.2 ρ-graduations. Extract of deviance profiles for positioning of two knots

	Deferred Period 4 Weeks				
	$z_1 = 5.5$	6.5	7.5	8.5	9.5
$z_2 = 15.5$	323.7	302.1	297.1	300.4	308.8
17.5	318.3	296.3	291.9	295.8	304.6
21.5	311.6	290.6	288.0	293.6	303.4
25.5	307.3	287.6	286.5	293.2	303.5
29.5	303.4	284.8	284.6	292.0	302.6
34.5	299.2	281.9	282.8	290.9	301.7
45.5	295.1	280.1	282.8	291.8	302.7
78.5	321.4	314.3	320.9	331.0	340.9

$$\log \rho_{xz} = \beta_0 + \beta_1 x + \beta_2 z + \beta_3 (z - z_1)_+ + \beta_4 (z - z_2)_+ + \beta_5 x (z - z_1)_+$$

based on the log link in combination with break-point predictor terms involving two knots in the effects of z are found to be effective in the case of the other deferred periods. The location of the optimum positions for the two knots is determined by scanning the deviance profile for positional choices for a pair of knots. We illustrate one such deviance profile in Table 3.2 for a deferred period of 4 weeks, resulting in the choice $z_1 = 6.5$ and $z_2 = 45.5$; while an extract of the corresponding predicted recovery intensities is presented in Table 3.3.

A similar relationship to this was also used to model v_{xz}, where it was

Table 3.3 Graduated sickness recovery intensities, deferred period 4 weeks

Duration	Age				
	20	30	40	50	60
4	4.66203	4.05554	3.52794	3.06899	2.66974
5	6.38265	5.55232	4.83001	4.20166	3.65506
6	8.73831	7.60152	6.61262	5.75237	5.00403
7	10.0191	8.68161	7.52263	6.51837	5.64818
8	9.62070	8.27129	7.11108	6.11361	5.25605
9	9.23829	7.88037	6.72204	5.73398	4.89115
10	8.87099	7.50792	6.35429	5.37792	4.55158
15	7.24227	5.89367	4.79619	3.90308	3.17627
20	5.91258	4.62649	3.62014	2.83269	2.21653
25	4.82703	3.63176	2.73246	2.05585	1.54678
30	3.94079	2.85091	2.06245	1.49205	1.07940
40	2.62656	1.75677	1.17501	0.78590	0.52565
50	2.12000	1.31096	0.81067	0.50130	0.30999
52.2	2.12924	1.29428	0.78678	0.47828	0.29074
104.4	2.35996	0.95265	0.38456	0.15524	0.06266
156.5	2.61580	0.70120	0.18796	0.05039	0.01351
208.7	2.89938	0.51611	0.09187	0.01635	0.00291
260.9	3.21380	0.37988	0.04490	0.00531	0.00063

Parameter estimates, standard errors:

$\hat{\beta}_0 = 0.5617\,(0.3252)$, $\hat{\beta}_1 = -0.01394\,(0.002991)$, $\hat{\beta}_2 = 0.3141\,(0.04870)$,

$\hat{\beta}_3 = -0.3390\,(0.05062)$, $\hat{\beta}_4 = 0.04254\,(0.005140)$, $\hat{\beta}_5 = -0.0007846\,(0.0001754)$.

necessary to combine data sets over all deferred periods before modelling, because of the paucity of data at individual deferred periods.

For the graduation of σ_x, we denote the reported number of transitions from healthy to sick by J_x with matching central exposures e_x. It is then necessary to allow for left censoring induced by the deferred period d in the model and so we formulate

$$m_x = E(J_x) = \pi_{xd} e_x \sigma_x, \quad V(J_x) = m_x, \quad \omega_x = 1$$

with scale parameter ϕ, where

$$\pi_{xt} = \exp\left[-\int_0^{t/52} (\rho_{x+u,u} + v_{x+u,u}) \, du \right]$$

denotes the probability that the time in state S (sick) exceeds t. Log-link formulae in combination with polynomial predictors in x are found to be appropriate. Further details can be found in Renshaw and Haberman (1995).

3.6 Risk classification

3.6.1 Excess mortality for assured lives and Cox's model

The Prudential Life Insurance & Impaired Lives Mortality study centres on an extensive data set, comprising information derived from well over half a million life insurance policies effected on medically impaired lives, i.e. persons identified as exhibiting one of a list of medical conditions at the time when they applied and were accepted for life insurance. The study began in January 1947 and is ongoing. Aspects of this data set have been investigated and modelled by Renshaw (1988), Haberman and Renshaw (1990), England (1993) and England and Haberman (1993), using a proportional hazards model (based on the seminal work of Cox, 1972)

$$\lambda(t, z_i) = \lambda^*(t) \exp \sum_{j=1}^p z_{ij} \beta_j$$

with known base-line hazard function λ^* and known covariate structure (z_{ij}) where t denotes time. The 'A 1967–70' assured lives standard life table, suitably adjusted for time trends, was used to construct the base-line hazard function λ^*. Then, if τ_{ik} denotes the time at entry into the study and t_{ik} denotes the time at exit from the study, either by censorship or by death, for each member k of a particular study cohort i, it is possible to compute the accumulated integrated base-line hazard

$$m_i = \sum_k \int_{\tau_{ik}}^{t_{ik}} \lambda^*(s) \, ds$$

and hence to model the number of deaths per cohort, D_i, as independent Poisson responses

$$\mu_i = E(D_i) = m_i \exp \sum_{j=1}^p z_{ij} \beta_j$$

Table 3.4   Prudential study: impairment of coronary arteries, parameter estimates

Overall mean	$\hat{\mu} = 1.910$		
Age at entry	$\hat{\alpha}_1 = 0$	$\hat{\alpha}_2 = -0.5668$	$\hat{\alpha}_3 = -1.354$
Policy duration	$\hat{\beta}_1 = 0$	$\hat{\beta}_2 = -0.4109$	$\hat{\beta}_3 = -0.3354$
Complications	$\hat{\gamma}_1 = 0$	$\hat{\gamma}_2 = 0.3359$	

subject to a variety of covariate structures determined by the various medical impairments.

The factors

$$\exp \sum_{j=1}^{p} z_{ij}\beta_j = \frac{\mu_i}{m_i} \tag{3.2}$$

are closely connected to the traditional, widely used actuarial mortality ratios (actual deaths/expected deaths: for further discussion see Berry, 1983; England and Haberman, 1993).

As an illustration of the technique, we consider the case of those lives with 'impairment of the coronary arteries': there are 3307 such entrants to the study and 297 deaths during the period 1947–87. The data are partitioned according to:

- age at entry (three levels i: 16 to 49 years, 50 to 59 years, 60 to 79 years);
- policy duration (three levels j: 0–2 years, 2–5 years, 5–8 years);
- complications indicator ($k = 1$ without 'complications', $k = 2$ with 'complications', with 'complications' defined as subsequent chest pain on exertion);

and the parameter estimates for the factor-based main effects model

$$\sum_{u=1}^{p} z_{uv}\beta_v = \mu + \alpha_i + \beta_j + \gamma_k$$

are presented in Table 3.4. These estimates quantify the impact of age at entry, policy duration and the presence of complications on the mortality ratios (see Equation (3.2)). Thus, close inspection reveals that the presence of complications leads to mortality ratios that are 1.4 times higher than when complications are absent. The negative values of α_i ($i = 2, 3$) and β_j ($j = 2, 3$) indicate that the mortality ratios decrease with increasing age at entry and with increasing policy duration.

3.6.2   Excess assured lives' mortality and smoking

Renshaw (1994a) has compared the mortality of assured lives classified as non-smoking and smoking, using deaths and exposures from policy counts in the two-year calendar period 1988–89, by modelling the recorded number of deaths as independent over-dispersed Poisson response variables in order to target the force of mortality μ_{ijkx}. In this study, the data are cross-classified by:

- gender ($i = 1$ female, $i = 2$ male)
- habit ($j = 1$ non-smoker, $j = 2$ smoker, $j = 3$ undifferentiated)

Table 3.5 Analysis of assured lives' mortality by age, gender and smoking habit (1988–89)

$$\hat{\alpha}_0 = -7.324\,(2.686 \times 10^{-1})$$

$\hat{\tau} = -5.031 \times 10^{-1}\,(8.794 \times 10^{-2})$	$\hat{\theta}_2 = 4.291 \times 10^{-2}\,(3.477 \times 10^{-3})$
$\hat{\psi}_2 = 6.456 \times 10^{-2}\,(6.418 \times 10^{-3})$	$\hat{\psi}_3 = 4.157 \times 10^{-2}\,(4.991 \times 10^{-3})$
$\hat{\beta} = 9.222 \times 10^{-2}\,(9.128 \times 10^{-3})$	$\hat{\gamma} = -2.819 \times 10^{-3}\,(3.019 \times 10^{-4})$

Deviance $= 332.52$ on 199 degrees of freedom from 206 observations, with scale parameter $\hat{\phi} = 1.671$.

- status ($k = 1$ medical, $k = 2$ non-medical, i.e. whether or not the life had a medical assessment at the time that the insurance policy commenced)
- age ($x = 11–15, 16–20, \ldots, 96–100$)

giving rise to $2 \times 3 \times 2 \times 18 = 216$ cells, a few of which are empty at the fringes of the data array. A thorough GLIM analysis of these data is supportive of the model structure

$$\log \mu_{ijkx} = \alpha + (\tau + \theta_i + \psi_j)x + \beta x^2 + \gamma x^3$$

comprising the log link in combination with a cubic in age effects and with two of the three factors impacting additively on just one of the coefficients. The effects of medical status are not statistically significant. Details of the fit are presented in Table 3.5. Comparisons age for age within gender reveal lower mortality for non-smokers compared with smokers, with the mortality of the undifferentiated category intermediate throughout. Further, while it is well established that mortality for females is lower than for males, age for age and allowing for comparable circumstances, here the predicted mortality for smoking females is higher than for non-smoking males, age for age.

3.6.3 Lapse rates

In life insurance, the premature termination of a policy is called a 'lapse'. It is important for an insurer to understand how the probability that a policy lapses varies across its portfolio (and through time). The Faculty of Actuaries Withdrawals Research Group collected data covering the lapse experience for the calendar year 1976 for seven insurance companies based in Scotland. The resulting policy lapse rates have been modelled using the empirical lapse rates as the binomial responses of a GLM to target the probability of a policy lapse: full details are given in Renshaw and Haberman (1986).

3.7   Premium rating

In non-life insurance, the premium charged to the policyholders can be broken down into the risk premium (representing the product of the claim frequency and the claim severity for each type of claim) and the components related to profit and the expenses of running the insurance company.

Two critical aspects of premium rating in non-life insurance involve the statistical modelling of claim frequency and claim severity patterns by rating factors (either existing or potential). We shall consider motor insurance and related data sets

$$n_u^t, e_u^t, x_u^t$$

for specific claim type t (e.g. insured vehicle, third party property or injury), where u denotes cross-classification by factors (e.g. policyholder's age, vehicle type, rating district) and where:

n_u^t = claim numbers;
e_u^t = exposure;
x_u^t = average claim size.

Then, given the expected claim rate λ_u^t and expected claim severity μ_u^t for each claim type t, it is possible to estimate the total risk premium

$$rp_u = \sum_t \lambda_u^t \mu_u^t$$

for each combination of rating factors u.

3.7.1 Modelling claim frequency

We consider a specific claim type t and denote the data by (n_u, e_u), comprising the claim numbers n_u accruing from matching exposures e_u, in a fixed observation window of typically 6 or 12 months duration. Then, a number of different possible modelling scenarios are of interest.

Foremost amongst these is the modelling of the claim numbers as independent Poisson responses N_u, for which

$$m_u = E(N_u) = e_u \lambda_u, \quad V(m_u) = m_u, \quad \omega_u = 1$$

with possibly a scale parameter $\phi > 1$ to represent within-cell heterogeneity. Used in combination with the log link function, for which

$$\eta_u = \log m_u = \log e_u + \log \lambda_u = \log e_u + \sum_j x_{uj} \beta_j$$

with offsets $\log e_u$, it gives rise to a multiplicative structure in the rating factors. A number of applications appear in the literature. Thus in marine insurance, McCullagh and Nelder (1983 and 1989), using data provided by the Lloyd's Register of Shipping concerning damage incidents caused to the forward section of cargo-carrying vessels, model the reported number of damage incidents classified by the three factors, namely ship type, year of construction and period of operation. Andrade e Silva (1989), Brockman and Wright (1992) and Boskov (1992) have each applied this same model to motor claims data using a variety of potential rating factors in the linear predictor. Centeno and Andrade e Silva (1991) discuss the case when there are certain fixed linear relationships between covariates in the predictor, while Stroinski and Currie (1989) discuss the selection of rating factors in automobile claim frequency modelling.

Historically, heterogeneity across risks has been modelled in the claim frequency setting by treating λ_u as a random variable. Let $Y_u = N_u/e_u$ for which $E(Y_u|\lambda_u) = \mathrm{Var}(Y_u|\lambda_u) = \lambda_u$, then

$$E(Y_u) = E\{E(Y_u|\lambda_u)\} = E(\lambda_u)$$

while

$$\mathrm{Var}(Y_u) = E\{\mathrm{Var}(Y_u|\lambda_u)\} + \mathrm{Var}\{E(Y_u|\lambda_u)\}$$

implies that

$$\mathrm{Var}(Y_u) = E(\lambda_u) + \mathrm{Var}(\lambda_u)$$

For the heterogeneous case, $\mathrm{Var}(Y_u) > E(Y_u)$ and the model has over-dispersion. If, in addition, λ_u has a gamma distribution with mean $E(\lambda_u)$ and variance $\tau\{E(\lambda_u)\}^2$, then, it is well known that the responses Y_u have the negative binomial distribution, for which

$$m_u = E(Y_u), \quad V(m_u) = m_u + \tau m_u^2, \quad \omega_u = e_u, \quad \phi = 1$$

Alternatively, if λ_u is assigned the inverse Gaussian distribution with mean $E(\lambda_u)$ and variance $\tau\{E(\lambda_u)\}^3$ for $\tau > 0$, the responses Y_u have the Poisson–inverse Gaussian distribution, for which

$$m_u = E(Y_u), \quad V(m_u) = m_u + \tau m_u^3, \quad \omega_u = e_u, \quad \phi = 1$$

Both cases revert to the Poisson distribution as $\tau \to 0$. Renshaw (1994b,c) has investigated both these cases in a motor insurance context.

The modelling of claim numbers as independent binomial responses N_u, for which

$$m_u = E(N_u) = e_u p_u, \quad V(m_u) = m_u\left(1 - \frac{m_u}{e_u}\right)$$

in combination with the log-odds link so that

$$\eta_u = \log\left(\frac{m_u}{e_u - m_u}\right) = \log\left(\frac{p_u}{1 - p_u}\right) = \sum_j x_{uj}\beta_j$$

to target p_u, the probability of a claim (or at least one claim), has received less attention in this context. An application of the targeting of the probability of at least one claim in the context of (Belgian) car insurance claims is given by Beirlant *et al.* (1991), while a number of researchers, including Coutts (1984), have used this predictor-link structure to target claim proportions but with estimation by weighted least squares.

3.7.2   Modelling claim severity

Again, we consider a specific claim type t and denote the data by (n_u, x_u), where x_u is the average claim size based on n_u claims, so that the total claim size is the product $n_u x_u$. It is well established that claim severity (or loss) distributions have positive support and are invariably positively skewed (Hogg and Klugman, 1984). Therefore, we consider the modelling of average claim sizes as the independent responses X_u of a GLM, with power variance function, subject to

$$\mu_u = E(X_u), \quad V(\mu_u) = \mu_u^\zeta(\zeta \geqslant 2), \quad \omega_u = n_u$$

and scale parameter ϕ. Jorgensen (1987) discusses the characteristics of the family of power variance distributions for general ζ, showing that, for $\zeta \geqslant 2$, the distribution has positive support and is positively skewed.

The case $\zeta = 2$, corresponding to a gamma claim severity distribution, is of special interest. Precedence for its use in this context is to be found in McCullagh and Nelder (1983, 1989) in which a re-analysis of the car insurance data of Baxter *et al.* (1980) is presented. There, the data are cross-classified by policyholder's age, car group and vehicle age, and the additive main effects in all three factors are established in the subsequent analysis of possible predictor structures η_u, using the reciprocal link function. This choice of link, a member of the parameterised family of power link functions:

$$\mu_u^{\gamma} = \eta_u$$

is justified on the basis of the deviance profile constructed by allowing for incremental changes in γ. Both Mack (1991) and Brockman and Wright (1992) make identical distributional assumptions but restrict the modelling to the log-link, the limiting form of the power link as γ tends to zero, in order to focus on a multiplicative structure. The detail is presented in terms of two rating factors so that $u \equiv (i, j)$ with predictor-link structure

$$\log \mu_{ij} = \mu + \alpha_i + \beta_j$$

The more general case of $\zeta \geqslant 2$ includes the inverse Gaussian distribution ($\zeta = 3$) and has the gamma distribution ($\zeta = 2$) as a limiting case. Coutts (1984) suggests the potential of these two specific cases in the claim severity modelling context. For the general case and a given predictor-link structure, the optimum value for $\zeta (\geqslant 2)$ is determined by scanning (minus twice) the extended quasi log-likelihood, namely

$$-2q^+ = \sum_{u=1}^{n} \frac{d_u}{\phi} + \sum_{u=1}^{n} \log\{\phi V(y_u)\}$$

for incremental changes in ζ. McCullagh and Nelder (1989) illustrate the extended quasi log-likelihood profile for the Baxter *et al.* (1980) car insurance data set, which is optimal in the vicinity of $\zeta = 2.4$. They also demonstrate how contour plots of the extended-quasi-likelihood determine the joint optimum position for the parameters (γ, ζ) when the parameterised power link function is used in combination with the parameterised power variance function.

Jorgensen and Paes de Souza (1994) model motor insurance claims data by focusing on the power variance function with $1 < \zeta < 2$ giving rise to continuous responses, with positive support, and a finite quantum of probability to allow for the possibility of a zero claim.

3.8 Claims reserving in non-life insurance

In insurance, the liability to pay a claim crystallises at the moment of time when the insured contingency occurs. But there are many factors which can lead to very considerable delays between occurrence and payment and final

Table 3.6 Incremental claims data (Source: Mack, T. (1994) *Historic Loss Development Study* (1991 edition) by Reinsurance Association of America, p. 96)

	$d=0$	$d=1$	$d=2$	$d=3$	$d=4$	$d=5$	$d=6$	$d=7$	$d=8$	$d=9$
$i=1$	5012	3257	2638	898	1734	2642	1828	599	54	172
$i=2$	106	4179	1111	5270	3116	1817	-103	673	535	
$i=3$	3410	5582	4881	2268	2594	3479	649	603		
$i=4$	5655	5900	4211	5500	2159	2658	984			
$i=5$	1092	8473	6271	6333	3786	225				
$i=6$	1513	4932	5257	1233	2917					
$i=7$	557	3463	6926	1368						
$i=8$	1351	5596	6165							
$i=9$	3133	2262								
$i=10$	2063									

settlement of a non-life insurance claim (for example, the time taken to decide that a liability exists, to determine the amount of damages and to process the claim and associated payments).

The prediction of outstanding claim amounts in non-life insurance is necessary to ensure that adequate reserves are maintained to meet future payments when delay is incurred in settling insurance claims. Typically, the format for the data is that of a triangle, see Table 3.6 for an illustration, in which the rows i $(= 1, 2, \ldots, r)$ denote accident years and the columns d $(= 0, 1, \ldots, c-1)$ delay or development years. The settlement or payment year is $k = i + d$. The entries in the body of the run-off triangle **D** are the (non-cumulative) claims amounts y_{id}, adjusted for both inflation and exposures where appropriate. The remit is to predict the future claims amounts in the incomplete south-east triangular region Δ. Numerous estimation and prediction methods have been proposed including some which fall within the GLM framework.

We define the responses of the GLM to be the set of augmented incremental claim amounts $\{Y_{id} : (i, d) \in \mathbf{D} \cup \Delta\}$ where $Y_{id} = y_{id}$ if $(i, d) \in \mathbf{D}$ and $Y_{id} = 0$ if $(i, d) \in \Delta$. Denote exposures, if present, by e_i. Then, Jung (1968) has investigated aspects of the claims reserving method based on Poisson responses

$$m_{id} = E(Y_{id}), \quad V(m_{id}) = m_{id}, \quad \omega_{id} = 1 \quad \text{if} \quad (i, d) \in \mathbf{D}, \quad \omega_{id} = 0 \quad \text{if} \quad (i, d) \in \Delta$$

and $\phi = 1$ in combination with the predictor-link

$$\eta_{id} = \log m_{id} = \log e_i + \mu + \alpha_i + \beta_d$$

while Mack (1991) discusses the equivalent case based on gamma responses, for which $V(m_{id}) = m_{id}^2$ with a scale parameter ϕ. Wright (1990) describes a claims reserving method with responses:

$$m_{id} = E(Y_{id}), \quad V(m_{id}) = m_{id}, \quad \omega_{id} = 1 \quad \text{if} \quad (i, d) \in \mathbf{D}, \quad \omega_{id} = 0 \quad \text{if} \quad (i, d) \in \Delta$$

and specially constructed non-constant scale parameters ϕ_{id}, in combination with the log link and a somewhat complex parameterised predictor structure which may be reparameterised (Renshaw, 1995) to read in essence as

$$\eta_{id} = \log m_{id} = \log e_i + \mu + \beta_{i1} + \beta_{i2} \log d + \beta_{i3} d$$

In addition, the Kalman-filter is applied by Wright (1990) to smooth the parameter estimates, while Renshaw (1995) describes how it is possible to generate the scale parameters ϕ_{id} using the joint GLM modelling technique (mentioned in Section 3.4.3).

3.9 Conclusions

In this chapter, we have attempted to demonstrate the versatility of generalised linear models (and the statistical package GLIM) for tackling a range of important, practical problems arising in actuarial science. We have considered the fitting of univariate and bivariate curves to hazard rates and transition intensities (Sections 3.4 and 3.5); multi-stage modelling to allow for varying numbers of insurance policies per person and the varying size of insurance cover (Sections 3.4.2 and 3.4.5); risk classification for life insurance applicants (Section 3.6); modelling of claim frequency and claim severity for the determination of premium rates (Section 3.7); and estimation of outstanding claims reserves (Section 3.8). For a more extensive review, readers are referred to Haberman and Renshaw (1996). We believe that this degree of versatility will mean that the collection of successful applications will be widened further in the future.

References

Andrade e Silva, J. M. (1989) An application of generalized linear modds to Portuguese motor insurance. *Proceedings of the XXI Astin Colloquium*, New York, pp. 633–649.

Baxter, L. A., Coutts, S. M. and Ross G. A. F. (1980) Applications of linear models in motor insurance. *Transactions of 21st International Congress of Actuaries*, **2**, 11–29.

Berry, G. (1983) The analysis of mortality by subject-years method. *Biometrics*, **39**, 173–184.

Beirlant, J., Derveaux, V., De Meyer, A. M., Goovaerts, M. J., Labie, E. and Maenhoudt, B. (1991) Statistical risk evaluation applied to (Belgian) car insurance. *Insurance: Mathematics and Economics*, **10**, 289–302.

Bennett, M. (1978) Models in motor insurance. *Journal of the Institute of Actuaries Students' Society*, **22**, 87–148.

Boskov, M. (1992) Private communications.

Bowers, N. L., Gerber, H. U., Jones, D., Hickman, J. C. and Nesbit, C. (1986) *Actuarial Mathematics*. Chicago: Society of Actuaries.

Brockman, M. J. and Wright, T. S. (1992) Statistical motor rating: making effective use of your data (with discussion). *Journal of the Institute of Actuaries*, **119**, 457–526.

Centeno, L. and Andrade e Silva, J. M. (1991) Generalized linear models under constraints. *Proceedings of the XXIII Astin Colloquium*, Stockholm.

Chadburn, R. G., Cooper, D. R. and Haberman, S. (1994) *Actuarial Mathematics*. Oxford: Institute and Faculty of Actuaries.

CMI Committee (1976) The graduation of pensioners' and of annuitants' mortality experience 1967–70, Continuous Mortality Investigation Report No. 2, p. 57. London: Institute and Faculty of Actuaries.

CMI Committee (1991) The analysis of permanent health insurance data. Continuous Mortality Investigation Report No. 12. London: Institute and Faculty of Actuaries.

Copas, J. B. and Haberman, S. (1983) Non-parametric graduation using kernel methods. *Journal of Institute of Actuaries*, **110**, 135–156.

Coutts, S. M. (1984) Motor insurance rating, an actuarial approach. *Journal of the Institute of Actuaries*, **111**, 87–148.

Cox, D. R. (1972) Regression models and life tables (with discussion). *Journal of the Royal Statistical Society Series B*, **34**, 187–220.

Currie, I. D. and Waters, H. R. (1991) On modelling select mortality. *Journal of the Institute of Actuaries*, **118**, 453–481.

England, P. D. (1993) Statistical modelling of excess mortality of medically impaired insured lives. PhD Thesis, City University, London.

England, P. D. and Haberman, S. (1993) A new approach to modelling excess mortality. *Journal of Actuarial Practice*, **1**, 85–117.

Francis, B., Green, M. and Payne, C. Eds. (1993) *The GLIM system, Release 4 Manual*. Oxford: Clarendon Press.

Forfar, D. O., McCutcheon, J. J. and Wilkie, A. D. (1988) On graduation by mathematical formula. *Journal of the Institute of Actuaries*, **115**, 1–135.

Gavin, J., Haberman, S. and Verrall, R. (1993) Moving weighted average graduation using kernel estimation. *Insurance: Mathematics and Economics*, **12**, 113–126.

Gavin, J., Haberman, S. and Verrall, R. (1994) On the choice of bandwidth for kernel graduation. *Journal of the Institute of Actuaries*, **121**, 119–134.

Gavin, J., Haberman, S. and Verrall, R. (1995) Variable kernel graduation with a boundary correction. *Transactions of the Society of Actuaries*, **47**, 173–209.

Grimes, T. (1971) Claim frequency analysis in motor insurance. *Journal of the Institute of Actuaries Students' Society*, **19**, 147–154.

Haberman, S. and Renshaw, A. E. (1990) Generalised linear models and excess mortality from peptic ulcers. *Insurance: Mathematics and Economics*, **9**, 21–32.

Haberman, S. and Renshaw, A. E. (1996) Generalised linear models and actuarial science. *The Statistician*, **45**, 407–436.

Hobcraft, J. B., Menken, J. and Preston, S. H. (1982) Age, period and cohort effects in demography: a review. *Population Index*, **48**, 4–43.

Hogg, R. V. and Klugman, S. A. (1984) *Loss Distributions*. New York: Wiley.

Johnson, P. D. and Hey, G. B. (1971) Statistical studies in motor insurance (with discussion). *Journal of the Institute of Actuaries*, **97**, 199–232.

Jorgensen, B. (1987) Exponential dispersion models (with discussion). *Journal of the Royal Statistical Society, Series B*, **49**, 127–162.

Jorgensen, B. and Paes de Souza, M. C. (1994) Fitting Tweedie's compound Poisson model to insurance claim data. *Scandinavian Actuarial Journal*, No. 1, 69–93.

Jung, J. (1968) On automobile insurance rate making. *Astin Bulletin*, **5**, 41–48.

London, D. (1985) *Graduation: The Revision of Estimates*. Winsted CT: Actex Publications.

Mack, T. (1991) A simple parameteric model for rating automobile insurance or estimating IBNR claims reserves. *Astin Bulletin*, **21**, 93–109.

Mack, T. (1994) Which stochastic model is underlying the chain ladder method? *Insurance: Mathematics, and Economics*, **15**, 133–138.

McCullagh, P. and Nelder, J. R. (1983) *Generalized Linear Models*. London: Chapman and Hall.

McCullagh, P. and Nelder, J. R. (1989) *Generalized Linear Models*. Second edition. London: Chapman and Hall.

Osmond, C. and Gardner, M. J. (1982) Age, period and cohort models applied to cancer mortality rates. *Statistics in Medicine*, **1**, 245–259.

Renshaw, A. E. (1988) Modelling excess mortality using GLIM. *Journal of the Institute of Actuaries*, **115**, 299–315.

Renshaw, A. E. (1991) Actuarial graduation practice and generalized linear and non-linear models. *Journal of the Institute of Actuaries*, **118**, 295–312.

Renshaw, A. E. (1992) Joint modelling for actuarial graduation and duplicate policies. *Journal of the Institute of Actuaries*, **119**, 69–85.

Renshaw, A. E. (1994a) A comparison between the mortality of non-smoking and smoking assured lives in the U.K. *Journal of the Institute of Actuaries*, **121**, 561–571.

Renshaw, A. E. (1994b) Modelling the claims process in the presence of covariates. *Astin Bulletin*, **24**, 265–285.

Renshaw, A. E. (1994c). A note on some practical aspects of modelling the claims process in the presence of covariates. *Proceedings of the XXV Astin Colloquium*, Cannes.

Renshaw, A. E. (1995) Claims reserving by joint modelling. *Proceedings of the XXVI Astin Colloquium*, Leuven.

Renshaw, A. E. and Haberman, S. (1986) Statistical analysis of life assurance lapses. *Journal of the Institute of Actuaries*, **112**, 459–497.

Renshaw, A. E. and Haberman, S. (1995) On the graduations associated with a multiple state model for permanent health insurance. *Insurance: Mathematics and Economics*, **17**, 1–17.

Renshaw, A. E. and Haberman, S. (1996) Dual modelling and select mortality. *Insurance: Mathematics and Economics*, **18**, 105–126.

Renshaw, A. E., Haberman, S. and Hatzopoulos, P. (1996) The modelling of recent mortality trends in U.K. male assured lives. *British Actuarial Journal*, **2**, 449–477.

Renshaw, A. E. and Hatzopoulos, P. (1996) On the graduation of 'amounts'. *British Actuarial Journal*, **2**, 185–205.

Royston, P. and Altman, D. G. (1994) Regression using fractional polynomials of continuous covariates: parsimonious modelling (with discussion). *Applied Statistics*, **43**, 429–467.

Stroinski, K. J. and Currie, I. D. (1989) Selection of variables for automobile insurance rating. *Insurance: Mathematics and Economics*, **8**, 35–46.

Wright, T. S. (1990) A stochastic method for claims reserving in general insurance. *Journal of the Institute of Actuaries*, **117**, 677–731.

Part II
CREDIT

4

Consumer Credit and Statistics

David J. Hand

4.1 Introduction

'Consumer credit' is a phrase used to describe the supply of goods or services to be paid for by an individual at some future time or times, along with interest. In its most basic form, 'credit scoring' is deciding where to put someone on a 'creditworthiness' scale – finding their 'score' on a latent variable or construct indicating the extent to which they can be trusted with credit. This scale can then be used to help decide whether or not to grant an applicant credit. This basic model has been extended in many directions, some of which we shall see below. First, however, by way of motivation for the remainder of this chapter, let us see what is happening to the credit industry.

To get a flavour for the figures involved, consider the following: in 1989, excluding mortgages, there were over 700 billion dollars of outstanding consumer credit in the US (Lewis, 1994); total UK consumer debt is about £500 billion; in 1994 in the UK some £36 billion of retail expenditure was made using credit cards; around 10 million British households currently have mortgages; credit card spending in the UK increased by 16% between January 1995 and January 1996; in December 1995, credit card borrowing in the UK increased by £276m, the largest increase recorded; in December 1995 the total amount of new consumer credit rose by £797m, the second largest monthly increase recorded. Walter Wriston nicely summed things up when he said, as far back as 1975: 'I see a trillion dollars in consumer savings out there, and I don't see any number that looks like that anywhere else' (Lawrence, 1984).

This credit comes from a wide variety of sources: mortgages, credit cards, bank loans, overdrafts, car finance schemes, retailer's cards, and so on, and this variety is indicative of what is happening in the industry. It is changing dramatically. Increased competition (at the time of writing, supermarket chains have just entered the financial services industry), rapidly growing databases holding details of all aspects of our lives, more demanding requirements from those seeking loans, the need for quicker loan decisions in a faster moving world, and new credit products and systems appearing daily, mean that the profile of the industry can change almost overnight. In such a market, speed is of the essence, and performance only a tiny bit better than a competitor's, when grossed up over the huge number of potential transactions, can mean the difference between profit and loss. Increasingly sophisticated mathematical and statistical models implemented as computer programs have become essential to the financial well-being of the various organisations involved. The numbers of records mean that the problems are intrinsically statistical, though often with a new slant – some of the data sets are vast.

Other changes are also occurring which interact with the above. For example, there is a significant shift towards the cashless society, with such innovations as the Mondex card, and the rate of unification of banks' activities is increasing, with the different databases (e.g. credit card usage, current account usage, mortgage, and so on) being combined.

In the light of all this, it is hardly surprising that the banks are tending to recruit more and more mathematically and statistically sophisticated staff to explore and develop credit scoring and credit control strategies. The aim of this chapter is to indicate some of the places where statistical models are used in such strategies.

4.2   The judgemental approach to granting credit

Up until a few decades ago, the decision as to whether or not to grant credit was made on the basis of personal knowledge. Success or failure in such an application would depend on how well the hopeful borrower was known to the potential lender, and how reliable the latter thought the former to be. In some situations this practice continues – for example in the form of a tab run up behind a bar. Over the course of time, however, as the trends noted in the opening section began to crystallise, so the practice was extended to a less personal, but still subjective judgemental assessment. This has been described as being based on the 'three Cs': *character* – how trustworthy the applicant is thought to be; *capacity* – how financially able they will be to repay (based on their income, job security, etc); and *collateral* – what resources they had that could, in the event of default, be claimed to pay off the loan.

From a modern perspective, the weaknesses of such a system will be evident. Primarily it is subjective: the loan decision is based on the opinion of the lender. Arbitrary and irrelevant prejudices can creep in. The decisions are not replicable and are subject to arbitrary outside influences: if the bank manager had had a row with his wife (and in those days bank managers were male) that morning then this may well have coloured his decisions. The decision making process could not be taught, except by an apprenticeship process. Certainly it could not be automated – so that it could not be applied to thousands, even millions, of transactions daily. And the process was slow. As we shall see, speed matters: a day's delay in reaching a decision may mean that the applicant will have gone elsewhere.

To overcome these weaknesses of the traditional judgemental approach, more formal and objective methods were needed. The human judge, with all his weaknesses and shortcomings, needed to be taken out of the loop.

4.3   What scoring can be used for

The motivation for credit scoring that we have provided above is couched in terms of identifying good and bad credit risks. However, this is a superficial conception of the objectives and it is worth spending some time examining more closely what we really want – or may really want – to do. Once again, the freedom to do this is only earned as a consequence of the scope given to us by the computer.

The ideal bottom line is that we should only grant credit to people who are going to be profitable for the company. 'Profitability', however, is a hard thing to gauge. It will depend on recruitment cost, the workload the customer generates, the interest rate charged, and the repayment period, as well as the customer's default risk. A good risk, in terms of default probability, may not be profitable (because, for example, they pay off their credit card bill fully each month, so incurring no interest charges). Conversely, a bad risk may be profitable, provided they have paid enough before they default. Default risk is thus just one aspect of profitability. However, it is a very important one, and often the main one on which a loan decision is taken. (The role of other variables is growing as more sophisticated computer models become possible.)

Even if risk is taken as the key measure, we have to decide how much risk is acceptable. Zero risk can be achieved – simply by accepting no customers – but this is not an acceptable option. There are many strategies which might be followed – and this is one reason why the market is so complex. Many organisations – especially the ones with larger market penetration – simply seek to accept the lower risk applicants. Other organisations aim at niche marketing, targeting the higher risk groups but setting their interest rates so that the higher default loss is covered. At the opposite extreme from accepting no-one we have the extreme of accepting everyone, with an interest rate appropriate for their estimated default risk or with a common interest rate which yields profit on average. But accepting everyone, with a common interest rate sufficient to ensure a profit overall, is not without its risks. Once the policy becomes known, one might reasonably expect more poor risk applicants to seek loans. This is an illustration of a more general phenomenon termed *population drift*, which describes how the population of applicants evolves over time. We have more to say about it below. Different organisations also target different sectors of the market, making loans for different purposes or with different repayment structures.

Even deciding on the definition of what should be regarded as a good or bad risk may be far from straightforward. It will, of course, depend on the nature of the loan: a credit card, an overdraft, a mortgage, or whatever. The definition may be based on slow repayments (but is one month overdue to be regarded as 'bad' or should it be two, or . . . ?), a combination of account balance below some level throughout the month and overdraft limit exceeded at some point, or some more sophisticated combination. Inevitably there is some arbitrariness in such definitions. There are underlying continua (of profitability or default risk, for example) and the lender is trying to split these at appropriate points – but the appropriate points change with changing circumstances.

The most common use of credit scoring is to decide whether or not to grant an applicant a loan – this is termed 'application scoring'. But predictive models can also be built for other purposes. 'Behavioural scoring' describes the process of constructing a predictive model for someone who has already been given a loan: to predict such things as imminent default or whether they should have their credit limits changed. In general, behavioural scores are more accurate predictors than applicant scores. Scoring systems have also been constructed for fraud detection (e.g. Leonard, 1993). For example, with credit cards one may seek unusual expenditure patterns, such as a large number of purchases in a short time, or have flags set by certain classes of purchases: those which are easy to dispose of, such as jewellery and electrical goods, are higher risk. If a card has not been used for a long time it may indicate that it has been lost. Yet another usage is to guide the collections strategy. If payments are overdue, a series of actions, of gradually increasing severity will take place, such as overprinting the regular statement, sending a mild reminder letter, a strong letter, telephoning, all the way up to legal action. An area which has been less explored, but which has considerable potential, especially with the growth of large and very detailed databases, is in marketing – identifying those potential customers who are more likely to take up an offer.

From a statistician's perspective, all of the above are merely predictive models, for which a large range of statistical techniques exist. Thus, the scope for statistical work is huge. Moreover, as we shall see below, simple predictive models do not exhaust the possibilities. As the complexity of the loan products grow, as the available information increases, as the need to have a slight edge becomes more important, and as the need to extend and deepen the relationship with the customer increases in value, so the opportunity for applying increasingly sophisticated statistical models becomes greater and greater.

4.4   What is a scorecard and how to construct one

Within the credit industry, the term 'scorecard' is often used to describe the numerical recipe for assigning a creditworthiness score to an applicant. In its most elementary form, for each applicant a scorecard assigns a numerical (usually integer) value to each level (the levels are called 'attributes' in the credit scoring industry) of each predictor variable (the variables are called 'characteristics'). These are added to produce an overall score signifying the 'creditworthiness' of the applicant. Note that the creditworthiness of an individual is not immutable – it can and does change in response to changing circumstances. This opens up possibilities for more sophisticated statistical modelling which are only beginning to be explored.

To produce the numbers, a whole range of statistical techniques has been investigated. Detailed discussions are given in the Chapter 5 of this book by Lyn Thomas and in Rosenberg and Gleit (1994) and Hand and Henley (1997). The methods used include linear regression, logistic regression, linear and quadratic discriminant analysis, recursive partitioning methods, neural networks, nearest neighbour methods, and others. Because of the competitiveness of the commercial environment, even a slight hint that some method might outperform others attracts attention.

Apart from mere predictive and classificatory performance, other aspects of the underlying problem need to be taken into account – and these impose constraints on the optimality of the statistical solution. Sometimes the numerical values assigned to the attributes may appear unreasonable (perhaps having an increasing and then decreasing relationship with ordered attributes of the characteristic). In this case, to preserve the common-sense interpretation, there may be advantages in adopting an apparently sub-optimal set of values. This sort of situation may also arise for legal reasons: in the United States the age of elderly people may not be used as a negative or derogatory characteristic.

It is clear that 'optimality' here has a more subtle and complex interpretation than merely that the solution minimises or maximises some simple well-defined statistical measure of fit, such as a sum of squared deviations. In general, small improvements may be swamped by the exigencies of reality. For example, optimising a prediction method on current data is all very well, but things, including the type of people who will apply for loans, change over time – as discussed in Section 4.7 below. Moreover, even if a small apparent improvement in predictive performance could be achieved in practice, it may not be commercially worthwhile. In an increasingly competitive environment, the limited lifetime of a financial product may mean that the cost of changing a method of scoring is not worthwhile – the costs will not be recouped before the entire system is replaced. Moreover, credit scoring models have been built, refined, and extended for some decades now so that highly sophisticated versions have been built. Workers in the credit scoring industry know their data well. It would be surprising, to say the least, if a new kind of predictive model were to come along which significantly outperformed existing methods. Real improvements are likely to come from extending the models and problems that are examined. There is plenty of scope for this as new types of loan instrument continue to appear on the market.

Not all statisticians have appreciated these points. To quote Lawrence (1984, p. 55) 'We have had poor results when we relied on outside "professional" mathematicians, typically academics not associated with a specialist firm, to develop scoring systems. While they are steeped in statistical mumbo-jumbo, these mathematicians tend to have little knowledge of the underlying business concepts of scoring. In one business, we waited more than 20 months for a professor of statistics to come up with the "Cadillac" of scoring systems, while all the business needed was a "Chevrolet" that would work.'

4.5   The data

Predictor variables for credit scoring models come from application forms, records of past behaviour (so an exemplary past financial life may imply difficulty in getting a loan if there is no previous track record of repayments), and perhaps from credit reference bureaux (see below). They include such things as time at present address, postcode, annual income, age, occupation, educational level, and so on. Most are categorical and standard practice is to categorise those (such as age) which are not, into a relatively few categories. Obviously the items will differ from product to product (the questions appropriate to a large mortgage will differ from those for a small hire purchase) and targeted population to population, but there are also national differences. For example, Lewis (1994) remarks that, in Germany, applicants may be asked if they have completed their military service, in Italy the province of birth and whether the marital contract calls for community property, in France and Belgium the country of origin, and in Japan whether the employer is publicly traded and its number of employees.

Data for credit scoring usually have the common characteristic of multivariate data, that there are missing values. Such values may be structurally missing (for example, questions which are only asked conditionally on the responses to previous questions) or randomly missing. For example, in one data set the author analysed, there were 3883 applicants in the design set with values recorded for 25 characteristics. Of these 25 characteristics, just 5 had no missing values and 2 had over 2000 missing values. Figure 4.1 shows the distribution of the numbers of missing values over the cases. It is worth noting that the fact that a value is missing may contain valuable predictive information, so it may be useful to regard missing as another category.

One of the anxieties often voiced by the public is that formal credit scoring systems are too inflexible. To allow for this, many systems permit unusual cases (for example, a lawyer with a good credit record who has just been sent to prison) to be flagged and a personal decision to be made. Sometimes this

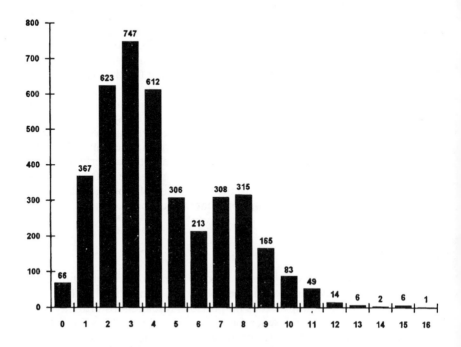

Figure 4.1 Numbers of missing values in 3883 applications for home improvement loans. There were 25 characteristics altogether. Only 66 applications had no missing values. One had 16.

a loan being withheld despite a score above the threshold. Such decisions are called *overrides*. The former is a *low-side override* and the latter is a *high-side override*. Caution has to be exercised to ensure that the scoring system is not ignored too often, or its predicted performance will suffer. *Policy overrides* describe a company's policy to make certain kinds of decisions even though the raw score may not justify it.

4.6   Reject inference

Statistical inference from a sample to a population is based on the assumption that the sample has been drawn in a random way in some sense. Without this condition being satisfied, no statistical inference can be made. In particular, it is assumed that the sample is not distorted and does not over-represent or under-represent parts of the population. Unfortunately, in the real world of consumer credit, this ideal seldom holds.

Consider, for example, personal loans. First, the population of potential borrowers has to be identified and sent an invitation to apply for a loan. A self-selecting subset of this population will respond to the invitation, completing an application form. The bank will score these applications and offer loans to some of them. Of the people offered a loan, only some will take it up. And, of those who take it up, some will default on their repayments, some will repay early (which may well mean a loss of interest revenue to the bank), and some will pay off according to the original plan. With such a complex selection process as this, it is clearly very important to be sure of the population to which the inference is to be made.

A similar problem of sample selection arises when new application scorecards are to be developed. Only very rarely will the available data be a random sample from the intended population. It is much more typical (for obvious reasons) that the available data will describe those people who were scored as acceptable in an earlier application process, using an already existing scorecard. The 'bad risks' will be underrepresented.

This problem is widely recognised within the industry, and many attempts to adjust the model by taking into account those applicants rejected by the previous scorecard have been explored. These attempts are collectively termed *reject inference*. Effective reject inference requires information about the way in which the previous reject decision was made. Unfortunately, much of the work in the industry has not recognised this and demonstrably ineffective models have sometimes been adopted. Further discussion of reject inference is given in Hand and Henley (1993/4) and Joanes (1993).

4.7   Population drift

Scorecards are built using records of past applicants and their subsequent behaviour. However, the objective is to predict the behaviour of new or future applicants. Unfortunately the distribution of applicants is unlikely to remain static. Changes arise from changing economic climate, changing population demographic characteristics (for example, the growth in one parent families and the increase in unmarried cohabiting couples), changing lifestyles due to technological progress (the increase in the number of telephones and cars), the dramatic increase in the number of people studying for university degrees in the UK, marketing campaigns bringing in different kinds of people, and other reasons. All of these influences lead to a degradation in predictive performance of a scorecard built on an earlier dataset. To allow for this, either the scorecard must be replaced, or it must be dynamically updated.

Replacement is expensive and one would want to avoid unnecessary replacement. It is necessary to monitor the population, and change the scorecard when some tracking measure of difference from the original development sample has been exceeded. Such population stability analysis can be based on a comparison of summary statistics of new applicants with the earlier ones, or can be based on performance statistics: when the scorecard performance degrades below a certain level it is replaced. In general, scorecards and other predictive models yield a creditworthiness continuum, and an applicant's score on this continuum is compared with some threshold to yield a decision. In practice, changes of the threshold alone are often sufficient, and these will be made before going to the expense of completely revising the scorecard.

To test whether to install a new scorecard (or, more generally, change the system in some other way) the old and new methods are run in parallel. Typically, the new method is applied to a randomly selected subset of applicants, the proportion selected being chosen on the basis of the perceived risk of the new approach relative to the existing one. This strategy, termed the 'Champion–Challenger strategy', can be extended to simultaneous comparison of more than two methods.

Bazley (1992) has argued for dynamic updating methods, and these have been explored to a limited extent. For example, the nearest neighbour method (Devijver and Kittler, 1982; Hand, 1997) permits earlier applicants to be removed from the database as new ones arrive, without requiring any re-calculations of the model.

4.8   Reject option

A commonly employed strategy for selecting customers in the credit industry is what is technically known as the *reject option*. This, a very different concept from reject inference above, makes the extent of investigation of an applicant depend on the degree of uncertainty about the decision. Applicants who are clearly good or bad risks can be dealt with quickly, with a minimum of investigation. These will be accepted or rejected on the basis of a simple questionnaire (for example, in the case of mail order, perhaps on the strength of their responses to the items on a short application form printed in a magazine). Applicants about whom the short questionnaire does not permit a confident decision require more careful consideration. These may be asked for more details, or perhaps may be checked out with a credit reference agency. (They are 'rejected' from the initial classification – hence the term 'reject option'.) By varying the proportion of applicants for whom the decision is deferred, one can control the classification errors in the selection. But, of course, this has to be balanced against the speed and ease of obtaining the credit by the customer. It is possible that an applicant, frustrated by having to complete yet another form, may go elsewhere. Yet again the simplistic statistical model has a great deal of scope for expansion and improved accuracy. Relatively little formal work has been carried out within the credit industry on the best divisions when implementing the reject option (partly because of difficulties such as the problem of defining profitability, noted above), but the statistical literature includes some detailed work (especially with nearest neighbour methods – see, for example, Devijver and Kittler, 1982).

4.9 Assessing performance

Assessing the performance of a scorecard is clearly of central importance. Not only will this allow us to choose between alternative scorecards, but it will also allow us to see if absolute performance is good enough and if performance degrades as time passes.

Essentially, there are two possible objectives. One may simply seek to compare the distribution of good risks with the distribution of bad risks – without imposing a level of risk below which an applicant is to be rejected. Or one may adopt a particular threshold and assess the resulting classification. The first of these allows an overall statement of how well the scorecard separates the two risk classes. Since the threshold risk level may be changed (see Section 4.7 above), this provides an overall measure which one may hope will be valid fairly generally. There is always a chance, however, that this general measure will be at odds with the performance of the classification induced when any particular threshold is chosen. Hence, if one is confident that a chosen threshold is a reasonable one, the second approach can be adopted, and performance assessed using measures of the proportions of good risks misclassified as bads and vice versa.

Examples of general measures, in the first class, are those based on the chance that a randomly selected good risk applicant will have a score higher than a randomly selected bad risk applicant. One such measure is the *Gini coefficient*. This has a convenient representation in terms of the area under a curve showing the proportion of good risks accepted plotted against the proportion of bad risks accepted, as the threshold risk level for acceptance varies. Such curves are called ROC curves or Lorenz diagrams. Details are given in Wilkie (1992) and Hand (1997).

The Gini coefficient does not take into account the relative sizes (the priors) of the good and bad risk classes, but simply compares the shapes of the distributions. A measure which does take this into account is the difference between the average estimated probability of belonging to the good class for applicants who did turn out to be good and the average estimated probability of belonging to the good class for applicants who turned out to be bad. This, of course, can be standardised in various ways.

Examples of measures falling into the second class, when an acceptance/rejection threshold has been chosen, are such things as overall misclassification rate, proportion of those accepted who turn out to be bad risks, proportion of bad risks who get accepted, and so on. One complication is that, if performance is to be assessed using data collected by applying the scorecard, normally only those scored as 'accepts' will have their true class known (see Section 4.6). Thus, one might focus one's attention solely on measures of bad rate amongst accepts. Again these sorts of issues are discussed in Hand (1997).

4.10   Legal issues

In most countries there are stringent legal restrictions on what may be used as a basis from which to obtain a credit score. Examples of such restrictions are the Equal Credit Opportunity Act of 1974 in the USA (which forbids discrimination on grounds of race, colour, religion, national origin, sex, marital status, or age) and the Consumer Credit Act (1974) in the UK. There is also often a legal obligation to give some explanation of the reason when credit is declined. In addition to such legislation aimed directly at the credit scoring process, much other legislation has bearing on credit. For example, in the UK, the Sex Discrimination Act, the Race Relations Act, the Fair Trading Acts, the Unfair Contracts Terms Act, and the Data Protection Act all have relevance. Statistical scoring and classification systems all have to work within the boundaries defined by such legislation. The problems are thus not merely ones of mathematical optimisation, and neither are the solutions entirely clear cut. There is, for example, an argument that if a scorecard is produced by objective statistical methods from a data set of known good/bad risks, and with all variables which might be related to risk used, then it cannot be discriminatory – by definition it will only use information which is relevant to the risk and not 'information' based on arbitrary prejudice. If one requires that, for non-statistical (legal) reasons, some variables which are relevant to risk should not be used, then inevitably discrimination on grounds not related to risk is introduced. (A lower risk group which would have been identified as such will now not be so identified. Indeed there is some evidence that current legislation discriminates against females by virtue of this effect, see Chandler and Ewert, 1976.) The only way around this is to include all variables which might have relevance and let the statistical method decide whether or not to include them. Of course, if on the other hand the aim is to ensure that certain groups have the same proportion of granted loans, regardless of their true default risk (that is, if 'non-discrimination' is defined in this way), then this can be achieved by building separate scorecards for the two groups and classifying the same proportions as bad risks.

Constraints on the information which may be used in constructing a credit classification rule are one class of legislative restrictions. Another relates to legal requirements for credit reference agencies to divulge information on individuals to those individuals on request. Credit reference agencies collect repayment behaviour information on individuals, pooling the information from a wide variety of banks and other financial organisations. Most adults in Britain have records with these bureaux. A brief extract from Section 159 of the Consumer Credit Act (1974) indicates the sorts of rights consumers have with reference to the files describing them:

2. If you think that anything in the file is wrong and you are likely to suffer as a result, you have the following rights.

3. If you think there is no basis at all for the entry, you may write to the agency requiring it to remove the entry.

4. If the entry is incorrect you may write to the agency requiring it to remove or amend the entry.

A leaflet explaining in laymen's terms how credit ratings in the UK work, and consumers' rights in relation to them (including how to get mistakes corrected) can be obtained from: Consumer Affairs Department, CCN Credit Systems, PO Box 40, Nottingham NG7 2SS, UK. The Office of Fair Trading has also published a leaflet (*'No Credit'*) explaining how to correct wrong information on a credit reference agency's files. This leaflet can be obtained from: OFT, PO Box 2, Central Way, Feltham, Middlesex, TW14 0TG, UK.

4.11 Consumer versus corporate loan

All of the above is concerned with consumer loans. A few words about corporate loans also seem appropriate. In general, objective scoring has been slower to take off in the corporate sector, and there are sound reasons for this. Statistical applications in the consumer sector have population size on their side – there are many more applications for consumer loans than for corporate loans. Also, consumer loans are typically for relatively small amounts, whereas corporate loans are for larger amounts. The size of a corporate loan will determine who approves it, with more senior managers being required for larger loans. In contrast, with consumer loans, the more senior managers will not be concerned with individual loans, but with developing and implementing new *ways* to lend money – new *products*. For example, a senior manager will approve a new car finance scheme, a new credit card, convenience cards, a second mortgage strategy, a system of personal loans, a pre-screened loan offer, and so on.

The information available for decision making in the two domains also differs. The decision about whether to grant a corporate loan will be based on the company's financial statement and business plan. In contrast, consumer loan decisions are based on demographic characteristics of the applicant along with any available information about past repayment behaviour. The review and monitoring of corporate loans, like the initial granting decision, tends to be on an individual basis. In contrast, as we have seen above, for consumer loans, threshold criteria are established using statistical methods and the monitoring simply checks to see if the thresholds are breached.

Despite these differences, statistical models are being built for the corporate sector and it is a ripe area for future investigations.

4.12 Conclusions

Statistical methods are widely applied in the consumer credit industry but the level of technical sophistication is not great. There is scope for refining the models and, more importantly, for developing and implementing more sophisticated statistical models which better match the problems. This is becoming more urgent as more complex financial products appear in the consumer credit market. Interestingly enough, in contrast to what this might seem to imply, the banks and other lenders do make an effort to keep abreast of new techniques, although the awareness is patchy. Thus, neural networks, expert systems, and genetic algorithms have all been the focus of interest, while less glamorous sounding, but arguably more pertinent techniques have attracted less interest.

It is essential that any statistical development should take place in the environment of the problem it is intended to address. Statistical tools and models developed without close contact between those developing them and those who need them to solve real problems are unlikely to be of value. The quotation at the end of Section 4.4 illustrates the dangers. The point is also illustrated in Hand and Henley (1997).

Although most emphasis has been on scoring and classification models, other classes of statistical technique have been used. For example, survival analysis has been proposed to model the proportion of applicants remaining 'good' as time passes (Narain, 1992). (Note, however, that the simple survival model has to be modified to allow for a proportion who will not go bad, no matter how long they remain in the system.) Conditional independence graphs have been proposed as a holistic model of applicants (Hand *et al.*, 1997). Markov chain models have been used to model the way customers move between levels of repayment status (for example, Cyert *et al.*, 1962; van Kuelen *et al.*, 1981; and Frydman *et al.*, 1985). Cluster analysis is a common technique for segmenting markets, but it has also been proposed for other uses, such as segmenting applicants for credit (Lundy, 1992) and to group month-to-month transitions (Edelman, 1992).

It will be clear from what has been said above that developing a credit scoring system involves input from a wide variety of people with special skills and knowledge. Wilkinson (1992) lists the steps to be taken. Collections of papers on credit scoring and credit control can be found in occasional issues of the *IMA Journal of Mathematics Applied in Business and Industry* as well as in Thomas *et al.* (1992) (these are selections from the papers presented at a regular conference on credit scoring and credit control which is held in Edinburgh, UK).

References

Bazley, G. (1992) Profit by the score. In *Credit Scoring and Credit Control*, L. C. Thomas, J. N. Crook and D. B. Edelman (Eds), Oxford: Clarendon Press, pp. 204–208.

Chandler, G. G. and Ewert, D. C. (1976) Discrimination on basis of sex under the Equal Credit Opportunity Act. Credit Research Centre, Purdue University, W. Layfayette, Indiana.

Cyert, R. M., Davidson, H. J. and Thompson, G. L. (1962) Estimation of the allowance for doubtful accounts by Markov chains. *Management Science*, April, 287–303.

Devijver, P. A. and Kittler, J. (1982) *Pattern Recognition: a Statistical Approach*. Englewood Cliffs, New Jersey: Prentice Hall.

Edelman, D. B. (1992) An application of cluster analysis in credit control. *IMA Journal of Mathematics Applied in Business and Industry*, **4**, 81–87.

Frydman, H., Kallberg, J. G. and Kao, D.-L. (1985) Testing the adequacy of Markov chains and mover–stayer models as representations of credit behaviour. *Operations Research*, **33**, 1203–1214.

Hand, D. J. (1997) *Construction and Assessment of Classification Rules*. Chichester: John Wiley and Sons.

Hand, D. J. and Henley, W. E. (1993/4) Can reject inference ever work? *IMA Journal of Mathematics Applied in Business and Industry*, **5**, 45–55.

Hand, D. J. and Henley, W. E. (1997) Statistical classification methods in consumer credit scoring: a review. *Journal of the Royal Statistical Society, Series A*, **160**, 523–541.

Hand, D. J., McConway, K. J. and Stanghellini, E. (1997) Graphical models of applicants for credit. *IMA Journal of Mathematics Applied in Business and Industry*, **8**, 143–155.

Joanes, D. N. (1993) Reject inference applied to logistic regression for credit scoring. *IMA Journal of Mathematics Applied in Business and Industry*, **5**, 35–43.

Lawrence, D. B. (1984) *Risk and Reward: the Craft of Consumer Lending*. New York: Citicorp.

Leonard, K. J. (1993) A fraud-alert model for credit cards during the authorization process. *IMA Journal of Mathematics Applied in Business and Industry*, **5**, 57–62.

Lewis, E. M. (1994) *An Introduction to Credit Scoring*. San Rafael, California: Athena Press.

Lundy, M. (1992) Cluster analysis in credit scoring. In *Credit Scoring and Credit Control*. L. C. Thomas, J. N. Crook and D. B. Edelman (Eds), Oxford: Clarendon Press, pp. 91–107.

Narain, B. (1992) Survival analysis and the credit granting decision. In *Credit Scoring and Credit Control*. L. C. Thomas, J. N. Crook and D. B. Edelman (Eds), Oxford: Clarendon Press, pp. 109–122.

Rosenberg, E. and Gleit, A. (1994) Quantitative methods in credit management: a survey. *Operations Research*, **42**, 589–613.

Thomas, L. C., Crook, J. N. and Edelman, D. B. (Eds) (1992) *Credit Scoring and Credit Control*. Oxford: Clarendon Press.

van Kuelen, J. A. M., Spronk, J. and Corcoran, A. W. (1981) On the Cyert–Davidson–Thompson doubtful accounts model. *Management Science*, **27**, 108–112.

Wilkie, A. D. (1992) Measures for comparing scoring systems. In *Credit Scoring and Credit Control*, L. C. Thomas, J. N. Crook and D. B. Edelman (Eds), Oxford: Clarendon Press, pp. 123–138.

Wilkinson, G. (1992) How credit scoring really works. In *Credit Scoring and Credit Control*, L. C. Thomas, J. N. Crook and D. B. Edelman (Eds), Oxford: Clarendon Press, pp. 141–159.

5

Methodologies for Classifying Applicants for Credit

Lyn C. Thomas

5.1 History of credit scoring

There are two types of decision situations that credit granters have to face. The first is whether to grant credit to a new applicant and the second is how to adjust the credit restrictions or the marketing effort directed at a current customer. Techniques that support the first type of decision are called *credit scoring*, while techniques that support the second type of decision – whether it be adjusting the credit limit, determining what debt recovery procedure to implement or which other product to cross-sell – are known as *behavioural scoring*. The situations are different because of the extra information available in the second case, namely the repayment and ordering history of the customer. This chapter will concentrate on classification techniques as they are used in the first of these situations – the credit scoring techniques.

The information available to help decide whether to grant credit to a new applicant comes from various sources. First, there is the information that the applicant gives on the application form. This information can include details on family, occupation, residence, banking relationships as well as some details of major sources of income and expenditure. Information can also be obtained from a credit reference bureau which, as well as confirming residence and electoral roll information, will record any county court judgments for default and information on other instances of default which have been identified by the bureau. Most important, there are the application form details of all previous applicants to the lender and the credit histories with the credit lender of those who were granted credit. From this the lender can determine for each previous customer whether the customers' repayment performance was satisfactory or not. Having classified previous customers in this way the aim of credit scoring is to derive a classification rule on a sample of the previous customers which can then be used to determine to which of the new applicants to give credit. Those classified as good are given credit and those classified as bad are rejected. Some authors (Hand, 1981) refer to the process of deriving the classification rule as *discrimination* and the process of applying it to new applicants as *classification*, but in this context credit scoring encompasses both processes.

Historically, credit lending decisions were a matter of the subjective judgment of credit granters based on properties of the applicant. Usually these properties were some subset of the five Cs – capital, collateral, character, capacity and condition. The first credit scoring system was developed by Durand (1941) for the US National Bureau of Economic Research to investigate instalment loans made by 37 firms. This idea was taken up by mail order companies during the 1950s as is described in Lewis's book (Lewis, 1992), but as Johnson (1992) indicates it was the advent of credit cards and the Equal Credit Opportunities Act (ECOA) of 1974 and 1976 that really gave the impetus to credit scoring. The former meant that the demand for credit cards far outstripped the ability of the industry to train and hire personnel to screen applicants, while the ECOA only allowed discrimination in the granting of credit if it could be statistically validated – essentially outlawing judgemental approaches to the process. Despite the criticisms levelled at credit scoring, as in Capon (1982), it has proved so much more successful than the previous approaches that it has become the standard decision support aid in almost all areas of credit granting now.

5.2   Credit scoring: the art and the objective

The outcome of the credit granting process is the choice between two actions – grant credit to the applicant or refuse them credit. Thus, it is natural to assume that the credit scoring system seeks to classify the applicants into two groups – one group of those who will be accepted and the other of those to be rejected. Sometimes there is a claim that there are other groups, such as those on whom further information is required or those who must be handled manually. However, these categories are more to do with the ways the organisations wish to organise their decision making, than being real final categories. It is odd to try and classify which applicants will be in a particular group when the decision of whether they are in that group or not is wholly that of the organisation. Thus it seems sensible to classify into two groups only: 'goods' and 'bads'.

If A is the set of all possible application form data, we want to split A into two subsets A_G and A_B so that classifying applicants in A_G as 'good' and those in A_B as 'bad' minimises errors. There are several ways of defining the error in this case. Let

L be the average lost profit of classifying a 'good' as a 'bad';
D be the average debt incurred by classifying a 'bad' as a 'good';
p_G (p_B) is the proportion of 'goods' ('bads') in the applicant population;
$p(x \mid G)(p(x \mid B))$ is the density function that a good (bad) will give application data x;
$q(G \mid x) \propto p(x \mid G)p_G$ is the probability that someone with application data x is a 'good'.

One way of minimising errors is to minimise the expected loss per application, which is given by

$$L \int_{x \in A_B} p(x \mid G) p_G \, dx + D \int_{x \in A_G} p(x \mid B) p_B \, dx$$

$$= L \int_{x \in A_B} q(G \mid x) \, dx + D \int_{x \in A_G} q(B \mid x) \, dx \quad (5.1)$$

This is a slight approximation because, in reality, the lost profit L and the debt D may vary from applicant to applicant whereas, in Equation (5.1), the average values are taken. Since $p(x \mid G) p_G / p(x \mid B) p_B = q(G \mid x)/q(B \mid x)$, the expected loss is minimised by the rule

$$A_B = \{ x \mid q(G \mid x)/q(B \mid x) < D/L \} \quad (5.2)$$

If $p(x \mid G)$ and $p(x \mid B)$ are multivariate normal distributions this rule translates into a linear rule $x.w < c$ $(x_1 w_1 + x_2 w_2 + \ldots x_n w_n < c$ where x_i is the ith component of the application data). In most classification problems one seeks to minimise the expected classification error which is the special case of (1) when $D = L = 1$. This seems inappropriate in this context since the cost of D, the defaulting error is much greater than L, the loss of profit error. In reality it is difficult to put a figure on D and especially L. Thus, the usual approach is to try and set one of the errors at a fixed value and minimise the other error. Usually firms decide on a given rate of acceptance of applicants, say a, so that one requires

$$\int_{A_G} (p(x \mid G) p_G + p(x \mid B) p_B) \, dx = a \quad (5.3)$$

and seeks to minimise the default rate $\int_{x \in A_G} p(x \mid B) p_B \, dx / a$. This can be achieved by a suitable choice of L and D in Equation (5.1) (equivalent to the Lagrange multiplier approach to constrained optimisation) and so the same optimal rule applies.

Fisher (1936) sought to find the linear surface that best separates two groups – the 'goods' and 'bads' in our case – by finding the direction that maximises the distance between the group means using the within group variances in that direction as the unit of measurement. Defining m_G, m_B to be the group means and S to be the (common) covariance matrix, then if x is a unit directional vector, the distance between the means in that direction is $x'(m_G - m_B)$. The standard deviation within each group in the x-direction is $(x'Sx)^{1/2}$ and hence Fisher's criterion seeks to maximise $x'(m_G - m_B)/(x'Sx)^{1/2}$. Differentiating with respect to x and setting the derivative equal to zero gives the necessary conditions that x must satisfy for the separation to be a maximum, namely

$$m_G - m_B = x'(m_G - m_B) Sx/(x'Sx)^{1/2} = \lambda Sx \quad (5.4)$$

where the point is that $\lambda = x'(m_G - m_B)/(x'Sx)^{1/2}$ is a scalar. Hence the optimal x is given by

$$x = \lambda^{-1} S^{-1} (m_G - m_B) \quad (5.5)$$

which is the Fisher linear discriminant function. Thus, this linear scorecard has appeared without any assumption of the form of the application data distribution. However, in the case of a multinormal distribution this is the same function as obtained via (5.2).

This underlying theoretical support might suggest that developing a scorecard is just a routine application of Fisher's discriminant analysis but there are a number of difficulties that immediately arise. First, what variables or characteristics should be used in the classification? These correspond to the questions asked on the application form, the publicly available data on the applicant and the credit reference bureau information on the applicant. There are often more than 100 such characteristics. This is too many both because it leads to collinearity and dependency among the variables with a consequent lack of robustness in the coefficients of the linear function obtained and also because it becomes difficult to explain to the users what is going on. Thus, part of the art of developing a credit scorecard (the linear classifying function) is to identify which variables to use and which to ignore. This can be done using a mixture of approaches – experience of what proved important in previous similar credit granting situations, stepwise applications of the discrimination techniques and weights of evidence. Stepwise discrimination can start by only having the variable that gives the biggest separation between the group means in the discriminant function and then adding one variable at a time. Each time choose the variables that has the greatest additional effect on this separation until the increase is below some prespecified limit. Alternatively, stepwise discrimination starts with a function involving all the variables and drops out one by one those that have the smallest impact on the distance between the means until again one comes to variables whose elimination will cause a change greater than some prespecified limit.

'Weights of evidence' is related to the logarithm of the likelihood ratios of goods to bads at each answer. If a variable has n possible values $i = 1, 2, \ldots, n$ and $p(i \mid G)(p(i \mid B))$ is the proportion of goods (bads) who take value i in the variable, then the weight of evidence is defined as $\Sigma_i(p(i \mid G) - p(i \mid B))\log(p(i \mid G)/p(i \mid B))$. If this value is large then the variables distinguish well between the goods and the bads, and any variable with a value greater than 0.1 is usually worth considering for inclusion.

Although some variables, like age and income are continuous, many of the variables in the application data are categorical, like residential status, phone ownership and employment category. The variables in the application data are called the characteristics of the applicant and the different answers that are possible for each variable are called the attributes of that characteristic. One needs to be able to deal with variables or characteristics that have a finite number of attributes, i.e. categorical characteristics. There are three obvious ways of introducing categorical characteristics into discriminant systems. One is to represent an n-attribute categorical variable by $n - 1$ dummy binary variables. This has the advantage of not imposing any relationship on the coefficients of the variables in the linear discriminant rule apart from that which comes from the statistical calculations, but the approach leads to a large number of variables. A second approach is the location model, suggested by Krzanowski (1975) in which a discriminant function on the continuous variables is built for every combination of attributes of the categorical characteristics. In the credit scoring context this leads to building a very large number of different discriminant functions. The third approach is to give appropriate numerical values to the N-attributes of the categorical variables. Let g_i be the number of goods who have attribute i of the variable and b_i be the number of bads who have that attribute. If $g = \Sigma_i g_i$ and $b = \Sigma_i b_i$, then let attribute i have a value which can be g_i/b_i, $(g_i/g_i + b_i)$, $g_i b/b_i g$, $\log(g_i/(g_i + b_i))$

or $\log(g_ib/b_ig)$. This approach gives the attribute values of the characteristics an ordering related to the odds of goods to bads among the sample of past applicants who have that attribute. The reason why this is the approach that is used in general is because it is appropriate for the continuous variables as well as the categorical variables. For example Fig. 5.1 describes the relationship of credit risk with age for a sample of credit card customers.

Since the relationship is not monotonic and the aim is to have a predictive system rather than an explanatory system, it may be better to split age into a number of age groups and think of it as a categorical variable with the attributes being these various age groups. This loses the smoothness that a continuous view of age guarantees, but allows one to include external information about the ages when there are life changes. By defining the attribute value for each age group using g_i and b_i as above, one can ensure that these values reflect the differences in credit risk.

There are two other general difficulties that arise in using credit scoring. The first is the inherent bias introduced because the classification is being done on a sample of accepted customers. There is no knowledge of the credit histories of those who were previously rejected by the organisation. Credit scorecard builders use the term *reject inference* for the ways they seek to compensate for this bias. One method first described by Hsia (1978) is 'augmentation' which initially builds a linear function to discriminate between those who are accepted and are rejected. For each score of this function the proportion of those with this score who are accepted is calculated and those accepted at that score are then weighted proportionally to the inverse of their probability of acceptance. The weighted accept sample is then used to build the linear function to discriminate between the goods and the bads. Hand and Henley (1993) point out that this and other methods used all have the assumption that $P(g \mid x)$ is the same for the sample of those accepted and those rejected. Unless this or some other relationship between clients who were accepted and clients

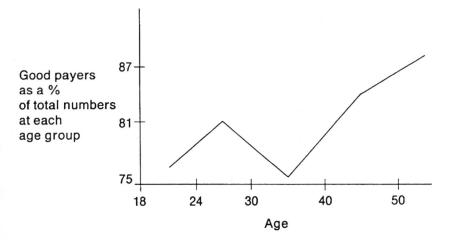

Figure 5.1 Relationship of credit risk with age

who were rejected can be assumed then there is no way the inherent bias can be overcome.

The second problem is to determine how effective the discrimination is likely to be. It is clear that applying the classifier function to the sample it was built on puts an optimistic bias on the misclassification error. So how can one obtain better estimates of how the function will perform on a future sample than this optimistic error, called the 'apparent error'. When data are limited, authors have suggested cross-validation and bootstrap methods for estimating the error rate of the discrimination function. The cross-validation (leave-out-one or Lachenbruch method, Lachenbruch, 1975) builds a classifier on all but one sample point, tests it on this last point and repeats this experiment for each point in the sample space. Although there is little bias it is expensive in computing time and has a relatively large variance. The bootstrap method assumes that if e_T is the true error rate, e_A is the apparent error rate and b is the expected optimism of the apparent error rate then $e_T = b + e_A$. The bootstrap approach estimates b by drawing random samples S^* from the original sample S to build a classifying function. The classification function error on S^* is compared with its error on S and the difference is taken as an estimate of b. This is repeated for other samples S^* and an estimate for b and hence e_T obtained, Srinivasan and Kim (1987) and Leonard (1988) have used this approach in calibrating credit scoring data. However, in almost all credit scoring problems the size of the population which can be used as a sample is very large. In that case the best and easiest thing to do is to split the data set into two subsamples, build the classifier on one subsample and test it on the other. This hold-out method gives a robust and reliable way of estimating the error and is the approach used in most credit scoring situations.

There is also the question of whether the scorecard should be built on the whole application population or if the population should be segmented into subpopulations and a different scorecard built on each. This could be considered as a variant of the location approach to discrimination (Krzanowski, 1975) mentioned earlier. What it means is that the discrimination involves the cross terms between the variable that the population was split on and all the other characteristics used. Thus, in principle it appears to be no different than using the standard approach but including more variables. One would expect such an approach always to give 'better' scorecards, and in many instances this is the case. One famous example was the work of Chandler and Ewert (1976) who looked at the effect of building a separate scorecard for males and females. They showed that doing this was advantageous to females since characteristics that are derogatory for males – part-time employment, etc – will not have the negative effect on female only scorecards that they have on a full scorecard. However, doing this is illegal since the Equal Credit Opportunity Act prohibits using sex in any way in a scorecard. One suspects this is an example of legislators not understanding the statistical aspects of what they were doing. Banasik *et al.* (1996) showed there is a negative side to segmentation and that unless the two subpopulations are sufficiently different, segmentation may result in a system that discriminates less well than if one built a scorecard on the whole population.

Now we will look in more detail at the different approaches that have been used in credit scoring. The literature on credit scoring is somewhat limited but there have been three surveys of these approaches by Thomas (1992), Rosenberg and Gleit (1994) and more recently by Hand and Henley (1996). There are also two books that look at credit scoring: Lewis (1992) describes the area from a practitioner's viewpoint while the book edited by Thomas, Crook and Edelman (1992) describes some of the practical problems and theoretical underpinning of the area as presented at an international conference held in 1989.

5.3 Regression and logistic regression approaches

The idea of building a credit scorecard – the function that classifies into goods and bads – was introduced by looking at the discriminant analysis approach. This is essentially the same as a linear regression approach to the problem, and if one defines a dependent variable by $Y = 1$ for a good and $Y = 0$ for a bad then the linear regression equation estimating Y as a linear function of the characteristics gives the same linear function as Fisher's discriminant analysis. Thus, it is no surprise that the fastest way of obtaining the coefficients of the linear function is to use the least squares approach of linear regression. Assume applicant i has application attributes $x_{i1}, x_{i2}, \ldots, x_{ip}$ in answer to the p characteristic variables. We identify the dependent variable y_i as 0 if bad and 1 if good. Alternatively if there are n_G goods and n_B bads in the sample population, one can define $y_i = n_G/(n_G + n_B)$ if applicant i is bad and $y_i = n_B/(n_G + n_B)$ if applicant i is good. We seek to find weights $w_1, w_2, \ldots, w_m$ so that $\Sigma_j w_j x_{ij}$ best approximates y_i. The objective is to

$$\text{minimise} \, (y_i - \Sigma_j w_j x_{ij})^2 \tag{5.6}$$

Thus, one can obtain w_i by using the usual mean-square regression formula.

The fact that this linear function is the best separator of the goods from the bads depends on the covariance matrix of the application attributes – the x_{ij} – being the same for the population of goods and bads. This will not be the case in reality and, if one assumed different covariance matrices for the goods and the bads, one would end up with a quadratic expression in the xs to separate A_G and A_B. However, practical experience shows that the benefits of this more accurate delineation are often lost because of the many more coefficients that have to be estimated. Consequently, normally linear scorecards are constructed.

Many of the difficulties, both in the theoretical justification and in the practical implementation of linear discriminant analysis for building credit scoring systems, are addressed in two papers by Eisenbeis (1977, 1978). These papers were based on investigations of statistical scoring techniques brought on by the Equal Credit Opportunities Act and, although there are criticisms of the approach, there are none which suggest it is not a useful and successful tool.

Although the linear regression approach works satisfactorily in credit scoring it is in spite of the fact that the data usually break all the assumptions being made about normality and common covariance. There are several other models, which have theoretically stronger underpinning, which can be used to classify binary response data with many categorical attributes. Logistic regression is the one that has found most favour in credit scoring. If p_i is the 'probability ' that the ith applicant is a good, linear regression can be thought of as assuming

$$p_i = \boldsymbol{w}.\boldsymbol{x}_i = w_0 + w_1 x_{1i} + w_2 x_{2i} + \ldots w_m x_{mi} \qquad (5.7)$$

while logistic regression assumes

$$\log(p_i/(1 - p_i)) = \boldsymbol{w}\boldsymbol{x}_i \qquad (5.8)$$

This is a much more general assumption in that it holds not just in the case when the attributes have a multi-normal distribution – the linear regression case – but also when they are independent binary. It seems much more satisfactory to have the left-hand side of Equation (5.8) being able to take any value between minus and plus infinity rather than just between 0 and 1 when one is trying to approximate it by a linear function. The difficulty that held up the use of logistic regression was essentially computational. There is no explicit formula for the maximum likelihood estimators of the w_i and so one must use Newton–Raphson approximations or some other estimation method. However, the power of standard computers in the 1990s makes this problem obsolete.

One might expect logistic regression to be much superior to discriminant analysis given its stronger theoretical justification, so it is surprising that the results are so similar. One of the most detailed published analyses was performed by Henley (1995). His results suggested that for fixed levels of acceptance, logistic regression would generally produce a slightly lower level of bads accepted – but it was never statistically significant, being at most a 0.5% drop in the default rate. The reason for so little change may be apparent from Fig. 5.2 which plots $\log(x/(1 - x))$ against x the left-hand sides of Equations (5.7) and (5.8). The fact that the log function is non-linear only starts showing itself if $x < 0.2$ or $x > 0.8$ and in these ranges of probabilities it should not be difficult to classify who is good and who is bad. Thus, over the critical area the logistic regression function acts very much as the linear regression function.

Wiginton (1980) also made the comparison of logistic regression and discriminant analysis on consumer credit scoring while Srinivasan and Kim (1987) did a similar analysis for lending to commercial customers. Both found a small advantage in logistic regression though Wiginton's results – that neither approach was really effective – is probably due to only three variables being used in the regression.

Two other alternatives of logistic regression have been used in credit scoring and related classification approaches. In probit analysis, if $N(x)$ is the cumulative normal distribution function so

$$N(x) = \frac{1}{\sqrt{2\pi}} \int_{-\infty}^{x} \exp(-y^2/2) \, dy$$

then the aim is to estimate $N^{-1}(p_i)$ as a linear function, so

$$N^{-1}(p_i) = \boldsymbol{w}\boldsymbol{x}_i = w_0 + w_1 x_{1i} + w_2 x_{2i} + \ldots + w_m x_{mi} \qquad (5.9)$$

Grablowsky and Talley (1981) compared a probit model with a linear

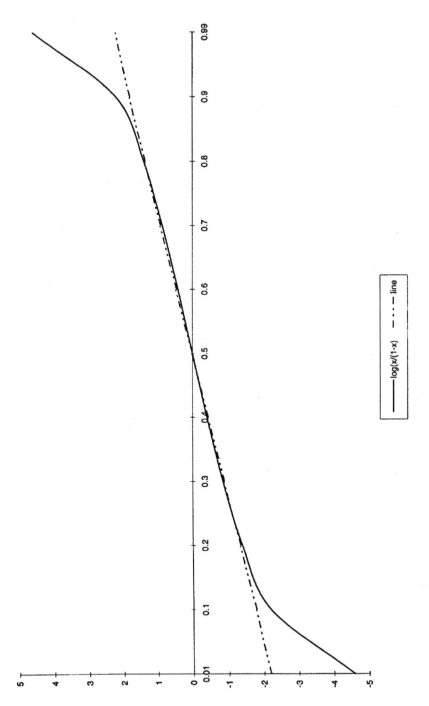

Figure 5.2 Approximation of $\log(x/(1-x))$ by straight line

discriminant model and concluded the former was slightly superior. Another variant is the tobit transformation which seeks to estimate p_i by

$$p_i = \max\{wx_i, 0\} = \max\{w_0 + w_1 x_{1i} + w_2 x_{2i} + \ldots + w_m x_{mi}, 0\} \qquad (5.10)$$

5.4   Other statistical approaches

There are other statistical approaches to classification which have been suggested for credit scoring. The first is the recursive partitioning algorithm (RPA) or classification tree approach. This was developed by Breiman and Friedman independently in 1973 and came to prominence with their book on the subject published in 1984 (Breiman *et al.*, 1984). The idea is to split the set of application data A into two subsets, each of which is more homogeneous in the credit-behaviour of their members than the original set. Each of these sets is then again split and the process repeated until the subsets that remain are considered to be terminal nodes of a tree. Each terminal node is then classified as a member of A_G or A_B and the whole process is represented as a classification tree. This tree depends on the splitting rule that divides the sets, the decision on when a subset is a terminal node and the way of classifying these terminal nodes as good or bad. This last decision is usually straightforward in that a terminal node is considered a member of A_G if the proportion of goods in that set in the sample exceeds either 50% or $D/(D+L)$.

At each node in the tree the splitting rule is applied to each variable in the application data in turn and the best of these splits used. Assume that the variable's attributes – its values – are ordered $s_1, s_2, \ldots s_n$ and that $F(s \mid G)$ is the cumulative distribution function of this variable among the goods and $F(s \mid B)$ is the same function over the bads. The myopic rule would be to split at a value s^*, i.e. $A_G = \{s \mid s > s^*\}$, which minimises the expected loss

$$LF(s^* \mid G)p_G + D(1 - F(s^* \mid B))p_B \qquad (5.11)$$

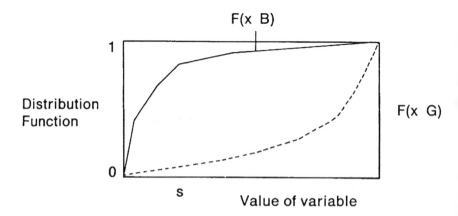

Figure 5.3   Kolmogorov–Smirnov distance

This is essentially choosing the s that maximises the Kolmogorov–Smirnov (K–S) distance between the two distributions (it is exactly that if $p_G L = p_B D$) as Fig. 5.3 shows

This chooses the best value to split on for a particular variable. One chooses the variable to split on as the one whose best split gives the smallest expected loss as defined in Equation (5.11). As Breiman *et al.* (1984) point out, one can take more sophisticated methods where one chooses the criterion not to be the expected loss after the next split but the one after r more generations of splits and so takes into account the likely future splits as well as the immediate one. Whichever approach is taken it results in a tree where the splits are usually on different variables at different levels and so seeks to capture the relationship between the variables. Fig. 5.4 gives an example of a typical classification tree.

The final decision is when to stop and leave a subset as a terminal node. This can be for one of two reasons – either because the number of sample points in the subset is too small or because, when one splits, the two daughter subsets are very similar – this shows itself in a small K–S distance. There is a temptation to keep growing the tree and to keep splitting more and more because eventually one will end up with a tree with only terminal nodes, each containing applicants who are either all good or all bad and this would be a perfect classifier of the data it is being built on. The problem is that the bigger the tree the less robust it is when applied to different data and so the final task in any tree classification is to prune back the tree. Breiman *et al.* (1984) discuss the various methods available for pruning these trees. Applications of the recursive partitioning algorithm are given in Coffman (1986) while Boyle *et al.* (1992) compare the recursive partitioning algorithm with discriminant

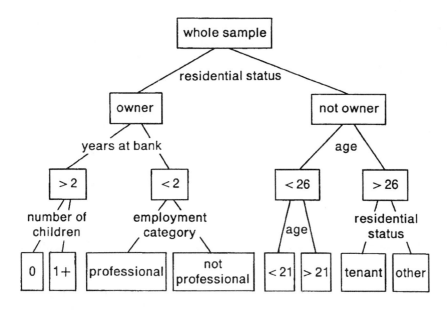

Figure 5.4   Classification tree

analysis. They suggest RPA is a competitive approach but it is also a useful way of identifying those variables which have strong interactions as they tend to be ones that are being split at the top levels of the tree. This may help regression methods determine which variables need to be combined together to allow for important second- and third-order effects.

The result of the regression trees approach is to divide A, the set of possible application data, into a number of cells and each cell is then labelled good or bad. Thus, there is no obvious scorecard or even a score for an applicant, though one can use the ratio of goods to bads in the sample who are in that cell as a surrogate score. However, it does have the advantage that it can deal with the non-linearities in the data which the linear methods can only cope with by introducing composite variables – e.g. $d(X, Y) = 0$ only if $X = 1$ and $Y = 1$, or $X = 0$ and $Y = 0$ and is 1 otherwise.

A second statistical approach is to use nearest neighbour ideas to classify new applicants. The idea is very simple. One chooses a metric on the space of application data to measure how far apart are two applicants. A new applicant is pronounced good if enough of its nearest neighbours measured by this metric are good. The three features needed for this approach are the metric, how many neighbours, k, to consider and what is enough of them. Normally one takes a simple majority as enough but it may be appropriate to require at least $D/(D + L)$ to be good. In their detailed investigation of a credit scoring application of the nearest neighbour method, Henley and Hand (1996) found any k between 100 and 1000 had broadly the same effect, though there were considerable local differences between using k and using $k + 1$. These were sufficient for them to suggest averaging over k. The vital question though is the metric. This is almost equivalent to finding the weights in the discriminant analysis approach and identifying which variables should be ignored altogether. In fact, Henley and Hand (1996) use a metric $d(x, y) = (x - y)(I + dw^\mathrm{T}w)(x - y)$ where w are the weights from discriminant analysis. This uses the Fisher idea that this is the direction that best separates the goods from the bads.

Henley and Hand (1996) show that, on data for credit scoring mail-order customers, the nearest neighbour methods slightly outperformed regression approaches. The nearest neighbour approach also has the intuitively appealing idea that one can constantly incorporate new data by introducing new points in the space as the credit behaviour of existing applicants becomes clear. Chatterjee and Barcun (1970) and Hand (1986) also looked at nearest neighbour approaches to credit scoring.

5.5   Mathematical programming

The mathematical programming approach to classification first reached prominence with the paper by Freed and Glover (1981a) who outlined how a two-group classification problem could be set up as a linear programming problem. This was quickly followed by their description of how to use mathematical programming in the n-group case (Freed and Glover, 1981b) and Hand's description of the ideas in his book on classification (Hand, 1981). This resulted in a number of authors investigating this approach to classification so that Joachimsthaler and Stam's review (1990) of the subject nine years later refers to over 70 papers in the area.

In the credit scoring context we assume that all dummy variables, partitioning of ranges and transformations of categorical variables to give quantitative values have been performed as was outlined in Section 5.2. Each applicant is represented by p characteristic values. Assume that in the sample on which the scorecard is to be built there are n_G goods labelled $i = 1, 2, \ldots n_G$ and the ith good has characteristic values $x_{i1}, x_{i2}, x_{i3}, \ldots x_{ip}$. Assume there are n_B bads labelled $i = n_G + 1, n_G + 2, \ldots n_G + n_B$ and their characteristic values are similarly labelled. Thus, a perfect discrimination requires that we find weights $w_j \ j = 1, 2, \ldots p$ and a cut-off value c so that for the goods $w_1 x_{i1} + w_2 x_{i2} + \ldots + w_p x_{ip} > c$ and for the bads $w_1 x_{i1} + w_2 x_{i2} + \ldots + w_p x_{ip} < c$. This would give the perfect scorecard. Usually this cannot happen and so we seek to minimise some function of the absolute errors by solving the linear program

Minimise $a_1 + a_2 + \ldots + a_{n_G + n_B}$

subject to

$$w_1 x_{i1} + w_2 x_{i2} + \ldots + w_p x_{ip} \geqslant c - a_i \quad 1 \leqslant i \leqslant n_G \qquad (5.12)$$
$$w_1 x_{i1} + w_2 x_{i2} + \ldots + w_p x_{ip} \leqslant c + a_i \quad n_G + 1 \leqslant i \leqslant n_G + n_B$$
$$a_i \geqslant 0 \quad 1 \leqslant i \leqslant n_G + n_B$$

This formulation minimises the sum of the absolute errors – which are the a_is – but if one replaced them all with the same variable a and just minimised a one would minimise the maximum absolute error. Linear programming requires that one changes the strict inequalities about the cut-off score to greater than or equals, and less than or equals. This is no problem provided one recognises that this allows a solution to the Linear Programming with $c = 0$ and all the weights 0. Thus, one needs to normalise c in some way to avoid the trivial solution.

Allowing integer variables, which tend to make the solution times considerably longer, one can develop scorecards that minimise the number of misclassifications in the sample as follows. Let $g_i = 1$ if the ith case is a good misclassified as a bad; 0 otherwise. Let $b_i = 1$ if the ith case is a bad misclassified as a good; 0 otherwise. Then minimising the number of misclassifications is obtained by the integer program

Minimise $g_1 + g_2 + \ldots + g_{n_G} + b_{n_G+1} + \ldots + b_{n_G+n_B}$

subject to

$$w_1 x_{i1} + w_2 x_{i2} + \ldots + w_p x_{ip} \geqslant c - M g_i \quad 1 \leqslant i \leqslant n_G \qquad (5.13)$$
$$w_1 x_{i1} + w_2 x_{i2} + \ldots + w_p x_{ip} \leqslant c + M b_i \quad n_G + 1 \leqslant i \leqslant n_G + n_B$$
$$b_{n_G+j}, \ g_i \geqslant 0 \quad 1 \leqslant i \leqslant n_G, \ 1 \leqslant j \leqslant n_B$$

There are a number of other objectives that one could use. Lam *et al.* (1996) for example set up a pair of linear programs that seek to minimise the deviation from the mean score for each of the two groups of goods and bads. They apply this methodology to credit scoring data.

The general consensus (see the review by Joachimsthaler and Stam, 1990) of those who have made comparative studies of the mathematical programming approaches and the statistical approaches to classification is that the statistical ones classify marginally better. Defenders of the statistical approaches also decry the mathematical programming approach because it may lead to the 'all weights are zero' solution if some of the x_{ij} are negative. This can be overcome by transforming the x's to make them all positive but the results of the linear program approach are not invariant under such a transformation. Glover *et al.* (1988) has presented a more complicated Linear Programming formulation which overcomes these difficulties but for almost all cases the approaches suggested above will be robust enough. The linear programming approach also has a lot going for it. First of all it can deal with large numbers of variables – hundreds of thousand if necessary – whereas the statistical methods do not cope as well with such large numbers of variables. In particular, it would be quite acceptable to represent each characteristic as a number of dummy variables. This would allow the program to choose the scores within a characteristic rather than have them forced on them by taking the good-to-bads odds ratio or the log odds ratio suggested earlier. Secondly, it allows one to require certain properties of the weights, for example that $w_1 > w_2$ which may describe a discrimination which the credit granter wishes to encourage. For example, one could ensure that the under 25s have a higher weight than the over 50s to try and get a young client base. This would be impossible to impose in the statistical approaches. Also, goal programming, where there is more than one objective to be aimed for, can help develop scorecards that try to discriminate on credit worthiness and also on other factors like profitability or card-usage.

One integer programming formulation of credit scoring is much quoted in the literature because of its surprising result. Kolesar and Showers (1985) reported the results of a scorecard they had built for AT&T to determine who should pay a deposit on getting a phone service. They reported that a rule which asked at most 10 yes/no questions and accepted those with at least k yes answers worked very well. This is an extremely simple system where all the weights are either 0 or 1 and they used an integer programming formulation to determine which weights to use. However, their choice of this system was constrained by the requirements of the telephone company and they do report that a linear discriminant-based scorecard would have performed better. The reason why this simple system works so well might well be the 'flat maximum' effect first reported by Lovie and Lovie (1986) which says that there is a very wide choice of the weights that give almost as good discrimination as the optimal weights. Overstreet *et al.* (1992) have exploited this effect in building generic scorecards on one population of creditors to determine who gets credit from a different organisation whose customers have different geographic and social variables. Again the scorecards built in this way work tolerably well.

5.6   Neural networks and expert systems

A neural net consists of a number of processing units each of which is a simple computation device. Unit i receives input signals from the other units,

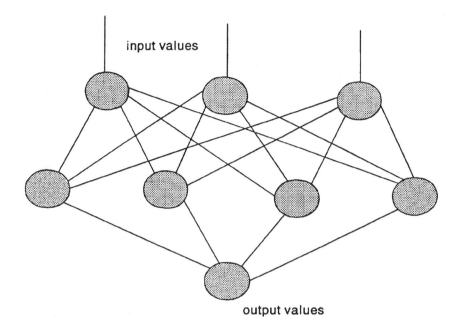

input values

output values

Figure 5.5   Topology of neural network

aggregates these signals based on an input function, I_i and generates an output signal based on an output or transfer function O_i. This output signal is then sent to other units as directed by the topology of the network. Although no assumptions apart from continuity are imposed on the input and output function by far the most common functions used are

$$I_i(w, O) = \Sigma_j w_{ij} o_j + c_j; \qquad O_i(I) = 1/(1 + e^{I_i}) \qquad (5.14)$$

where I_i is the input of unit i, O_i is the output of unit i, w_{ij} is the connection weight between unit i and unit j and c_i is the 'bias' of unit i. The most common topology is a series of layers of units with connections between all the units in one layer and those in the next layer as in Fig. 5.5.

The neural network is presented with a training sample of data. Each data point has a set of input values and corresponding output values. The objective is for the neural network to find the weights w_{ij} and c_i so that it can replicate the data set results as closely as possible. The most common learning algorithm to calculate the weights is back propagation (Freeman and Skapura, 1991). The back propagation algorithm has two phases. First, forward propagation calculates the error on the training sample given the present value of the weights. Suppose the sample has s data points with data point r having input vector x_r and output vector d_r. If the output from the neural net when x_r is input is y_r then the sample error is $E = \Sigma_r \Sigma_j (y_{rj} - d_{rj})^2/2$. E is a function of the weights and we try to minimise the error by adjusting these weights using the steepest descent method. The direction and magnitude of the change in w_{ij}, denoted Dw_{ij} is calculated by

$$Dw_{ij} = -\frac{\partial E}{\partial w_{ij}} \varepsilon \qquad (5.15)$$

where ε is a parameter controlling the convergence rate of the algorithm. The error $E(w)$ is propagated back layer by layer from the output units to the input units. The adjustment in weight w_{ij}, where i and j are in intermediate layers, is obtained by using the chain rule

$$\frac{\partial E}{\partial w_{ij}} = \frac{\partial E}{\partial O_i} \frac{\partial O_i}{\partial I_i} \frac{\partial I_i}{\partial w_{ij}} \qquad (5.16)$$

The weights can either be updated after each data point is tested or after the whole set of sample points has been taken through the neural net. The procedure is repeated until E converges, although there is no guarantee it will find the weights that give the minimum error.

It is not surprising that this methodology has been used in credit scoring. The input variables are the application characteristics and the output is the credit performance – good or bad – of the sample creditor. Tam and Kiang (1992) applied it to identifying bank-bankruptcy cases, Altman *et al.* (1994) looked at identifying corporate bankruptcy in Italy, while Carter and Catlett (1987) used it in accepting consumers for credit cards. More recently, Desai *et al.* (1996) reported on a comparison of neural net, linear discriminant and genetic algorithm methods. In almost all cases the authors report results that suggest that neural networks do very well in classifying compared with other approaches.

Care has to be taken with these results in that much of the corporate lending decisions use sample sets of less than 200 with the number of variables also of this order of magnitude. Consumer credit usually has data sets in the hundreds of thousands and we have been surprised by the length of time it takes the net to learn in such cases. Also, one has to examine the stopping criteria used in the analysis because the error function, E, has the tendency to start increasing after a certain number of back propagation iterations through the data.

With only two layers of the network it is well known that the neural net can only develop a linear relationship between the input and output data and so theoretically cannot do better than linear discriminant analysis. Thus, the power of neural nets must be in the ability of the 'hidden layers' to tease out some of the non-linearities in the relationships. In exchange, it has to give up the statistical underpinning and strength of evidence tests which surround the statistical procedures.

Ignizio and Soltys (1996) gave an alternative approach to the use of neural networks in bank classification. They used the idea of masking functions to cover the whole of a region of one type of credit risk – good or bad – by saying items of that type have to be in one of a number of simpler regions whose boundaries are defined by masking functions. The coefficients of these masking functions can be obtained using linear programming approaches. The result is a methodology similar to a neural network where the inputs to the input layer are the values the variables take. At the next layer these are combined to calculate the masking function values for these input values and at the final layer there is a check to see which masks, or regions, the input values fall into. Again, with the much smaller data sets of commercial lending, this approach gave very competitive results.

Neural networks are sometimes considered to be expert systems – machines behaving as human experts because of their learning ability. True expert systems, which seek to embody the knowledge of human experts in a set of rules and then use an inference engine to infer what these rules suggest about a set of data presented to it, have also been used in credit scoring situations. The applications have tended to be in commercial loan situations (Zocco, 1985; Leonard, 1993a) and fraud detection (Leonard, 1993b). Davis *et al.* (1992) take a very preliminary look at using a Bayes expert system to classify credit card applicants and compare the results with a simple neural net. It does seem that expert systems are much more successful at detecting fraudulent transactions and so are used at the authorisation stage of a credit card purchase rather than at determining credit worthiness initially, while statistically based methods are the reverse of this. The reason is the size of the two groups one is trying to discriminate between in the two cases. There are enough poor credit risks to allow statistical methods a chance of recognising the attributes of each group successfully, while there are so few fraudulent transactions compared with non-fraudulent ones that statistical methods cannot find the needle in the haystack. However, since fraudsters tend to have a *modus operandi* (buy power tools which are easily sold on stolen credit cards for example), expert systems are able to look out for these clues by having them embodied in their rules. The expert system *Authorizer's Assistant* developed at American Express for this purpose was one of the first of its kind and has proven very successful (Davis, 1987).

5.7 Genetic algorithms

Genetic algorithms, like neural networks, are motivated by using biological analogies to develop effective optimisation approaches. In the case of genetic algorithms the idea is to try and mimic the 'survival of the fittest' rule of genetic mutation to develop optimisation algorithms. One starts with a population of potential solutions to a problem and a way of measuring the fitness or value of each solution. One then produces a new generation of solutions by allowing existing solutions to mutate (change slightly) or crossover (two solutions combine to produce a new solution with aspects of both). The choice of which solutions to mutate and crossover is related to their fitness so better solutions are more likely to be chosen for both operations. In this way it is hoped to produce new generations of solutions that have higher values. Mutation is a form of random hill climbing in optimisation terms while crossovers are some form of hyperplane sampling – jumping to other solutions with certain common properties.

The solutions in the credit scoring context are potential scorecards and so in order to apply this approach it is necessary to find a way of representing scorecards, and then a way of mutating and crossing over these representations. One way suggested by Albright (1994) is to define the scorecard by the coefficients and the powers on the variables it uses. Thus, if the score is obtained from the variables $x_1, x_2, \ldots x_p$ by

$$\text{score} = a_{11}x_1 + a_{12}x_1^2 + \ldots + a_{1m}x_1^m + a_{21}x_2 + \ldots + a_{pm}x_p^m \quad (5.17)$$

then an obvious representation of the scorecard is $(a_{11}, a_{12}, \ldots a_{pm})$. Mutations can be obtained by allowing each component of the vector to have a probability of changing its value. This is often done by representing the numbers in binary form and giving a probability for each digit to change. Crossovers could be represented by cutting the string at some arbitrary point and taking the first half of the *a*s from one scorecard and the second half of the *a*s from another scorecard. This involves a considerable amount of computing although not very intensive calculation and so is particularly suitable for use on parallel computers. Preliminary results on the application of this approach to credit scoring are given in Fogarty and Ireson (1993) and Albright (1994). The result of the comparison made by Desai *et al.* (1997) using credit union data is somewhat disappointing for genetic algorithms. Overall, the statistically based methods outperformed both neural nets and genetic algorithms, but whereas there were types of behaviour – customers who were tardy in their repayments but did not default – that neural nets were able to predict better than the standard methods, there were no examples of genetic algorithms performing better than the other methods.

5.8 Conclusions

One of the interesting aspects of credit scoring is the diversity of techniques that can be applied to the problem as has been outlined in the previous sections. It is therefore somewhat surprising that the recurring theme is how little difference there is in the classification accuracy of the different approaches. The reason for this is the 'flat maximum effect' mentioned earlier. Experimentation suggests that there is a wide region of different linear weights which give very similar classification errors. 'All wolves are grey in the dark' could be the way to summarise the results. Closer investigation though shows there are differences and this is an area where even a small difference in classification accuracy can lead to significant savings. A credit card company with eight million accounts, which on average loses a thousand pounds on each account that defaults, can save twenty million pounds if it can cut its default rate from 5% to 4.75%. That is worth doing a lot of research to try and achieve.

The other point to make is that the various methods strengthen or extend the basic credit scoring methodology in different ways. The regression techniques allow one to do hypothesis testing to see which variables make significant contributions to the classification. Recursive partitioning and neural nets are able to deal with complex dependencies between the variables automatically without the analyst having to recognise such dependencies. Mathematical programming techniques allow large numbers of variables and can incorporate most of the biases management want to introduce into the scorecard. It is these added features as much as pure classifying ability that determines which technique is used in practice. In the UK most scorecard developers would use several of the techniques in developing their product but the basis of the scorecard is usually a logistic regression or a mathematical programming methodology.

What are likely to be the future developments in scoring techniques? The important change will be in the objective of the scorecard. Up to now that has been to identify the new customers who are at highest risk of defaulting. Almost all financial organisations would like to change to a scorecard that identifies the profitability of a new customer rather than just his or her default risk. This is a much harder criterion to satisfy, for as well as estimating the default risk one needs to estimate the usage of the credit line by the customer and the likely repayment behaviour. This may necessitate using ordinary regression approaches or possibly building Markov chain models. On top of that, profit will be strongly related to how long the consumer stays with that credit granter. This suggests that one will need to develop methodologies that have aspects of survival analysis in them. There is still plenty of work to be done in developing credit scoring techniques!

References

Albright, H. T. (1994) Construction of a polynomial classifier for consumer loan applications using genetic algorithms. Working Paper, Department of Systems Engineering, University of Virginia.

Altman, E. I., Marco, G. and Varetto, F. (1994) Corporate distress diagnosis; comparisons using linear discriminant analysis and neural networks (the Italian experience). *Journal of Banking and Finance*, **18**, 505–529.

Banasik, J., Crook, J. N. and Thomas, L. C. (1996) Does scoring a subpopulation make a difference? *International Review of Retail, Distribution and Consumer Research*, **6**, 180–195.

Boyle, M., Crook, J. N., Hamilton, R. and Thomas, L. C. (1992) Methods for credit scoring applied to slow payers. In *Credit Scoring and Credit Control*, L. C. Thomas, J. N. Crook and D. B. Edelman (Eds), Oxford: Oxford University Press, pp. 75–90.

Breiman, L., Friedman, J. H., Olshen R. A. and Stone, C. J. (1984) *Classification and Regression Trees*, Belmont, California: Wadsworth.

Capon, N. (1982) Credit scoring systems: a critical analysis, *Journal of Marketing*, **46**, 82–91.

Carter, C. and Catlett, J. (1987) Assessing credit card applications using machine learning, *IEEE Expert*, **2**, 71–79.

Chandler, G. G. and Ewert, D. C. (1976) Discrimination on basis of sex under the Equal Credit Opportunity Act, Indiana: Credit Research Centre, Purdue University.

Chatterjee, S. and Barcun, S. (1970) A nonparametric approach to credit screening, *Journal of the American Statistical Association*, **65**, 150–154.

Coffman, J. Y. (1986) The proper role of tree analysis in forecasting the risk behaviour of borrowers. Atlanta: MDS Reports, Management Decision Systems, 3, 4, 7 and 9.

Davis, D. B. (1987) Artificial Intelligence goes to work. *High Technology*, April 16–17.

Davis, R. H., Edelman, D. B. and Gammerman, A. J. (1992) Machine-learning algorithms for credit-card applications, *IMA Journal of Mathematics Applied in Business and Industry*, **4**, 43–52.

Desai, V. S., Crook, J. N. and Overstreet, G. A. (1996) A comparison of neural networks and linear scoring models in the credit environment. *European Journal of Operational Research*, **95**, 24–37.

Desai, V. S., Conway, D. G., Crook, J. N. and Overstreet, G. A. (1997) Credit scoring models in the credit union environment using neural networks and genetic algorithms. *IMA Journal of Mathematics Applied in Business and Industry*, **8**, 323–346.

Durand, D. (1941) *Risk Elements in Consumer Instalment Financing*. New York: National Bureau of Economic Research.

Eisenbeis, R. A. (1977) Pitfalls in the application of discriminant analysis in business, finance and economics, *Journal of Finance*, **32**, 875–900.

Eisenbeis, R. A. (1978) Problems in applying discriminant analysis in credit scoring models, *Journal of Banking and Finance*, **2**, 205–219.

Fisher, R. A. (1936) The use of multiple measurements in taxonomic problems, *Annals of Eugenics*, **7**, 179–188.

Fogarty, T. C. and Ireson, N. S. (1993) Evolving Bayesian classifiers for credit control – a comparison with other machine learning methods. *IMA Journal of Mathematics Applied in Business and Industry*, **5**, 63–76.

Freed, N. and Glover, F. (1981a) A linear programming approach to the discriminant problem. *Decision Sciences*, **12**, 68–74.

Freed, N. and Glover, F. (1981b) Simple but powerful goal programming formulation for the discriminant problem. *European Journal of Operational Research*, **7**, 44–60.

Freeman, J. A. and Skapura, D. M. (1991) *Neural Networks, Algorithms, Applications and Programming Techniques*, Reading, MA: Addison-Wesley.

Glover, F., Keene, S. and Duea, B. (1988) A new class of models for the discriminant problem. *Decision Sciences*, **19**, 269–280.

Grablowsky, B. J. and Talley, W. K. (1981) Probit and discriminant functions for classifying credit applicants; a comparison. *Journal of Economics and Business*, **33**, 254–261.

Hand, D. J. (1981) *Discrimination and Classification*. Chichester: Wiley.

Hand, D. J. (1986) New instruments for identifying good and bad credit risks; a feasibility study. Report prepared for the Trustee Savings Bank.

Hand, D. J. and Henley, W. E. (1993) Can reject inference ever work? *IMA Journal of Mathematics Applied in Business and Industry*, **5**, 45–55.

Hand, D. J. and Henley, W. E. (1997) Statistical classification methods in consumer credit. *Journal of the Royal Statistical Society, Series A*, **160**, 523–541.

Henley, W. E. (1995) Statistical aspects of credit scoring, PhD Thesis, Open University.

Henley, W. E. and Hand, D. J. (1996) A k-NN classifier for assessing consumer credit risk, *The Statistician*, **65**, 77–95.

Hsia, D. C. (1978) Credit scoring and the Equal Credit Opportunity Act. *The Hastings Law Journal*, **30**, 371–448.

Ignizio, J. P. and Soltys, J. R. (1996) An ontogenic neural network bankruptcy classification tool. *IMA Journal of Mathematics Applied in Business and Industry*, 7, 313–326.

Joachimsthaler, E. A. and Stam, A. (1990) Mathematical programming approaches for the classification problem in two-group discriminant analysis. *Multivariate Behavioral Research*, 25, 427–454.

Johnson, R. W. (1992) Legal, social and economic issues implementing scoring in the US. In *Credit Scoring and Credit Control*, L. C. Thomas, J. N. Crook and D. B. Edelman (Eds), Oxford: Oxford University Press, pp. 19–32.

Kolesar, P. and Showers, J. L. (1985) A robust credit screening model using categorical data. *Management Science*, 31, 123–133.

Krzanowski, W. J. (1975) Discrimination and classification using both binary and continuous variables, *Journal of American Statistical Association*, 70, 782–790.

Lachenbruch, P. A. (1975) *Discriminant Analysis*. New York: Hafner Press.

Lam, K. F., Choo, E. U. and Moy, J. W. (1996) Minimizing deviations from the group mean: a new linear programming approach for the two-group classification problem. *European Journal of Operational Research*, 88, 358–367.

Leonard, K. J. (1988) Credit scoring via logistic models with random parameters, PhD Dissertation, Concordia University, Montreal, Canada.

Leonard, K. J. (1993a) Empirical Bayes analysis of the commercial loan evaluation process. *Statistics and Probability Letters*, 18, 289–296.

Leonard, K. J. (1993b) Detecting credit card fraud using expert systems. *Computers and Industrial Engineering* 25, 103–106.

Lewis, E. M. (1992) *An Introduction to Credit Scoring*. San Rafael, California: Athena Press.

Lovie, A. D. and Lovie, P. (1986) The flat maximum effect and linear scoring models for prediction. *Journal of Forecasting*, 5, 159–186.

Overstreet, G. A., Bradley, E. L. and Kemp, R. S. (1992) The flat maximum effect and generic linear scoring models: a test. *IMA Journal of Mathematics Applied in Business and Industry*, 4, 97–110.

Rosenberg, E. and Gleit, A. (1994) Quantitative methods in credit management: a survey. *Operations Research*, 42, 589–613.

Srinivasan, V. and Kim, Y. H. (1987) Credit granting: a comparative analysis of classification procedures. *Journal of Finance*, 42, 665–683.

Tam, K. Y. and Kiang, M. Y. (1992) Managerial applications of neural networks: the case of bank failure predictions, *Management Science*, 38, 926–947.

Thomas, L. C. (1992) *Financial Risk Management Models in Risk Analysis, Assessment and Management*. J. Ansell, F. Wharton (Eds), Chichester: Wiley.

Thomas, L. C., Crook, J. N. and Edelman, D. B. (1992) *Credit Scoring and Credit Control*, Oxford: Oxford University Press.

Wiginton, J. C. (1980) A note on the comparison of logit and discriminant models of consumer credit behaviour. *Journal of Financial and Quantitative Analysis*, 15, 757–770.

Zocco, D. P. (1985) A framework for expert systems in bank loan management, *Journal of Commercial Bank Lending*, 67, 47–54.

6

Credit Scoring and Quality Management

Kevin J. Leonard

6.1 Risk management

In the statistical literature, 'risk' is referred to as a measurement of uncertainty – usually quantified with probabilities or percentages. There is a
0.9 probability of rain tomorrow; there is a 50% chance that the investment
will be profitable. The more risky a decision is, the more uncertainty
surrounds it. For example, in the context of finance, the risk-free return rate
refers to the rate of return on an investment that is guaranteed – e.g. Canada
Savings Bonds. Canada Savings Bonds (or CSBs) are guaranteed by the
Government of Canada and are virtually risk-free. The investor knows in
advance with certainty the value of his or her investment upon maturity.
Hence, there is no need for a risk premium. On the other hand, investment in
Mutual Funds, for example, does incur inherent risk. As a result, the rate of
return for a risky investment must be higher than the rate associated with an
instrument that is risk-free. If not, then why would anyone invest in the riskier
alternative? The level of uncertainty must be compensated by a higher
expected return.

When there is uncertainty therefore, possible outcomes are assigned
probabilities; the outcomes and the probabilities are then combined to
calculate a weighted average or expected outcome. It is these expected values
that form the basis of risk assessment. (Over the years, the term 'risk' has
gained negative connotations – i.e. the term 'risky loan' is employed to mean
a 'bad loan'. Strictly speaking – risk is neither good nor bad: it just means that
the outcome is uncertain. The loan could turn out to be a good or a bad. If we
know that a loan will be written off with *certainty* – then it is not a risky loan:
it is simply a 'bad' loan.)

Risk Management involves both the quantification and the evaluation of the
probability of financial return compared with the expectation of financial loss
pertaining to the introduction or the continuance of a business venture. This
presumes, first, the ability to identify the financial gains and losses for a
prospective investment and, second, the capacity to measure and manage this

evaluation process. Without this management component, the above description would be better classified as 'risk evaluation'.

The measurement of risk usually involves the application of sophisticated mathematical and/or probability models. One such set of models pertains to the area of credit or loan application evaluation – called *Credit Scoring Models*. It is this very specific area of Risk Management, including all the issues contained therein, that this chapter seeks to address.

Banks must evaluate the ability of their customers to repay their financial obligations according to the agreement established between the respective parties. This evaluation process can be carried out using credit scoring models. (Credit evaluation can be done in a number of ways; however, the emphasis here is solely on credit scoring.) We discuss here decision making models that aid in the prediction of the outcome of these credit applications. For the most part, this outcome relates to whether or not a customer turns out be a good account – payment conditions have been satisfied.

6.2   What is credit scoring?

The process of modelling creditworthiness is referred to as *credit scoring*. This modelling process, which often uses statistical methodology (such as discriminant analysis or log-linear models) is carried out by banks and other financial institutions. Based on the statistical analyses of historical data, certain financial variables are determined to be important in the evaluation process of the customer's financial stability and strength. Subsequently, information on these variables is obtained for new bank customers. This information is summarised and a point system is used where the different variables have different weights. An overall score is produced by adding these weighted scores. If this overall score is above a predetermined cut-off point, the credit applicant receives a certain line of credit. If not, the applicant is denied credit.

The primary objective in credit scoring is to develop an effective scoring model that contains only a small number of predictor variables. These scoring algorithms or 'scorecards' are then used to evaluate all credit applicants in the future. This allows for consistency in credit evaluation and efficiency in processing. When implemented effectively, the scorecard should be able to rank order the entire relevant population of applicants by risk. That is, every applicant or account is assigned a score. This score then equates to an odds quote, likelihood level or probability estimate, which provides an estimate of payment performance or loan outcome – an estimate that an individual account will turn out, for example, to be a 'good' account. However, probabilities only make sense over the long run. Therefore, even though all individual accounts are assigned a score (and an odds quote), the probabilities only provide insight for all accounts – as a block – at a certain score. When considered in this manner, odds quotes can be very useful in the establishment and maintenance of the break-even or cut-off score.

Applying statistical methods to the credit decision is not new and the area of credit scoring for consumer credit has developed into a multi-million dollar industry. In fact, every major bank and trust company in North America presently employs some form of credit scoring. Further, credit scoring for consumer lending is well documented in the academic literature (see Reichert *et al.*, 1983, for example).

The reasons for the creation of a financial credit scoring model can be summarised as follows:

(i)   to quantify the mechanical procedures involved in credit scoring and gain the efficiencies of application processing that come through automation;

(ii)   to gain control and consistency in lending practices for the entire credit portfolio;

(iii)   to identify the variables that are important in the credit evaluation process;

(iv)   to improve delinquency statistics while maintaining desired approval rates.

6.2.1 Performance Definitions

The most pertinent question regarding credit scoring is: 'what am I trying to predict?' The objectives of the organisation that are motivating the introduction of scoring models will indicate, to a large degree, what procedures are appropriate. In general, however, there are usually two different types of scoring models:

(i)   predict the outcome of the decision – whether the credit is approved or a declined;

(ii)   predict the outcome of the credit – whether the customer turns out to be a good or a bad.

Most credit scorecards are built to address the latter good/bad decision. Client financial institutions are interested in identifying what makes a good loan and in learning what is indicative of a more risky loan (possibly a bad). Very seldom are banks, for example, interested in using credit scoring models to find out why their managers approved or declined a particular loan in the past.

(One area of interest has been the creation of profitability models. Unfortunately, profit models are very difficult to build because very little data is captured regarding the components of costs and revenue. For example, for revenue, data would have to be identified revealing spending per account, interest revenue, and fees charged; for costs, proper model building would require allocation of costs of collection, costs of capital, and write-offs/loan losses. To be able to obtain information on these two outcome groups – profitable and non-profitable – the credit industry as a whole will have to move to finding out more information regarding what makes money and what loses money.)

In order to build a scorecard that rank orders individuals based on the odds of being a good or a bad customer, one must first define these two principal sets. A good account is an account that has, for the most part, paid the financial obligations in full and on time. A good account is not necessarily an account that has never been delinquent. There are times when accounts go into arrears for a brief period of time (one cycle late) over a long evaluation period and this account should still be classified as a good. This is due to the fact that, as a whole, the customer has exhibited 'good' payment behaviour. Basically, the rule of thumb for a good account is: if you had to do it over again, you would.

Contrarily, a 'bad account' is an account that if you had the opportunity to do it over: you would not. This is an account that typically has been delinquent many times during a short time period; or perhaps is an account that is currently 90 days past due. Also, accounts that have been put into a non-performing category or have been written-off are classified as bads – these are, in fact, the worst of all the bads. It is important to note, however, that a bad is not necessarily an account that has been written-off or put into collections. Some bads are just accounts that have been such a nuisance that it costs more to keep them current than the revenue gained from interest charges and late fees.

A third category that is sometimes an issue is the indeterminate set. These are accounts that cannot be grouped or classified as either a good or a bad – they are not bad enough to be classified as a bad yet not good enough to be a good. In an effort to ease the scorecard development process, the account is not classified as either a good or a bad and, as a result, is not sampled and would not be represented in the development process.

A cut-off score can be effectively established in one of two ways. First, since a scorecard rank orders all applicants by risk (the documentation accompanying the scorecard at completion of analysis provides detailed information concerning the score breakdown of the development sample), a cut-off score can be set by referring to the cumulative statistics in an attempt to match previous acceptance rates. Then if we wish to accept (say) 70% of all applicants in the future, a cut-off score is set at the 30th percentile – we expect 70% of all applicants to score at or above this score (all other things being constant).

A second method for cut-off score establishment involves the calculation of a break-even score. This requires a fairly good estimate of the number of goods (revenue generation) it takes to cover the losses from one bad – often in the range of 5 to 1 (i.e. the revenue from five goods roughly approximates the loss from a single, average bad). If this is true, then the cut-off score should be set at the score equivalent to an odds quote of 5 to 1. In this way, at the margin, we are breaking even for all accounts in the score range at the cut-off. Any accounts booked above this score will be profit generating, on average. Remember, the scorecard rank orders by risk – therefore, higher scores relate to higher 'probability of good', or more profitable, accounts as a whole.

In some scorecard building programs, the two principal sets are defined as bankrupt versus non-bankrupt accounts. The definitions of these two groups are much easier to understand. However, one has to be very cautious regarding the interpretation of the scores. A high score (for most scorecard vendors – the higher the score the better) represents an account that is not likely to be bankrupt – which is not necessarily the same as 'not likely to be a bad'. If accounts that do not go bankrupt still cost you money (an account can be written off even though the customer was not bankrupt), then it is possible that you will be looking favourably on accounts that could be severely delinquent (yet not bankrupt).

It should be noted that there is an opposing argument to be made. All non-bankrupt accounts can be worked (i.e. can be placed in collection) and, based on the effectiveness of this attention, can result in being profitable accounts. If this is true, then perhaps the bankrupt versus non-bankrupt is the best of all possible principal set definitions, because it has allowed for the identification of the one type of account you wish to avoid – bankruptcies.

One last question is whether the creation of a profitability model is the ultimate goal. Instead of deciding a cut-off level based on probabilities of being good or non-bankrupt, we should investigate the possibility of developing models that will target profitable accounts. However, the key task in this case, which has been a very difficult obstacle to overcome, is that the account manager or the scorecard vendor must be able to identify which components of the business comprise a profitable account and what makes an account not-profitable. Only then can a scorecard be built with these components. Otherwise, the primary objective of being able to distinguish between profitable versus non-profitable accounts becomes untenable. Unfortunately, this identification has been so challenging that an effective profitability model has yet to be developed within the credit scoring industry.

6.2.2   Sampling issues

There are a number of subsidiary sampling questions that must be addressed when implementing credit scoring models.

6.2.2.1.   Is the past like the present?

A very important assumption is that the population that I analyse in the future will be similar to the population or sample (i.e. sample of customers) that I evaluated in the past.

The key underlying factor, of course, is that we believe that there is a significant statistical relationship among three components: the type of customer, the type of product and the actual specific criterion that predicts the outcome of the credit or, if you prefer, the payment performance. Once we have clearly determined who we wish to target, then a sample can be selected that is representative of the new target population that will be applying for credit for that product. We then proceed to examine the statistical (predictive) relationship. It is only then that we see if a relationship does indeed exist and, if so, exactly what form it takes. This statistical relationship may be similar to relationships in other portfolios or it may be unique.

Let us assume that we wish to build a scorecard that will help evaluate and rank order by risk a set of customers that apply for a $10 000 unsecured personal line of credit. However, this product will only be offered to customers who have, or are now applying for, a mortgage. Thus, we have now identified a very specific type of customer – a mortgage holder who is interested in a line of credit product. When we sample from past customers to do the analysis, we must select customers who came to the same financial institution in the past and who applied for a line of credit. Additionally, to ensure that we are dealing with the same customers, we want to sample only people that had or were planning to obtain a mortgage as well. In this way, we have replicated a very close approximation of the type of customer in our sample (that we are going to analyse from the past and develop the model on) that we wish to target, score and approve in the future.

A relevant question is: who cares if they have a mortgage? Why don't we just sample from the people that have applied for a personal line of credit in the past? The important point to remember is that we do not know *a priori* what is predictive and what is not, although we may have a good idea based on past experience. Therefore, in order to ensure the best possible model, as many factors as possible should be controlled for. This means replicating the future with past data as much as possible, including information on product (analysing line of credit customers with mortgages only), geographic location, seasonality and whatever else contributes, or may even remotely contribute, to the scorecard's effectiveness.

Perhaps, people who apply for a personal line of credit perform differently than people who apply for this same product and have a previous banking relationship with the same institution. (As an aside, research in this area is quite conclusive that banking relationships – such as a Retirement Savings Plan (RSP account, protected from income tax) – are highly correlated with payment performance.) If this is true, then we only want to analyse people that represent the same type of customer that we will score with the scorecard after it has been developed. If we do not, then other factors may affect the power of the scorecard and it may not perform as well and for as long as would otherwise be the case.

To summarise, we must remember the key point: the past must be representative of the future.

That being said, we should note that final scorecards (due to the types of statistical analyses performed) are very robust and often do a good job of rank ordering by risk the populations (or other segments of populations) that were never originally intended to be scored. However, one must be very careful when considering the use of scorecards outside the target population due to the reasons already discussed pertaining to approval rates, probabilities of performance and odds quotes. Scoring experts should be consulted before any implementation is considered.

It should be mentioned that, before any analysis, we are unaware of what the actual relationships are. It could be that, as in our above example, having a mortgage is an irrelevant piece of information, that it does not impact on performance. If this is the case, then it will not enter the scorecard as a characteristic. As a result, we may have been able to include all types of customers and not just those that had a banking relationship. However, this was not known at the outset, and it is always better to be safe – and to control for as much variability as possible. It does no harm to exclude accounts that you are questionable about, but it does seriously raise the risk level if we include these accounts in the analysis, because they could change the basis of the entire predictive relationship.

6.2.2.2 *What is the optimal sample size?*

Often, the number of cases, applications or amount of data captured is so large, a small 'representative sample' must be selected so that analysis is feasible. There are a number of different sampling techniques, but there are two key issues – randomness and representativeness. The sample must be drawn randomly to provide an environment that will promote representativeness and unbiased estimates of parameters.

If a type of customer occurs 5% of the time in the population, then a random sample will ensure that this is the case in our development sample, as long as the sample is large enough. (Most scorecards are built with about 1000 to 1500 goods, the same number of bads and about half this number of declines as a strict minimum.) To illustrate, suppose that 5% of the customers in a population are from Region A in the population. Random sampling from the entire population is expected to result in about the same percentage occurring in the sample. If it is the objective of the scorecard to score individuals at the same rate as before (the past is representative of the future) then the people in Region A are expected to be represented correctly by simply adhering to basic simple random sampling principles. If, on the other hand, the people from Region A, for whatever reason, in the future, will begin to constitute 50% of the population, then steps must be taken to adjust the sampling. This is the representativeness component. Simple programming controls can be put in place that will now adjust the expected total such that approximately 50% (given a large enough sample) of the final sample comes from Region A, thus ensuring representativeness while retaining the randomness component.

6.2.2.3 *What are the main problems with trying to predict the future?*

In an attempt to improve the accuracy of statistical inference, we try to control for as many extraneous influencing factors as possible. For example, it would be impossible to analyse every account that came through the door (accepted for credit) in the past. Therefore, for purely economic reasons, sampling is necessary.

In addition, application and performance data from accounts booked in the future are not available (at present) to review and analyse so, even if we knew perfectly the current predictive relationships, there would still be the need to infer present information for future time periods. Therefore, because of this time effect, sampling is always a necessity when forecasting.

It should be noted that there are many possible extraneous factors that could affect the sampling process and the effectiveness of the resulting scorecards. These can be categorised into three groups:

(i)   implementation of new credit policies;
(ii) marketing initiatives (solicitation programs);
(iii) economic conditions.

First, credit policies can address which accounts we would like to attract and score in the future. For example, we may believe that we should implement an income requirement. This may mean people that are below the specified income level will not be scored – their credit application will be turned down as a 'policy decline'. As a result, all our analysis should reflect this policy: previous applicants with income below this specified level should not be analysed.

Secondly, marketing programs can change the profile of the customer that 'comes to your door' applying for credit. If this profile changes substantially, the whole premise of the scorecard – what is predictive and to what degree of performance – may come into question. The reason is that the past is no longer precisely representative of future applicants.

Thirdly, economic conditions can change the effectiveness of the scorecard to a tremendous degree, even if all other factors are kept constant, such as policies and marketing programs. Should a geographic region encounter a recessionary-type effect, then individual profiles on a credit bureau report may change, even though performance by these specific individuals may not change at all. Also, the whole rank ordering by risk may change if the economic changes are not evenly distributed across all cross-sections of society – be they industries, geographic sections or even occupation types and levels. Once again, the future applicant is no longer well represented by what was analysed during the scorecard development process.

6.2.3   Reject Inference Problem

As stated, credit scoring is the process of modelling creditworthiness. Based on statistical analysis of historical data of the approved accounts (the good customers and the bad), certain financial variables are determined to be important in the evaluation process of a credit applicant's financial stability and strength. When implemented effectively, the scorecard should be able to rank order the entire population of applicants by risk.

One primary drawback related to model development is that only accounts that were approved in the past have any performance data. That is, once an account is opened, and a measurement time-window established, an account can be evaluated by observable payment behaviour criteria to judge whether it turned out to be good or not. However, those accounts that were declined have no such performance history and thus cannot be evaluated regarding eventual outcome status – either a good or bad. If these accounts are thrown out of the sample before the analysis takes place due to the lack of a value for the dependent variable, then all applicants that came to the door applying for credit are not being considered, i.e. the declined applicants are not sampled and not represented. Therefore, in a sense, the statistical analysis would be performed on a group of accounts that were already pre-screened. This pre-screening causes a statistical bias if the resulting scoring instrument is used to evaluate all credit applicants in the future. Consequently, if declines are not analysed during model development, then they may not be handled well or evaluated effectively during future credit application evaluation.

In order to address this shortcoming, a number of different statistical techniques have been introduced to solve this problem of 'Reject Inference'. The problem can be summarised as one of trying to estimate the payment performance of applicants who were previously declined credit so that they can be considered in the statistical analysis and represented in the subsequent scoring instrument. Knowledge of the outcome of these declined accounts is critical if these scoring models are to evaluate correctly all applicants in the future.

In the credit scorecard industry, there are many solution procedures that are employed. The first alternative is to assign a 'bad' loan status as the outcome for the value of the dependent variable for each of the declines. Then the statistical analysis is performed on all the applicants (with either observed or inferred behaviour) considered. The main problem with this approach is that the resulting scorecard only reinforces the screening process that is currently in place. If there were 'good' applicants that were being rejected before, a new scorecard built with this methodology would reinforce previous lending practices and these 'good' accounts would continue to be declined because they were represented as 'bad' accounts in the analysis. As a result, the new scoring tool will not improve overall portfolio bad rate.

A second alternative is to find, by a number of different methodologies, approved accounts that look very similar in profile to each individual rejected account. The behaviour of the approved similar account is then inferred or assigned to the rejected applicant. Again the statistical analysis to build the scorecard is performed with all applicants considered. The actual determination of what constitutes a 'similar account' can prove to be very difficult and, hence, makes this process almost unfeasible.

A final method of reject inference, which is the one favoured by the industry, is effectively to build two separate models. The first scoring model is built on the approved applicants only. Once this is created, the rejects are scored out and, based on their score and the probability or likelihood of being a good or bad, a good or bad outcome is estimated and assigned to the reject. Then all this information is combined (the observed and the assigned outcomes) to build an overall good/bad scorecard. Once again, the scorecards are created based on statistical analysis of the entire applicant population. Although there are shortcomings to this methodology as well, it has provided consistent estimates that result in fairly robust scoring instruments. The main concern with this approach is that the assignment process is primarily all or nothing (either assigned a good or a bad behaviour) where a level of degree (high probability of a good or a bad) may be better suited.

It is easy to see that the performance of the scorecard is really only as good as the 'reject inference' procedure and the corresponding results. There has been much debate within the credit scorecard industry concerning how reject inference should be performed and which method of reject inference is the most efficient and effective approach. At present, it is safe to say that there is no consensus as to the best methodology to be employed.

6.3 Using the scorecards

As we have discussed, the specific type of accounts that were studied as part of the development analysis of the scorecard dictates who should actually be scored with the card in the future. Remember the first principle: what happened in the past must be representative of what will happen in the future. Although scorecards built on accounts from a credit card portfolio may be very similar to the types of people that applied to an instalment loan portfolio (let's say), there is no certainty that what is predictive of a good account is the same in both. Even though the characteristics themselves could be the same, the weights may be entirely different. Consequently, after the analysis is complete, one person could be scored out for both products independently and be approved for one product and declined for a second.

This result is perfectly consistent with the statistical theory. One individual may be rank ordered in an odds group of an acceptable (or profitable) level for one product but not for the other. In some detail, we will consider the above example of a credit card portfolio and an instalment loan portfolio, where both portfolios have separate dedicated scorecards. An applicant may be ranked in the top 60% of risk for the credit card portfolio and in the top 70% of the instalment loan portfolio. If the cut-off score is established such that a 65% approval rate is expected for both portfolios, it is quite consistent to see a customer approved for one credit product and not the other. In scoring, not only is the applicant's credit profile relevant, but so too is the profile of all the remaining people that have applied (or will apply) for the portfolio's product.

Additionally, the amount of exposure and number of inquiries at the bureau may have some effect. If the same person applies for the two products at different times, then it is possible that the risk exposure for both combined is too much and therefore, the second application should be turned down.

Another concern related to the introduction of credit scoring models is how credit decisions should be made with incomplete data. The scorecard has a number of different characteristics (some of which are highly correlated) due to the fact that sometimes certain data are missing. In other words, redundancy is built into scorecards to ensure a robust decision tool. For example, applicant age and time at occupation are highly correlated but often both are found in scorecards. This is due to the fact that often both may not be supplied by the applicant. When a small number of characteristics are missing (at most two), the scorecard still has strong predictive power because of the fact that each of the characteristics are predictive in and of themselves. (When more than two are missing, many institutions decline the application, often encouraging the applicant to re-apply when more information is available.) This is especially true of credit bureaux report information. In some instances, such as the high risk or the extremely low risk cases, the information from a bureau report will not add much predictive power. That being said, many lenders will not lend money or approve an application unless a bureau report is purchased (this often avails the institution of information that is not available in other sources – such as legal disputes, or number of major derogatories at other Financial Institutions). It is safe to say that the actual decisions that must be made regarding an organisation's implementation of scoring tools will rely heavily on its strategic plans, objectives and policies and procedures.

6.3.1   Exceptions, overrides and branch recourse

Exceptions are accounts that are *not* scored. Due to whatever reason, certain accounts are either approved or declined before they are scored. VIP accounts are examples of accounts that are exceptions.

An *override* to scoring occurs when the scorecard decision is reversed after the score for an applicant has been achieved and compared with the cut-off score. A *high-side override* is an override to decline (i.e. a score is above the cut-off but was overridden and the applicant was denied credit). A *low-side override* is an override to approve (below cut-off score but application is booked).

Overriding the score is something that must be done occasionally (usually a maximum of 5% of the time on either side) because of the fact that there are times when there are valid extraneous factors. When this occurs and an experienced loan officer is aware of additional information, then the branch would be best served by including this information in the decision process and monitoring the override reason for validity and effectiveness. For the override reason to be valid, the additional information must have been un-available (or not consistently available) during the sample time frame and subsequently not analysed during scorecard development.

When a decision is made to decline an application by a credit granting centre and this decision is different from the recommendation from a branch office, then the centre often forces the branch to approve the application (if they still desire to do so) under the auspices of a 'branch recourse'. This then amounts to charging back the losses to the branch if the customer ever defaults. This is one method of making branches responsible for their decisions and their overrides. An important point to remember is that it is imperative for the branch staff to be aware of how scoring works. This will do more to implement scoring positively than forcing the branch to cover losses that are charged back to their branch. If their override decisions and, more importantly, override reasons are not effective, then the best solution is to educate the decision makers on how scoring works and how they can improve their decisions.

This can be best accomplished by designing reports that illustrate the various reasons that the branches use for overrides. These overrides are then matched with performance over time. For the low-side overrides, tracking performance is quite straightforward. For high-side overrides (where accounts that should be booked are declined), management must agree to sample some of these accounts and book them (in fact, override the override). These accounts should then be tracked for performance in the same manner as the low-side overrides. Once the branch staff see which reasons for overriding are successful and which are not, then the case can be made with the supporting evidence. In fact, in certain areas, overriding should be encouraged. Why not? If delinquency rates go down as a result of using additional information, then it should be incorporated. The argument is of course: is the time-consuming process associated with the application by application review, which then becomes necessary, cost effective?

Once again, the important point is to try and factor in all the relevant components into the scoring system. If this is achieved, then overriding should not exist. However, being realistic, this is never the case. The scorecard can incorporate only what is captured. There is always new information that can be added and analysed. Thus, the data capture could always be expanded. Further, some information is captured but not consistently entered. This means that the coding could be interpreted differently and, as a result, the predictability could decrease. For all these reasons, the scorecards will never include all possible scoring variables or components: thus, overriding will continue to exist.

6.3.2   Policies and procedures

Policy decisions are decisions that are made outside the pre-established scoring parameters. First, they can take place before a score is calculated. For example, the policy of minimum income is often used and extremely effective. The incorporation of this particular policy may be necessary in order to replicate the cases analysed in the development sample. If people below a minimum income were not approved before, then, in an attempt to replicate historical acceptance rates, they should not be approved again. The effectiveness of this, like all other policies, should be tracked by comparing actual score distributions with expectations.

Secondly, a policy can impact on the decision after an application has been scored. For instance, there may be the policy that no one with a major derogatory is approved. Therefore, a policy to allow for high-side overrides (score is calculated and then the decision is made to decline after a credit bureau report has been produced) has been created. Once again, the effectiveness of this override reason must be monitored.

Is it good to double count the negative factors – such as using derogatory information on a bureau report – as both a credit scorecard characteristic and an override reason? In effect, once the analysis has been completed and the correct weight for negative bureau activity has been determined, the creation of an override decision based solely on this same derogatory information is questionable. However, the best response is: be consistent, insist that policies be directly incorporated into the entire scoring process and monitor all deviations to the credit scoring decision process.

6.4   Examination of detailed reports

This section contains descriptions of management information reports that provide very insightful perspectives. Additionally, we create a set of quality measures that can be used to evaluate the effectiveness of credit scoring models.

6.4.1   Management information

Normally, in credit scoring, the following statistics are generated in an attempt to evaluate the performance of scoring models. These statistics are created in an effort to measure portfolio efficiency or output:

- acceptance rate based on approvals of application volume;
- adherence to expected score distributions;
- frequency of reversing score decision (overrides);
- bad rate (number of accounts) as a percentage of the total number of accounts;
- loan losses versus profitability.

While these statistics are valid measures, they only describe how well the scoring algorithms are working within a very narrow focus. For instance, acceptance levels or approval rates are a direct result of cut-off score. To control approval rate, an adjustment of cut-off score can be performed. Since the scorecard is a rank-orderer of risk, the higher the cut-off score, the fewer the number of applications (and higher quality) that will be accepted. Measurement of approval rates is well established within the industry.

As a second example, the bad rate can be directly adjusted as well (through scoring) by reducing the approval rate. Fewer accounts approved will eventually result in lower write-off levels. Once again, this is a valid measure for evaluating scorecard efficiency. If the scoring instrument is not working properly, then these changes would not produce the expected results. Although these statistics indicate the performance of scorecards in the short term, more information is often required. For example, these measures do not really measure the true impact on the portfolio brought about by the introduction of scoring. The quality measures introduced in the second part of this section will attempt to measure the true effectiveness derived from scoring.

6.4.1.1 Score distributions or final score report

The final score report gives the first insight into the performance of the scorecards. After only 30 days, delinquency information is not available. However, part of the benefit of scoring is the ability to set approval rates (and delinquency expectation) based on how accounts were scored in the development sample (when the scorecards were created). The hypothesis of the past representing the future can be tested based on how we see future accounts scored out by the scorecard.

If the people score similarly to what was observed in the development sample, then this is not a guarantee that it ranks orders by risk. However, it is a good sign that at least the first expectation has been reached. The rest of the model's effectiveness – the ability to reduce delinquency – will have to be tested as we progress through time. This is why it is so important to monitor the scorecards and regularly track performance.

On the other hand, if people do not score similar to the development sample, we have a good very early warning that the scoring model may not produce the desired effect. If the scores are not as expected, then there is a significant possibility that the delinquency numbers will not correspond to the development sample as well. The only way to measure actual performance statistics, unfortunately, is to track scorecard results over time.

6.4.1.2 Override reports

One of the most common reasons for differences from development is due to rampant overriding. If at a score of, let's say, 200, we expect 67% of the population to score above, and after the first month we have booked 80% of all applications, then this might signal that an override problem exists.

One possible problem is that accounts that are below cut-off are not being declined. Usually at the inception of scoring at an institution there is resistance to using the scorecard for all decisions. In fact, a certain level of overriding is expected (as stated earlier, around 5% on either side of cut-off). Some of the time additional information is available and should be incorporated into the decision. This often results in reversing the score decision. However, just as often, there is resistance to scoring and overriding takes place based solely on 'gut intuition'. This is not recommended. This results in the same types of decisions being made that were in place before the introduction of scoring and, consequently, the benefits from scoring will not be realised.

An override report must be generated which will first enumerate all the score overrides, and later, when the information is available, match this with delinquency information. This will allow for an educated response to the question of overriding. What is often seen is that some reasons are good and some are not. The delinquency information will allow for the good ones to be kept and the poor ones to be stopped. Delinquency information can provide a great deal of influence when interpreted correctly.

6.4.1.3   Dynamic delinquency reports

Dynamic delinquency reports are designed to give insight into how the scorecards are performing over time regarding booking accounts and their subsequent performance. The reports actually break the scorecard down by intervals (preferably intervals of the same range) and delinquency by number of accounts is shown for each interval. Often there is a break that illustrates accounts both below and above cut-off and as well as exceptions (those accounts that were booked by exception or policy, i.e. not scored).

The problem with the dynamic delinquency reports is that there is no control for 'time on books'. To help remedy this situation, accounts are often shown within a specific time interval or time on books. This allows one to see trends in the accounts booked and their performance over time.

6.4.1.4   Vintage analysis

Vintage analysis reports show accounts with the same maturity as a whole and the performance over all the accounts booked. This allows equal comparison of accounts and usually shows trends through comparison as early as within the first few months (if enough accounts are booked). The only drawback is that it does not show score interval breakdowns – this would have to be realised through the dynamic delinquency reports or a creative combination of the two report types.

6.4.1.5   Account management or behaviour score reports

Finally, perhaps the most effective type of scoring is the account management or behaviour score. This scoring involves predicting future payment with actual past payment history at that specific institution. This does not involve application information nor does it incorporate any credit bureau information. All that is needed is access to the monthly masterfile or billing file. The past history is evaluated and a score is produced. Account management strategies are then driven by the account's most recent behaviour or B-score.

These account management scores are evaluated by reports that are generated every month to validate both the distribution of B-scores and the effectiveness of the different strategies that are driven by these scores. After approximately 12 months of history, strategies can be re-evaluated and the best or superior (defined by a number of different criteria) strategies can be retained. At that time, new and improved strategies (challengers) can be tested against the new or reigning champion.

6.4.2 Quality measures

In the current study of management practices, it is becoming accepted that Information Systems (IS) theory and Total Quality Management (TQM) theory are very much integrated. On one level, in the development of information systems, many TQM principles hold. For instance, user participation is critical in systems development. A comparable component of TQM is the fact that it is imperative to know and satisfy the demands of the customer. Doing it right the first time is preached by leaders in both TQM and IS as a method of getting users/customers over the critical implementation period and in a position to adopt the new technology willingly. On a second level, TQM principles often need an information system to become feasible, due to the overwhelming need to process a wealth of data almost instantly. A good example is in the area of continuous improvement and benchmarking, where many inter-firm and intra-firm comparisons are necessary to update reports and potentially modify objectives (Leonard, 1994).

The measurement of the effectiveness and the success of scoring initiatives have not received widespread implementation. We recommend measures that describe the scoring impact on the rest of the organisation. In fact, the emphasis is often on the efficiency of scoring algorithms (lowering loan losses) rather than on the effectiveness derived from scoring models (lowering application approval time or increasing customer dollars spent per account). As an example, consider the average time it takes to approve a credit application. If the introduction of scoring initiatives at a credit operation has been effective, then approval time should decrease substantially. More applications will be able to be processed in a shorter amount of time resulting in customers having access to their credit faster. Further, accuracy levels should begin to increase due to the automated nature of the adjudication process. In all, the customer, as well as the employee, will, over time, begin to experience a higher level of quality. These are factors that will not change quickly but rather improve slowly yet steadily over time – being truly reflective of the impact from scoring (Leonard, 1995).

Below, we list ten criteria that could be evaluated.

1. Approval time (each application for credit).
2. Approval accuracy.
3. Authorisation time (each transaction).
4. Override level.
5. Average dollars spent per account.
6. Interest revenue per account.
7. Current balance as a percentage of outstandings.

8. Dollars 30 day delinquent to outstandings.
9. Write-off dollars per account.
10. Collection time per account.

These ten components will provide some insight into the magnitude of the impact from scoring throughout the organisation. Below, we discuss the contribution of each of these ten criteria.

6.4.2.1   Approval time

The approval time is greatly affected by scoring. Scoring helps to streamline the whole approval process and allows for rapid decisions when the answer is clear-cut. When the applicant scores in the grey area around cut-off, then more time can be allocated for review. As a whole, however, scoring will bring the average approval time down.

6.4.2.2   Approval accuracy

Secondly, the accuracy levels corresponding to the application evaluation and subsequent approvals will be affected. Due to the automation of much of the process, routine applications will have very few errors.

6.4.2.3   Authorisation time

Authorisation time is another area that can be greatly enhanced by scoring. Whether accounts are scored based on performance information (on an on-going basis) or not, proper scoring can influence what characteristics are investigated at the time of any particular transaction. The more streamlined the decision process, the quicker the authorisation is approved. Quality will be perceived by the customer when the bank responds to his/her purchase request in a shorter interval of time.

6.4.2.4   Override level

Overrides are a particular interesting component to scoring. Even when the scorecards are working at peak efficiency, there is still a significant number of overrides occurring. This is due to the fact that no scorecard can incorporate every single possible factor that will influence payment behaviour. There will always be times when there are factors that have to be considered in addition to the scorecard characteristics. Hopefully, with an effective scoring algorithm these exceptions are limited to about 5%.

6.4.2.5   Average dollars spent per account

Improving the quality of the customer base will eventually result in more outstanding balances being current and less dollars outstanding being delinquent or forced to be written-off. This and the other 'percentage of outstanding dollars' measures that follow are critical individual measures of effectiveness.

6.4.2.6   *Interest revenue per account*

The more spending and current dollars, the more the interest revenue that will accrue.

6.4.2.7   *Current balance as a percentage of dollars outstanding*

The importance of this measure pertains to the objective to match each customer with the appropriate credit limit. Further, if collection initiatives are targeted effectively, the incidence of delinquent accounts should decrease, resulting in more current balances.

6.4.2.8   *Dollars 2+cycles delinquent as a percentage of outstandings*

Not all accounts can be monitored in the collection department. Correct management of a collection department through scoring can reduce workloads and reduce delinquency.

6.4.2.9   *Dollars written off as a percentage of outstandings*

If the criteria discussed in 6.4.2.5, 6, 7 and 8 above are all managed effectively, then this factor should also show improvement. However, due to the fact that write-off values often reflect accounts that were booked some two years previous, this is usually the last quality measure to be affected by a scoring initiative.

6.4.2.10   *Collection time spent per account*

Once again, this can be a direct result of the portfolio improvement realised from scoring. Less accounts are in collection, and, as a result, less time is spent on the phone chasing customers to honour their payment agreement.

6.4.2.11   *Total scoring quality index*

This measure reflects the overall or combined effectiveness of the scoring program and provides a direct assessment of performance. In order to be truly effective, this scoring quality index should be tied to whatever criteria with which the entire portfolio is eventually evaluated by senior management. For example, if profitability is the key success driver, then each factor should be weighted by its affect on profitability and then combined. Otherwise, a single quality number of this type will confuse more than enlighten the scoring participants.

6.5   Total quality management in information systems (TQMIS)

The focal point of a TQMIS model (Leonard, 1994) is the strategic objectives of the organisation or functional area. Once the long term goals have been clearly defined and established, the design of an information system to address these objectives can begin. Subsequently, once designed, the development process is started. It is important to note that there is feedback built into the model representing flow of information from the systems developers to the designers, perhaps occurring in the case of infeasible design specifications. Further, feedback is necessary between developers and strategic level management in the scenario where objectives may have to be modified due to unrealistic initial goals.

Once the systems have been developed, implementation can proceed. An important point in the TQMIS model is the need for an effective, well-coordinated training program. Implementation of new technology requires tremendous support and training both for the people leading the implementation and those who will be impacted from the new development. If this training and support has not been well established and developed, effective implementation may never be accomplished.

Once the system is implemented, then on-going management and monitoring can take place. This requires the development of reports and presentation materials that will quickly communicate pertinent ideas and information. This transformation of data to information is very critical – and will form the basis of the external evaluation of any new program.

Training and education play a significant role in the development of management reports and summary documents. This education and support initiative is the key as it interacts directly with the perceptions and expectations of all the constituents regarding communication of new ideas and the ability to affect change in the organisation. The constituents contain both internal and external groups as well areas that are responsible for:

- research
- consultants
- suppliers
- legal
- government
- social
- environment
- community
- competition
- customers
- shareholders
- management
- labour/employees.

The philosophy of total quality brings the whole model together, where benchmarking allows the framework for continuous improvement. Whether the comparison audience is the same group (on a historical basis), a different department within the same firm or an organisation within a different industry, the objectives are the same – determine the placing relative to the 'best of breed'. This gives targets for improvement and feeling of achievement as minor goals are accomplished. Benchmarking can be effective throughout the systems development – from setting the goals through design, development, and implementation right up to on-going management and tracking.

In the TQMIS model, it is important to stress the presence of the feedback loops and the interface between all the constituents and the system. Successful implementation will only be completed when all of the feedback loops are identified and explored, and supported through extensive training and education programs.

The majority of the emphasis in this field to date has been on finding the best statistical model. For example, everything from decision trees to discriminant analysis to neural networks has been examined (see Banks and Leonard, 1995, for a discussion). However, what the industry is now demanding, quite independent of the current research orientation, is information resources concerning the use of these models. Practitioners are constantly combing the literature for guidance on such everyday issues as implementation, training and support, and on-going management and tracking. Even the credit vendors have done surprisingly little work in this area. The calculation process of creating the statistical model, i.e. the transformation of data into information, is only a small component, yet is receiving a disproportionate percentage of the research attention. For reasons unknown, other areas have been ignored.

In the credit field, the starting point must be the strategic objectives: in this case, to reduce loan losses and delinquency while expanding the customer base and market share. This can be achieved through the application of credit scoring technology to a credit portfolio. The scorecards are then designed and developed, which leads to implementation. This change in the 'everyday way' of doing things brings about need for support and education. This is where the application of TQM principles (training, continuous improvement and benchmarking) becomes crucial. First, training must be in place to effect this change efficiently. Secondly, proper education and training to support managers in the design of the reports, and the development of insightful statistics reflecting the effectiveness of credit scoring, will help ease the implementation process and foster an environment of continuous improvement. Finally, benchmarking further creates the necessary short and long term goals that are both realistic and achievable.

6.6   The need for change in continuous improvement

The adoption of new technology (credit scoring models) is not unique to the credit field. Management theorists have been documenting the effect of change on organisations for years. In general, after the introduction of any new technology to an organisation, there is resistance to the impending change. In an effort to overcome this resistance, employees must be made aware of the impact of this change, and aware of any additions or revisions to their responsibilities and their performance expectations. Communication concerning these factors is critical to ensure successful implementation.

Specifically, all the relevant issues (training, continuous improvement, benchmarking initiatives) stem from the strategic objectives of the organisation. In the credit industry, at this time, the main strategic objective is to gain consistency, efficiency and effectiveness in their lending practices. This is best achieved through the application of credit scoring models. These objectives will *not* be achieved unless the TQM programs are formalised and goal-oriented.

As stated by Nadler (1981), 'effective change requires an understanding of both what the change should be and how it should be implemented'. A possible training solution is outlined by Stanislao and Stanislao (1983): 'Failure to understand the new method or policy arouses suspicion or an insecure feeling rapidly, but feelings are hidden from immediate supervisors or other management personnel. Training programs to teach employees necessary skills should be started well before the change.'

> An education and communication program can be ideal when resistance is based on inadequate or inaccurate information and analysis, especially if the initiators need the resistors' help in implementing the change. But some managers overlook the fact that a program of this sort requires a good relationship between initiators and resistors or that the latter may not believe what they hear. It also requires time and effort, particularly if a lot of people are involved. (Kotter and Schlesinger, 1979)

Tichy (1982) presents a Human Resource System where the key action item for implementing change effectively is training. The training forms the basis of two-way communication and gets the 'buy in' from those that are actually involved. If this acceptance to new technology is not achieved, then successful implementation could be 'road-blocked' by a number of possible factors – anything from 'innocent ignorance' ranging all the way up to 'full intentional sabotage.'

In an effort to bring TQM principles to the credit scoring industry, as stated, the training and continuous improvement programs must be goal-oriented. In order to assist in the establishment of these goals, benchmarking must be introduced. Unfortunately, this then requires that information be shared between institutions or departments to be truly effective. This may sound like an insurmountable hurdle due to the fact no institution wants to share information on performance; however, this may not be as impossible as it first sounds.

Due to the technical sophistication of the statistical analysis involved in the creation and implementation of scoring models and the fact that there has been a scarcity of resources to turn to for adequate training and research, the credit industry in Canada has created an association of credit scoring practitioners. The principal mandate of this association is to address the training and implementation issues which continually arise within the credit scoring community. This association, the Canadian Credit Risk Management Association (CCRMA), more than anything else, provides a forum where member institutions can meet and discuss with others in the industry. This sharing of information has proven to be very beneficial due to the fact that scoring problems (and solutions) are often shared by all credit organisations.

One argument against the feasibility of inter-firm and intra-firm benchmarking in the credit field is the confidentiality factor. However, other industries have been faced with similar challenges and have overcome them so as to benefit all involved. The automobile industry overlooked many competitive issues to help bring such safety items as anti-lock braking systems and air bags to the market as soon as possible.

Consequently, what must be done is to convince banks and other financial institutions that if one institution benefits, they all benefit – be it with initiatives regarding accepted practices such as credit scoring or fraud detection, or with new areas such as collections scoring or relationship scoring. The customer is then well served and consequently receives products that they can afford. There will be less loan losses as the credit institutions loan money responsibly, and then less write-offs. In the end, much like with the fraud detection models, savings can be passed on to customers in the form of lower interest rates and better customer service.

For businesses in general, and the credit scoring industry in particular, the need for training and education is becoming more and more apparent. It is only through TQM and the creation of effective training and education, and benchmarking programs, that individuals will best be able to face the challenges of future career development and organisational change.

In short, the credit scoring industry has been concerned with the efficiency of scoring – what models are good?; what is our bad rate?; how much are our loan losses? Very little concentration however has been on the effectiveness of scoring – are we using the models right?; can we use these models to satisfy our customers?; who is using the models better?; what level of effectiveness has been reached by competition? Oddly enough, returning to the TQM principles, it would appear that the industry is concentrating solely on the 'data/ processing/information' component, to the exclusion of all else. These other components are just as important in the goal to improve effectiveness. Our challenge is to start thinking about our real objectives (total quality management issues) – and the best way to achieve them.

Further Reading

Bierman, H. and Hausman, W. H. (1970) The credit granting decision. *Management Science*, **16**, 519–532.

Leonard, K. J. (1993) *Credit Scoring Manual: A Comprehensive Guide to the Improved Selection, Creation, Implementation and Management of Credit Scorecards*, Chapter Four, Canadian Credit Institute, Consumer Credit Management, offered through University of Toronto, School of Continuing Studies, pp. 43–99.

Leonard, K. J. (1996) Information systems and benchmarking in the credit scoring industry', *International Journal of Benchmarking for Quality Management and Technology*, **3**(1), 38–44.

Leonard, K. J. and Banks, W. J. (1994) 'Automating the credit decision process', *Journal of Retail Banking, Spring*, pp. 39–44.

Lewis, E. (1990) *An Introduction to Credit Scoring*. San Rafael, California: Athena Press.

Platt, H. D. and Platt, M. B. (1990) Development of a class of stable predictive variables: the case of bankruptcy prediction. *Journal of Business Finance and Accounting*, **17**(1).

St. John, C. (1990) Are two scores better than one? Combining credit bureau scores with custom scores. *The Journal of Consumer Lending*, Fall, pp. 30–34.

References

Banks, W. J. and Leonard, K. J. (1995) 'Credit scoring and mathematical models'. *Credit and Financial Management Review*, **1**, 10–14.

Kotter, J. P. and Schlesinger, L. A. (1979) Choosing strategies for change. *Harvard Business Review*, March–April, pp. 451–459.

Leonard, K. J. (1994) The relationship between information systems and TQM in credit scoring. *Credit Research Digest*, October, pp. 1–4.

Leonard, K. J. (1995) The development of credit scoring quality measures for consumer credit applications. *International Journal of Quality and Reliability*, **12**(4), 79–85.

Nadler, D. A. (1981) Managing organizational change: an integrative perspective. *Journal of Behavioral Science*, **17**, 191–211.

Reichert, A. K., Cho, C. C. and Wagner, G. M. (1983) An examination of the conceptual issues involved in developing credit-scoring models. *Journal of Business and Economic Statistics*, **1**(2), 101–114.

Stanislao, J. and Stanislao, B. C. (1983) Dealing with resistance to change. *Business Horizons*, July–August, pp. 74–78.

Tichy, N. M. (1982) Managing change strategically: the technical, political and, cultural keys. *Organizational Dynamics*, Autumn, 59–80.

7

Consumer Credit and Business Cycles

Jonathan Crook

7.1 Introduction and summary

The aim of this chapter is to survey the literature which relates consumer credit to economic cycles. The literature on this topic is vast and this chapter can give only a very brief introduction to the relationships involved. However, before we proceed, a note on modelling and the role of statistics in this chapter. In each of the subsequent sections the first parts describe theoretical economic models and the second parts describe stochastic models. This is because we are interested in understanding the relationship between business cycles and consumer credit. We proceed by making various assumptions and deducing behavioural relationships. Different assumptions result in different hypothesised relationships. Statistical methods are then applied to data to try to test the validity of these relationships. If the hypothesised relationships are observed in the data, the behavioural assumptions and relationships are retained. If the hypothesised relationships are not observed in the data we have less confidence in them. However using statistical methods for this purpose is fraught with difficulties as we will reveal in our discussions of the empirical results.

Because models of the relationship between net credit extended and the output of the economy are complex, this chapter begins with a summary of the arguments that will be made.

Cycles in real net credit extended (new credit extended minus repayments) lead those in Gross Domestic Product of the UK economy by an average of three quarters (i.e. nine months). The cyclical movements of consumer debt outstanding as a proportion of personal income lags changes in output by a considerably longer period than has been found in the US.

One measure of the money supply is the value of notes and coin in circulation outside banks plus sight deposits (deposits which can be withdrawn without giving notice) which people deposit into banks. If time deposits (deposits which can be withdrawn only after a minimum period of notice is given) and building society deposits are added, we derive another measure of the money supply. Banks issue loans which are spent and redeposited into the banking system. The deposits are re-lent and so the volume of deposits, and therefore the money supply, increases. Deposits are part of the liabilities which banks have and loans are part of their assets. As a simplification to considering the effect of increased consumer debt outstanding on the output of an economy we assume that an increase in consumer loans would increase deposits and so the money supply. We then analyse the effects of changes in the money supply on output as proposed by different macroeconomic schools of thought.

Different macroeconomic models give different conclusions as to whether changes in the money supply do effect changes in output. Traditional Keynesian theory predicts that changes in the money supply will directly effect changes in the total output. To understand other theories of how changes in the money supply effect changes in output we explain that the equilibrium level of aggregate output occurs when aggregate demand for goods and services equals aggregate supply. We explain that the aggregate demand for output is negatively related to the average level of prices and the aggregate supply of output depends on a long run sustainable level of output, the 'natural rate of output' and the difference between expected and actual prices. In the rational expectations version of the New Classical model, increases in the money supply are predicted by workers when they bargain for a new wage, and so they bargain for a wage which leads to the same volume of labour being used as before a money supply increase. The aggregate supply curve shifts in the same direction as aggregate demand and by the same amount such that they intersect at the same level of output. Hence output does not change. On the other hand, real business cycle models assume that changes in aggregate output occur because of changes in fiscal policy or productivity, not due to changes in the nominal money supply and so are not due to changes in the money supply. A further group of models, the New Keynesian models, assume that prices and wages do not instantly adjust to equate supply and demand in the labour market. An implication of this is that changes in aggregate demand as caused by changes in the money supply will affect output.

We then turn to the statistical evidence concerning the relationship between the money supply and the level of output in an economy. If changes in output can be explained by unexpected monetary or aggregate demand shocks the deviation of output from the natural rate would be temporary: output would be trend stationary. If a change in technology leads to a permanent change in output, output would be difference stationary. Various studies have tested output for these properties. Other types of statistical tests are evaluated.

This chapter progresses to consider work which has separated the effects of changes in credit from changes in the money supply, on output. One model might be called the credit shock hypothesis: that shocks to the supply of credit, such as changes in credit worthiness or credit controls, which may be independent of changes in narrow definitions of the money supply, effect changes in output. A second model includes the demand and supply for credit separately from the demand and supply of money and the market for goods. It predicts that increases in the supply of loans at each loan interest rate would increase output (and that an increase in the demand for loans would reduce output). The effect of credit as opposed to that of money has been thought by some to be responsible for downturns in the US economy. The validity of these models has been tested using vector autoregression and linear regression models. The evidence suggests that, when considered separately, both changes in credit *and* changes in the money supply effect changes in output.

We then consider the microeconomics of the loan market. We explain that in the life-cycle/permanent income hypothesis of consumer behaviour, the volume of credit demanded by a person depends on lagged values of previous income levels (aggregate income of households equals the value of total output in the economy) and of interest rates. The effects of output on the supply of credit are not well developed. However, we would expect the demand for credit to exceed its supply; that is, credit constraints apply. The existence of these constraints implies that the standard macroeconomic models of economies need to be amended. Econometric models, which estimate the effect of income on the demand for credit, try to take the magnitude of such constraints into account. Estimates of the elasticity of demand with respect to income vary, but all suggest that demand is procyclical with income, and so with output. However there are a number of difficulties associated with these studies which suggests more research on this topic is needed.

There are five further sections in this chapter. Section 7.2 presents time series data which describe movements in various measures of consumer credit over time in the UK economy. Section 7.3 reviews causal relationships from credit to economic cycles. Section 7.4 reviews the reverse relationship, and Section 7.5 concludes.

7.2   Descriptive trends

The aim of this section is to illustrate any cyclical variation in aggregate consumer credit data which may exist for the UK economy. We leave analyses of causes of the observed patterns until the following sections. Figure 7.1 shows time series plots of indices (1990 = 100) of Gross Domestic Product (GDP), a measure of the total output of the UK economy, and net consumer credit extended. Figure 7.2 shows plots of GDP and consumer debt outstanding. All three series are seasonally adjusted and in 1990 prices. The series for GDP which we have used is the 'strike adjusted' series which is used by the Office of National Statistics (ONS) as the 'reference cycle' for the UK economy. That is, the cyclical behaviour of this series is taken as indicative of cycles in the domestic output of the UK economy. The credit variables relate to the following sectors: banks, building societies, non-bank credit grantors (finance houses, check traders, money lenders etc), insurance companies and retailers. Net credit is the quarterly change in debt outstanding (and from 1987 these changes were corrected for write-offs). The debt outstanding figures are therefore the cumulated sum of the net credit extended series (up until 1987). Figure 7.1 shows that the volume of net credit extended in the economy grew at a much faster rate between 1963Q1 and 1988Q3 than did GDP. However, between 1988Q3 and 1992Q2, the volume of net credit extended plummeted as the economy moved into recession. Since 1992 the increase in the volume of net credit extended has increased very substantially. Figure 7.2, which covers the period 1976Q3 to 1995Q3, shows that between late 1976

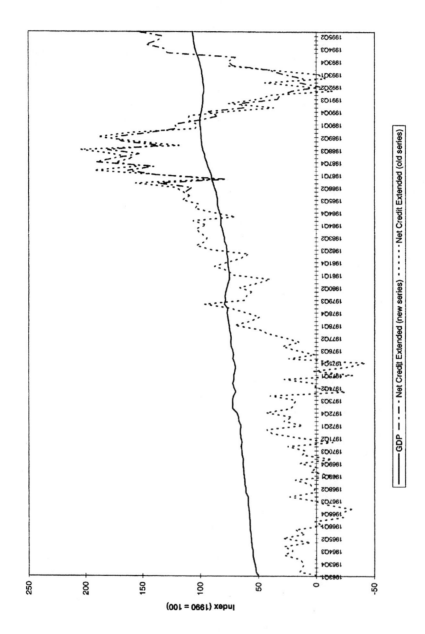

Figure 7.1 Net credit extended and GDP

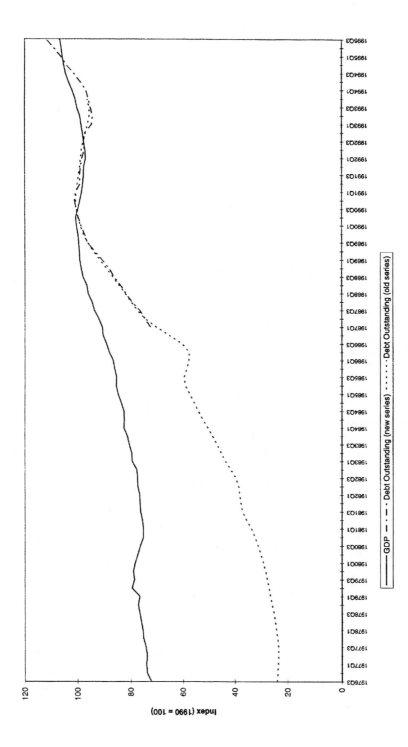

Figure 7.2 Debt outstanding and GDP

Table 7.1   Peaks and troughs in GDP and credit variables

Peak/trough	GDP	Net credit extended	Peak/trough	Debt outstanding
Trough	67Q3	66Q4		na
Peak	68Q4	68Q1		na
Trough	72Q1	69Q4		na
Peak	73Q2	73Q1		na
Trough	75Q3	75Q3		na
Peak	79Q3	78Q1		
Trough	81Q1	80Q3	Trough	81Q3
Peak	84Q1	83Q2	Peak	81Q3
Trough	85Q4	84Q4	Trough	82Q3
Peak	88Q4	89Q2	Peak	84Q3
			Trough	86Q2
			Peak	89Q4

and early 1990 the volume of outstandings has increased at a much faster rate than has GDP, but between then and late 1992 they both remained fairly constant and after early 1993 the growth rate of debt outstanding far exceeded that of output.

However, to identify *cyclical* behaviour in these series, the trend must be removed. Figure 7.3 shows detrended values of all three series. In all three cases the trend was estimated as (the antilogarithm of) a centred five year moving average of the logarithms of the (seasonally adjusted) values.[1] The deviation of the observed values from the estimated corresponding trend value, as a proportion of the trend value, was then calculated and then smoothed. The smoothing process was carried out using centred moving averages of lengths which were thought appropriate to the behaviour of the detrended series. This procedure is exactly that followed by the ONS when it estimates the reference cycle, and so the GDP series in the figure identifies the same peaks and troughs in output for the economy as have been identified by the ONS. The series for net credit extended was discontinued from 1993Q4 and for debt outstanding from 1994Q4. New series began in 1986 and 1987, respectively. Since the identification of cycles is very sensitive to the data used we have not tried to splice the new series onto the old to extend the old series. Therefore, after the filtering calculations the credit extended series ends in 1990Q2 and that for outstandings ends in 1991Q1.

The data shown in Fig. 7.3 suggest a number of conclusions. First, turning points in net credit extended do seem to lead those in GDP. For example it does seem possible to match peaks and troughs of the two series as shown in Table 7.1. Peaks and troughs were identified as major turning points in the series. Between 1966Q2 and 1990Q2 the number of quarters by which credit extended leads GDP varies between +9 (the 1969Q4 trough in extensions leading the 1972Q1 trough in GDP), and −2 (the 1989Q2 peak in extensions succeeding the 1988Q4 peak in GDP), with the mean lead being 2.9 quarters. Thus, for example, the 1967Q3 trough in GDP is preceded by the 1966Q4

[1] Twice the minimum value was added to the net credit extended series to make it positive in all quarters.

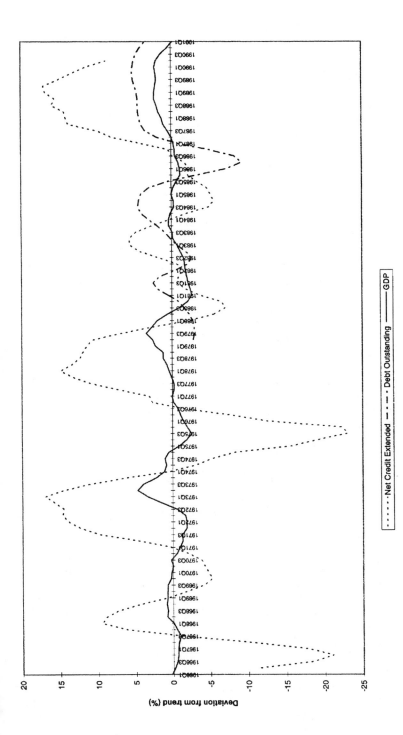

Figure 7.3 Cyclical behaviour of GDP, net credit extended and debt outstanding

trough in net credit extended and the 1979Q3 peak in GDP is preceded by the 1978Q1 peak in extensions.

These patterns suggest that net credit extended grows faster than its trend, from about three quarters, before output does likewise. This rapid rise in net extensions is due to a much greater growth rate in gross credit extended than in repayments. On average, net credit grows more slowly than its trend from a point around three quarters, before output grows slower than its trend. That is, repayments grow much faster than gross credit extended. In short, just before and during an upturn, gross credit extended grows faster than repayments, and from just before and during a downturn, people's repayments grow much faster than the gross credit they take.

Turning to the period, or length, of the cycle, between 1966Q2 and 1991Q2 the average length of the GDP cycle was 18.67 quarters. In the case of net credit extended, assuming that the first observed cycle begins in 1966Q1, the average length between 1966Q1 and 1984Q1 (the end of the last discernible cycle in the data period) was 18.25 quarters, almost the same as for the reference cycle. This adds strength to the claim that the cycles in both series appear to match each other. Notice also that although the smoothing process for net credit extended involved a longer moving average than was used for GDP and debt outstanding, (a 4 × 4 quarters MA compared with a simple 1 × 3 quarter MA or the latter two series) the proportionate deviation from trend for net extensions is far greater than for either GDP or debt outstanding. (Given that debt outstanding is the cumulated sum of quarterly net credit extended this is not a surprising result for these variables.) A 1% deviation from trend in net credit extended is followed by a much smaller percentage deviation from trend in GDP. Of course, since other variables such as interest rates, expected future income and the supply of credit may also change over the output cycle, we cannot, purely from these figures, ascribe output elasticities to the credit series. The explanations for changes in the credit series are discussed in the next section of this chapter.

Figure 7.3 also shows cyclical variation in debt outstanding, albeit for the shorter period of 1979Q1–1991Q1. In theory these values are the cumulated sum of the net credit extended series until 1987, after which both series are adjusted for write-offs. The debt series is difficult to interpret because it is difficult to match cycles in GDP and in debt. It would appear that, before 1983Q2, debt varied countercyclically with GDP; there being a trough in debt, but a peak in GDP in the same quarter: 1979Q3, and a peak in debt in 1981Q3, just following a trough in GDP in 1981Q1. But from 1983Q2, when both series were at their trend values, the two series move together, with GDP leading debt outstanding by about two quarters. These movements suggest that when the economy is expanding, people increase their debt faster than the trend rate of growth, and when the economy falls into recession, people expand their debt at a slower rate than the trend.

Figure 7.4 shows detrended values of GDP, net credit extended as a proportion of consumers' expenditure on durables, and debt outstanding as a proportion of personal disposable income. Net credit extended as a proportion of expenditure on durables shows almost exactly the same time series variation relative to GDP as does net credit extended alone. This is consistent with the argument that shortly before and during an upturn in the economy people

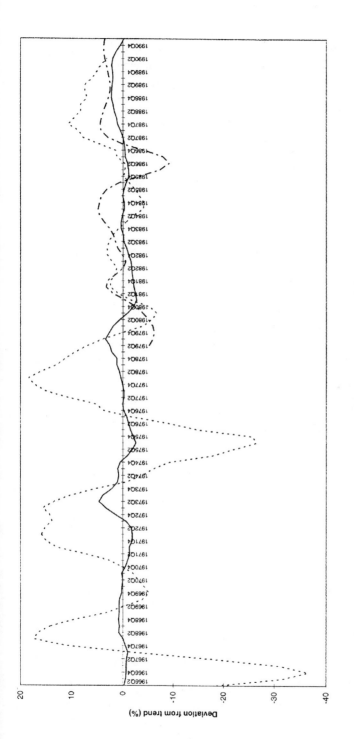

Figure 7.4 Debt outstanding as a proportion of income and credit extended as a proportion of expenditure on durables

increase the proportion of expenditure on durables, which takes the form of credit, more than the trend proportion, whilst shortly before and during a downturn in GDP relative to trend people decrease their proportion of durable expenditure, which they finance by taking credit relative to the trend. But we cannot be sure of this because we do not know the proportion of net credit extended that is used to pay for durables rather than non-durable goods and services. The variation in net credit extended relative to its trend values is greater than that of expenditure on durables. The cyclical behaviour of the series for debt outstanding as a proportion of PDI is also extremely similar to that of debt outstanding alone. From late 1982, debt as a fraction of income was procyclical, rising as the economy expanded and decreasing as the economy turned into a recession.

These patterns can be compared with those for the US as described by Sherman (1991). He calculated the ratio of consumer instalment debt outstanding to personal income relative to the average value (given a value of 100) over three output cycles between 1970 and 1982. He averaged these relative values over three cycles. He found that the average relative value of the debt income ratio fell from 98.8 to 97.7 between stages 1 and 2 (of a nine-stage cycle), then rose consistently to a value of 103 in stage 6 before declining to 98.9 in stage 9. In comparison, the relative values for the coincident indicator began at a low value of stage 1, peaked in stage 5 and decreased to stage 9. In short, changes in the debt to income ratio lagged output by about one stage. Since the duration of the stages varied slightly between the three cycles considered, the lag was between one and two quarters. This is a different pattern compared with that shown for the UK in Fig. 7.4. There we find that the peak of the debt income ratio lags output by around six quarters and the trough by about four quarters.

Unfortunately, there are no available data that allow us to observe long term patterns in changes in market shares of the sectors over several cycles. Data on sector shares are available only from 1987Q1 onwards during which time there has been only one peak and one trough in GDP. The only discernible patterns in this time are that, from 1991Q4, the share of the banks has consistently fallen as has that of insurance companies since 1993Q4. The share lost by banks appears to have been taken by other credit grantors such as finance houses, money lenders etc.

7.3   Credit and business cycle theories

7.3.1   Credit and the money supply

There are several definitions of the money supply. One measure, M1, is the volume of notes and coins in circulation outside banks plus sight deposits. When time deposits and deposits with building societies are added to M1 we gain another measure, M3.

There are several methods which a central bank· (like the Bank of England) can use to alter the money supply. One method is for the central bank to alter the reserve ratios of members of the monetary sector; that is, alter the minimum ratio of cash to sight deposits that banks create.

Secondly, it may alter the discount rate (i.e. interest rate) at which it lends to the monetary sector. Thirdly, it may sell or buy government bonds, etc. But in each case the method used results in a commercial judgement by banks as to whether, and if so by how much, to grant additional loans (or, in principle, withdraw debt) to all sectors of the economy including consumers. In addition, the banks themselves may wish to lend more, if for some reason they wish to lower their cash to deposits ratio. Suppose banks wish to make additional loans to firms and consumers and these loans are made. They are spent and most of the funds redeposited into the banking system. Banks then lend a proportion of these new deposits, which are subsequently spent and redeposited again. The cycle continues until the ratio of cash reserves to deposits which the banks choose, is reached. However, as more funds are lent and deposited, total deposits increase and so the money supply increases. The loans which the banks make are assets for the banks and the deposits which bank customers make are liabilities for the banks.[2]

Many macroeconomic models predict that changes in the money supply effect changes in the level of total output of the economy, aggregate output. Thus, changes in the volume of *consumer credit* outstanding are an intermediate mechanism between changes in monetary policy instruments and changes in aggregate output. However, almost none of the *standard* models of the economy separate the effects on aggregate output of changes in the money supply from the effects of changes in consumer (or other) credit outstanding. Nevertheless, there is a growing theoretical literature on the relationship between credit and economic cycles although the majority of it relates to business credit (Friedman and Kuttner, 1993; Blinder, 1987) rather than consumer credit. In this section we begin by explaining some 'standard' models of the economy making the assumption that changes in debt outstanding can be analysed as changes in the money supply. Subsequently we consider the implications of models which have separated the effects of changes in *credit outstanding* from those of the *money supply*.

7.3.2 A traditional Keynesian-type model

The traditional Keynesian view is that the government can reduce the amplitude of cycles in output and employment by adopting countercyclical monetary and fiscal policy. To explain this model, we shall assume, for simplicity, a closed economy consisting of households, firms and government. We will also assume that all prices are fixed. Planned expenditure, their total demand for goods and services, by these sectors consists of planned consumption (C), planned investment (I) and government expenditure (G). To

---

[2] The second, and much smaller (in terms of the stock of consumer debt outstanding) sector than banks, is the finance house sector. However, increases in the value of debt outstanding to consumers by this sector would normally be financed by Bills (promises to repay after, typically three to twelve months) or borrowing from banks. When banks make such loans to finance houses, and these loans are ultimately spent by consumers, the funds are redeposited back into banks and so the money supply increases. In addition, banks are one of the main buyers of bills. Our analysis therefore encompasses credit extended by finance houses, because much of it is ultimately borrowed from banks.

avoid ever increasing inventories or excessive demand, *planned* expenditure $(C + I + G)$ must equal *actual* expenditure, the value of *actual* output (Y). The market for goods and services is then in equilibrium. Thus, we can write:

$$Y = C + I + G \qquad \text{equilibrium condition} \qquad (7.1)$$

Assume that a businessman will add to his stock of capital if the net present value of this addition is positive. Assume also that the cost of holding inventories is higher when the real interest rate is high. Thus, the aggregate amount of additions to capital stock plus the volume of inventories which businessmen plan to hold is all negatively related to the real interest rate. If we assume a linear relationship then:

$$I = I_a - br \qquad (7.2)$$

where r is the real interest rate and I_a is the volume of investment which is independent of the real interest rate and b is a constant.

Planned consumption is assumed to depend linearly on the value of post-tax income which households have. Thus

$$C = C_a + c(Y - T) \qquad (7.3)$$

where C_a is the volume of planned consumption which is independent of income, Y denotes income and T denotes the total amount of tax paid by households, and c is a constant. The value of C_a will be affected by, amongst other things, the volume of new consumer credit taken. If this is high then consumers can plan to buy more consumption goods from a given income than if they take less credit. Substituting Equations (7.2) and (7.3) into (7.1) we derive, after some manipulation:

$$Y = \frac{C_a + I_a}{(1 - c)} + \frac{1}{(1 - c)} G - \frac{c}{(1 - c)} T - \frac{b}{(1 - c)} r$$

Given values for C_a, I_a, G, T and c, Equation (1) can be written as:

$$Y = d - er \qquad (7.4)$$

where

$$d = \frac{C_a + I_a}{(1 - c)} + \frac{1}{(1 - c)} G - \frac{c}{(1 - c)} T$$

$$e = \frac{b}{(1 - c)}$$

Equation (7.4) is known as the IS curve and shows combinations of output and interest rates where planned expenditure, total demand, equals actual output. (It is called the IS curve because the condition that planned expenditure equals actual output can be shown to imply that when there are no government or foreign sectors, planned *investment* equals planned *savings*).

Equilibrium in an economy also requires that the demand for money equals the supply of it. Keynes argued there were three demands for real money balances, (M/P), where M is the stock of nominal money (notes and coins in circulation plus deposits at current prices) and P is the price level (a weighted average of all prices in the current year relative to that in a base year). The demands are the transactions demand, the precautionary demand and the asset demand. In the first case we require real money balances to buy goods and services between dates on which we are paid. Given the level of nominal interest rates, the greater the level of output of an economy, the greater the number of transactions made and so the greater the demand for money to facilitate these transactions. Secondly, the precautionary motive refers to our holding of money in case we might need it to pay unexpected bills. Thirdly, if we hold money instead of interest earning assets ('bonds') we forego the nominal rate of interest the asset would have given us. Thus we can write:

$$L_d = l(Y, r_n)$$

where L_d denotes the demand for real money balances, Y denotes aggregate income and r_n denotes the nominal interest rate. If the demand for money is linear in both Y and r_n this becomes:

$$L_d = aY - br_n \qquad (7.5)$$

The supply of real money balances may be written as

$$L_s = \frac{\overline{M}}{P} \qquad (7.6)$$

where L_s denotes the supply of real money balances and $\overline{M}$ denotes the fixed stock of nominal money (controlled to some extent by the central bank). Since, in equilibrium, the demand will equal supply we can equate Equations (7.5) and (7.6) to give:

$$\frac{\overline{M}}{P} = l(Y, r_n) \Rightarrow r_n = \left(\frac{a}{b}\right)Y - \left(\frac{1}{b} \cdot \frac{\overline{M}}{P}\right) \qquad (7.7)$$

Equation (7.7) is the called the LM curve because it concerns liquidity and money.

Given that prices are fixed, the nominal interest rate equals the real interest rate, $r_n = r$. Then Equation (7.4) representing the equilibrium aggregate output, and Equation (7.7) representing equilibrium in the money market, can be solved to give a value of r and of Y at which both the product and money markets are in equilibrium. This is shown in Fig. 7.5 by r_0 and Y_0, given the IS curve labelled IS and the LM curve labelled LM_0.

If, for some reason, the economy was in a recession Keynesians argued that the real money supply could be increased. Suppose banks wish to, and succeed in, lending more funds to consumers and firms. At current interest rates and output this would lead to an excess supply of money, an excess demand for bonds, a rise in the price of bonds and so a reduction in their yield. A lower interest rate occurs at the original level of output. This argument could be repeated at each level of output and so the LM curve has shifted downwards. The LM curve cuts the IS curve at a lower interest rate, but at a higher output. The reduction in the interest rate has increased the demand for investment goods by firms because it has increased the net present value of each additional piece of plant etc. The reduction in the interest rate has also increased the demand for consumer credit, and so consumer goods, by households. This higher total demand would cause firms to increase output because if they did not there would be an unplanned fall in inventories.

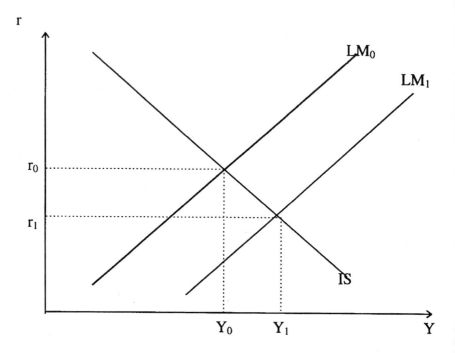

Figure 7.5 IS and LM curves

We should also add that the increasing use of credit cards has reduced people's transactions and precautionary demands for money at each level of income.

7.3.3   The new classical model

An alternative model of the economy, called the new classical model involves three groups of assumptions: rational expectations, market clearing and imperfect information. To explain this model we need to explore the concepts of aggregate demand and particularly aggregate supply further. We now relax the assumption of constant prices.

Suppose the price level, P, the weighted average level of all prices, falls. The real money supply, M/P, will rise. Given that the demand for real money balances is negatively related to the interest rate, and given the level of output and so the volume of transactions, the demand for money will equal its supply if the interest rate decreases. This is true whatever the original output level. Therefore, the LM curve will shift downwards such as from LM_0 to LM_1. The equilibrium output has increased at the lower level of prices. If this argument is repeated it can be seen that the aggregate output demanded in an economy is negatively related to the price level.

We now turn to aggregate supply. A number of different aggregate supply models have been proposed in the literature. An early such model was proposed by Friedman (1968) and is variously called the 'worker mis-perception model' (Mankiw, 1994) or the 'imperfect information model' (Gordon, 1987). In this model it is assumed that the amount of labour firms demand is negatively related to the actual real wage rate (W/P) and that firms are aware of this rate when negotiating with labour. On the other hand the amount of labour supplied by workers is positively related to the real wage they expect to occur (W/P^e), where P^e is the expected price level. Unlike firms, workers do not know the actual price level. Thus:

$$N_d = n_d(W/P) \tag{7.8}$$

$$N_s = n_s(W/P^e) = n_s\left(\frac{W}{P} \cdot \frac{P}{P^e}\right) \tag{7.9}$$

where N_d and N_s denote, respectively, the demand and supply of labour. Figure 7.6 shows equilibrium in the labour market.

Nominal wages and prices are assumed to be fully flexible and wages adjust to equate the demand and supply of labour. Now suppose the price level rises but workers did not expect this, so $P \neq P^e$. At the original level of W, (W/P) is lower than before the price level rose. Since firms know this they demand more labour. But workers believe the real wage is (W/P^e) which is higher than (W/P). So at the new (W/P) level they supply more labour than if their supply depended on (W/P), i.e. the N^s curve shifts to the right. At the original (W/P) level, there is excess supply of labour, so the real wage falls to $(W/P)_2$ to remove the excess.

Now suppose that the output of the economy depends on the amount of labour (N) and capital (K) used, where the latter is fixed (which is indicated by a bar over K): $Y = F(N, \overline{K})$. Thus, the increase in labour input from N_1 to N_2 will increase output. Suppose N_1 is the amount of labour used to produce the

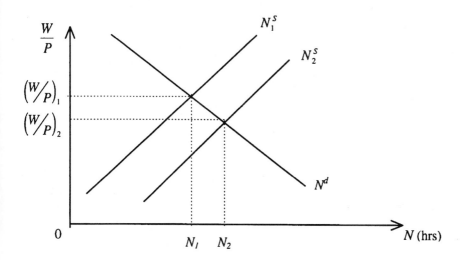

Figure 7.6   Equilibrium in the labour market

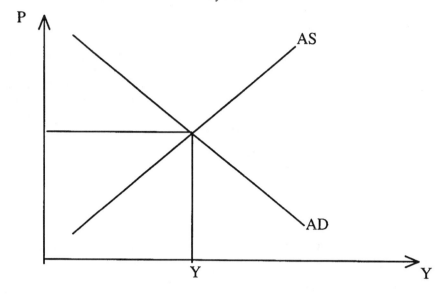

Figure 7.7　Equilibrium output in a closed economy

long run sustainable level of output – called the 'natural level' of output – Y^*. Then the difference between the actual and natural levels of output depends on the magnitude by which workers fail to anticipate future price levels:

$$Y = Y^* + \alpha(P - P^e) \tag{7.10}$$

This equation is the Friedman (or Lucas – see below) aggregate supply equation and shows the relationship between output supplied and the price level, given P^e and Y^*. The aggregate demand and aggregate supply functions are shown in Fig. 7.7.

However, Gordon (1987) and others have criticised Friedman's model arguing that workers are not likely to misanticipate prices over anything other than trivially short periods of time. Workers can observe the prices of goods they buy regularly. So Friedman's argument that workers adapt their expectations of future prices very slowly over weeks or months – the *assumption of adaptive expectations* – is unlikely to be valid. In its place Lucas (1975) proposed that every economic agent makes the most accurate forecast they can make given all the information they have at a point in time – the so-called *assumption of rational expectations*. Given this assumption, workers would be expected to notice that increases in employment followed instances where the actual price is higher than expected.

Lucas (1973) has derived an aggregate supply function for a firm, similar to Equation (7.10), without Friedman's assumption of asymmetry of information between firms and workers. Lucas' argument is as follows. Suppose an economy consists of many markets, each on a separate island. On each island there are many suppliers, and prices are competitively determined. Each supplier knows the current price on his island, but not that on other islands. It is assumed that each supplier will change his output if the price on his island, relative to that which he expects to hold for the economy as a whole, changes. If the actual price on his island exceeds (is lower than) the price he expects to hold for the economy as a whole, he will increase (reduce) his output above (below) the normal or natural level. He will do this because if all prices rise by the same proportion, including those of inputs, the output which maximises real profits is unchanged, whereas the output which maximises real profits will change if relative price changes. If the actual price equals the expected economy wide price, no output change from the equilibrium level occurs. Hence Equation (7.10) is deduced.

Notice that each supplier uses all available information to form his expectation as to the economy wide price, i.e. a *rational expectations* mechanism holds. Because each supplier does not know whether his observed price change is, or is not, a relative price change, he is said to face a 'signal extraction problem'. Given the assumption of rational expectations, Sargent and Wallace (1975) proposed the 'policy ineffectiveness proposition' (PIP) that perfectly anticipated changes in the money supply will not change output in a predictable way. To explain this, suppose the central bank increases the remaining money supply. At the original price level real money supply increases, so at the original output rate the nominal interest rate falls. This would be true whatever the original output rate so the LM curve shifts to the right. The equality of the IS and LM curves at the original price level occurs at a higher output. This would be true at each possible level of price, so the AD curve shifts to the right. This is shown in Fig. 7.8 where the initial equilibrium is at A.

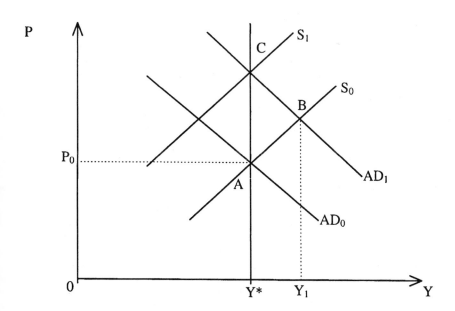

Figure 7.8   The policy ineffectiveness proposition

If the increase in the money supply was unexpected the shift in the AD curve would have been unexpected and so would the rise in price as the equilibrium moved from point A to B. Equation (7.10) would predict that output would exceed the natural level, that is Y_1 is greater than Y^*. However, if the increase in the money supply was *perfectly expected*, then the shift in the AD curve would be expected so the consequent increase in P would be expected. Since each aggregate supply curve is drawn for a given expected price level, a new and higher expected price level means the aggregate supply curve has shifted up. Given Equation (7.10), the aggregate supply curve shifts up by exactly the magnitude by which the AD curve shifts up, so output remains at the natural level, Y^*.

7.3.3.1   Monetary business cycles

Modigliani (1977) and others have criticised the Lucas aggregate supply function and rational expectations hypothesis by arguing that they cannot explain sustained deviations of output from the natural rate, yet such sustained deviations are what makes up the business cycle. According to Friedman and Lucas, output differs from the natural rate only if AD fluctuates randomly. That is, only if the actual price level differs from the expected price level and, given rational expectations, this will happen only randomly. To be exact, Lucas (1973) assumed that Equation (7.10) contained a lagged output term. Thus

$$Y_t = Y_t^* + \alpha(P_t - P_t^e) + \lambda(Y_t - Y_t^*) \tag{7.11}$$

But whilst the lagged term results in persistence, it has no theoretical rationale and is an *ad hoc* addition.

However, a number of propagation mechanisms have been proposed to change uncorrelated AD shifts into autocorrelated ones. Attfield *et al.* (1991) give an example of such a mechanism whereby firms hold inventories and respond to a random increase in AD by partially increasing output and partially reducing stocks. In the following periods, output also exceeds the natural level to replenish the depleted inventories.[3]

Therefore, the new classical model does offer an explanation of business cycles and PIP. In Lucas' model the cause of fluctuations in output is unexpected changes in the money supply. As explained earlier, changes in the money supply could be caused by changes in consumer debt outstanding.

7.3.3.2   Real business cycles

A different group of theorists from within the new classical school propose alternative sources of output fluctuations. These models, called real business cycle models, argue that the sources of fluctuations are 'real' factors such as

---

[3] Other propagation mechanisms include the serial correlation of the money supply (Lucas, 1975). Any unexpected increase in the money supply may occur. Prices rise but producers face a signal extraction problem. They believe their relative price has increased and produce more. Further serially correlated increases in the money supply, whilst not predicted, lead to the same change in prices and rational agents improve their ability to assess changes in relative prices, i.e. to see that relative prices have not changed. So in future periods output returns to the natural level.

changes in technology, preferences or fiscal policy.[4] Examples of these models are Kydland and Prescott (1982), Long and Plosser (1983) and King and Plosser (1984).

The important point for this chapter is the assumption of real business cycle models, in which changes in the nominal money supply, which may have been partly generated by an increase in debt outstanding to consumers, do not cause short run fluctuations in output. In many models the money supply is excluded altogether. In others, such as King and Plosser (1984), changes in the money supply are *a result* of changes in output. A fuller explanation of real business cycle theory is given in the Appendix.

7.3.4 The New Keynesian model

A further approach to explain short run output fluctuations, the New Keynesian School, assumes that prices and wages are 'sticky' and do not instantly adjust to clear labour and product markets. In these models, changes in credit and hence the money supply do have an effect on output. Here, we follow Fischer (1977). These models also predict that the aggregate supply function is a variant of Equation (7.10).

Assume, unlike Lucas and the new classical school, that the wage rate and price level do not completely adjust immediately to a change in the nominal money supply (some reasons for this will be given later), therefore such a change would lead to a change in output. This is explained in Fig. 7.9. Suppose nominal wages and prices are fixed. If aggregate demand decreases for some reason from AD^0 to AD^1, firms will reduce output because demand is less, and prices and wages cannot adjust. In the labour market firms will wish to hire less labour at the current real wage, although neither the supply nor demand for labour functions has moved.

In this model, nominal wages are fixed and so cannot be bid down by workers in response to unemployment, and prices are also fixed and so cannot be bid down by firms. In fact, over many time periods both the nominal wage and prices *would* actually decrease to increase firms demand for labour and output back to Y^*. However, before that is complete the

---

[4] Assume rational expectations. Assume wages in period $t + 1$, W_{t+1} are agreed based on workers' expected prices in period $t + 1$, P^e_{t+1}.

$$W_{t+1} = \delta P^e_{t+1} \tag{i}$$

where δ is the real wage workers expect to receive in period $t + 1$.

Assume the demand for labour is a negative function of the real wage rate and so, therefore, is aggregate supply, Y^s. Thus

$$Y^s_t = \alpha \left(\frac{P_t}{\overline{W_t}} \right)^\beta \tag{ii}$$

where α and β are positive constants and the bar denotes that W_t is given.

Lagging Equation (i) one period and substituting it into Equation (ii) we derive

$$Y^s_t = \frac{\alpha}{\delta} \left(\frac{P_t}{P^e_t} \right)^\beta \tag{iii}$$

This is a variant of the Lucas aggregate supply function which was given by Equation (7.10). (If α and β are scaled so that δ equals 1, and $\log \alpha$ in Equation (iii) is the natural rate of output, then the log of Equation (iii) has the same form as Equation (7.10).)

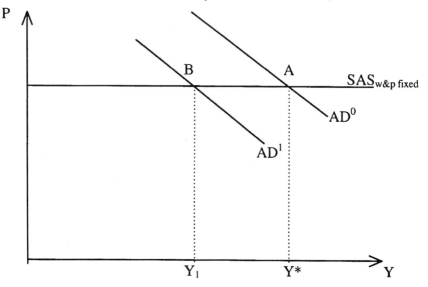

Figure 7.9   The New Keynesian model

government would increase the money supply leading to a rightward shift in the AD curve back to AD^0.

In the new classical model this increase in the nominal money supply could be *expected*, so the output would never reduce below the natural rate. However, in the new Keynesian model, because, for example, wages are set at t for periods t to $t+j$, the change in the price level in $t+j$ cannot be expected with perfect accuracy in period t. Thus, P_{t+1}^e will not equal P_t and from Equation (7.10) output will deviate from the natural rate.

New Keynesians give several reasons why wages and/or prices may not be completely flexible. For example wage rates are often negotiated for two or three year periods (Fischer, 1977).[5]

[5] Contracts for different groups may be renegotiated at different points in time (Taylor, 1979). So if contracts are negotiated every 2 years, but half of these alternate every year, only half of the contracts could be negotiated to incorporate changes in prices over the last year. The other half are not being renegotiated now. Prices may not be completely flexible because they are set by applying a mark up proportion to variable costs. If wage rates and input prices are constant, so are variable costs, and hence so are prices. Prices may also be inflexible because of so-called 'menu costs': the costs of informing customers of the new prices (Mankiw, 1985). Cooper and John (1988) argue that prices are inflexible because of failures by firms to coordinate prices. For example, suppose in a two-firm economy aggregate demand decreases due to a reduction in the nominal money supply, and a reduction in the price by both firms will shift the short run aggregate supply curve down sufficiently to restore Y^*. If the firms coordinated their price cuts, the pay-off to each would exceed those resulting from other actions. But if one cuts prices and the other did not, a recession would follow and the price cutter would make less profits than the other firm. Therefore, both firms will keep prices high to avoid being relatively worse off. Finally, arguments have been made as to why nominal wages relative to prices or other wages remain relatively constant. One of these is the efficiency wage model. It is assumed that a higher wage increases the efficiency of a worker, but non-linearly, such that there is a wage which corresponds to a minimum wage per unit of efficiency. Even if output, and hence the demand for labour by firms, decreases, firms will not wish to reduce the wage rate because this would increase the wage per unit of efficiency.

7.3.5   Empirical studies on the relationship between the money supply and output

Several empirical methods have been used to investigate the plausibility of monetary and real business cycles. One method is to test the time series data on real Gross National Product (GNP) (the output of an economy's factors of production) for trend or difference stationarities. The idea is that if business cycles can be explained by Equation (7.10) with unexpected monetary or aggregate demand shocks then the deviation of output from the trended natural rate would only be temporary. This is because, even if the unexpected monetary shock was serially correlated, firms and consumers would learn and solve their signal extraction problem, and because the persistence mechanism operates. So a shock would lead to a sudden rise and gradual return of output to its long term trend output, and would be trend stationary: its deviation around trend is stationary and with a constant variance. This may be represented as:

$$Y_t = \alpha + \alpha t + \varepsilon_t$$

where Y_t = real GNP in period t and ε_t is a random error which takes on values that may represent surprise changes in the money supply.

On the other hand, a change in, for example, technology could be expected to result in a permanently greater level of output with the same number of factors now being used more productively. The natural rate of output has risen and may continue to increase by the same amount per time period as before the shock, but from a higher level. This may be represented by the difference stationary model:

$$Y_t - Y_{t-1} = \beta + U_t \tag{7.12}$$

where U_t is a random error.

Nelson and Plosser (1982) have used a test devised by Dickey and Fuller (1979) to investigate whether real GNP and other variables coincident with business cycles are trend or difference stationary. To explain this method suppose we estimate

$$Y_t = \rho_1 + \rho_2 t + \rho_3 Y_{t-1} + \varepsilon_t \tag{7.13}$$

where $\varepsilon_t \sim iid(o, \sigma^2)$.

Trend stationarity implies $\rho_2 \neq 0$, $\rho_3 = 0$ and difference stationarity implies $\rho_2 = 0$, $\rho_3 = 1$. But if $\rho_3 \neq 0$ the OLS estimate of ρ_3 may be biased, and when Y_t is non-stationary little is known about the distribution of the student t-statistic. In addition, Equation (7.13) assumes ε_t is not autocorrelated. This may be false. Equation (7.13) is therefore reparameterised to be

$$\Delta y_t = \delta_0 + \delta_1 t + \delta_2 y_{t-1} + \sum_{j=1}^{n} \gamma_j \Delta y_{t-j} + \varepsilon_t \tag{7.14}$$

where $\varepsilon \sim iid(o, \sigma^2)$ and the null hypothesis $\delta_2 = 0$ in Equation (7.14) replaces that of $\rho_3 = 1$ in Equation (7.13). The t statistic for δ_0, δ_1 and δ_2 may then be compared with critical values calculated in Fuller (1976), Dickey and Fuller (1981) and/or MacKinnon (1990a,b). Joint tests are also possible. The lagged first difference terms are added to represent omitted autocorrelated variables since the error term is assumed to be white noise, which it may not be if these terms are omitted. Testing the null hypothesis that $\delta_2 = 0$ is to test for the existence of a unit root in Y_t.

Nelson and Plosser (1982) estimated Equation (7.14) for logs of a number of coincident indicators for the US economy using data for 1909–70 and concluded that they were difference stationary rather than trend stationary. They concluded that these, together with other findings, indicate that changes in real factors are an 'essential element' in models of business cycles. Using various ARMA non-parametric and unobserved components, approaches for seven major countries (Canada, France, Germany, Japan, Italy, UK and US), Campbell and Mankiw (1987, 1989) find that, except for the UK, shocks to output result in persistently different output over the foreseeable future. In contrast, using data for the UK for 1956 to 1991, Mullineux *et al.* (1993) found that GDP was difference stationary. That is, shocks in output did result in persistently different output. This has been found by many other authors although when seasonality is taken into account both seasonal stochastic and deterministic variation is observed (Osborn, 1990).

One of the first attempts to test the validity of the rational expectations hypothesis was by Lucas (1973). Recall Lucas's explanation of Equation (7.10). Suppose there is an aggregate demand shock in each of two economies, A and B. Suppose in economy A, in the past, the relative changes in a firm's demand have been less than those in aggregate demand. Given the assumption of rational expectations the firm will use these past relative magnitudes to predict future ones, so an unexpectedly higher (or lower) price in an A-type economy would be ascribed to a high (or low) relative price, rather than a high (or low) aggregate price; in B-type economies the reverse would be true. Such a shock in the A-type economy will imply a difference between actual and expected prices so there will be a change in output, but this is not so for the B-type economy. Therefore, in A-type economies, where aggregate demand is more volatile than B-type economies, a shock to aggregate demand would be expected to lead to a greater change in output than in B-type economies.

This argument may be represented as (see Attfield *et al.*, 1991):

$$\hat{Y}_{it} = Y_{it} - Y_i^* = \gamma_{ij}(\sigma_{vi}^2, \sigma_{\varepsilon i}^2)v_{it} \qquad (7.15)$$

where Y_{it} is the actual output for country i period t, Y_i^* is the natural output rate country i, period t, σ_{vi}^2 is the aggregate demand shock and $\sigma_{\varepsilon i}^2$ is the variance of relative demand changes (all in logs). Lucas (1973) adds a propagation term to Equation (7.15), assumes $\gamma_{ij}(\sigma_{vi}^2, \delta_{\varepsilon i}^2)$ is constant over time, and proxies v_{it} by the actual growth rate of nominal spending in country i, period t, $g_i t$, and also its mean value over the sample period, $\bar{g}_{it}$. Unexpected aggregate demand growth is assumed to be the difference between these two variables. Parameters for the regression of $\hat{Y}_{it}$ on $g_{it}, \bar{g}_{it}$ and $\hat{Y}_{it-1}$ are estimated for each of 18 countries for the period 1953–67. The theory makes two predictions: a positive coefficient on g_{it} and second, a low value of this coefficient for countries with the greatest volatility of aggregate demand. Both predictions were found. A similar result has been found by Kormendi and Meguire (1984).

Barro (1977, 1978, 1979; Barro and Rush, 1980) has tested the hypothesis that only unexpected changes in aggregate demand result in changes in output from the natural rate. These papers assume the cause of the unanticipated change in aggregate demand is an unexpected change in the money supply. A regression model explaining changes in the money supply is estimated and economic agents assumed to behave as if they used such a model to predict future money stock values. The residuals, representing unexpected changes in the money supply, were used as explanatory variables to explain changes in unemployment or output. The results using US data for various periods within the range 1946–77 indicated that unanticipated changes in the money supply did explain changes in output and that expected changes in the money supply did not add to the explanation. Similar results for the UK were gained by Attfield *et al.* (1981).

However, the empirical studies by Lucas and Barro have been subject to a considerable number of criticisms which are reviewed by Attfield *et al.* (1991). Attfield *et al.* argue that the aggregate demand variable is erroneously measured and show that the coefficient on g_{it} (see above) could be negatively correlated with the volatility of aggregate demand because the coefficient is biased, not because of the rational expectations hypothesis. The same error – an incorrect negative correlation between the coefficient on g_{it} and the volatility of aggregate demand – could also result if, as seems likely, an important variable was omitted from Lucas's empirical model of how unexpected changes in aggregate demand are determined.

In following Barro's papers, Mishkin (1982a, 1982b) included expected and unexpected money supply as explanatory variables of output changes with many more lags than in Barro's model and also estimated a two-equation simultaneous equation model to explain aggregate demand and output. Using US data for 1954–76 he found that the PIP hypotheses and that of rational expectations could be separately rejected. This suggests that expected changes in the money stock *do* affect output.

Further empirical tests of monetary versus real business cycle theories took the form of investigating whether money 'Granger causes' output. Variable X Granger causes variable Y if, in a regression of Y_t on lagged values of Y_t and lagged values of X_t, the estimated coefficients on the lagged Xs are collectively significant, i.e. they contribute to an explanation of Y_t over that explanation given by past values of Y (see Granger, 1969).

One way to investigate such a causal relationship between variables X and Y is to estimate the following regression equation:

$$Y_t = \sum_{i=1}^{k} \alpha_i S_{it} + \sum_{i=1}^{n} \beta_i Y_{t-i} + \sum_{i=1}^{n} \gamma_i X_{t-i} + \varepsilon_t$$

where S_{it} is deterministic variable i: trend, constant etc.

If the null hypothesis: $\gamma_i = 0$ $\forall i$ cannot be rejected then X does not Grange cause Y; if this null hypothesis can be rejected then X does Granger cause Y. Using a variant of this equation and US data for 1948 to 1968, Sims (1972) found that the level of the nominal money supply determined the level of nominal GNP and there was no reverse causation. However, the Lucas aggregate supply curve predicts that *unexpected* changes in the money supply determine deviations of output from the natural rate. In a later paper, Sims

(1980) used forecast error variances and innovation in (the logs of) money supply, industrial production, wholesale prices, and interest rates. He found that 37% of the forecast error variance in industrial production was explained by innovations in the money supply after the war (66% pre-war) when the interest rate was not included, but only 4% (58% pre-war) when the interest rate was included. This does not appear consistent with the rational explanations view of the monetary business cycle.

However, Hoover (1988) and others (see McCallum, 1986) have suggested that Sim's evidence is not necessarily against the monetary business cycle theory. McCallum argues that the monetary authorities in the US may have changed interest rates to effect changes in the monetary supply, so that surprise changes in both variables measure the same thing: unexpected changes in monetary policy. Only if both were unrelated to output would the evidence be against the Lucas model. Sims found that innovations in the money supply and in the interest rate explained 34% of the forecast error variance in industrial production. Secondly, Hoover (1988) notes that even if surprises in the money supply were shown not to be causally related to output, this would not necessarily be evidence in favour of real business cycle theories because certain Keynesian models would also be supported.

Recent empirical work has built on the Sims' (1980) vector autoregressive methodology. This methodology is explained more fully in the next section. Using this approach, and data relating to the US for 1965–86, Blanchard (1989) found that shocks to the supply or demand for money had no significant effect on the variance of forecast errors for output predicted by the model. Using data for the UK for 1955–89, Turner (1993) found that one-off increases in the supply or demand for money resulted in greater output over several periods and that 23% of the variance in forecast error in output two years in advance resulted from shocks to the demand or supply of money. Similar results for the UK using a model of an open economy and data for 1963–91 were obtained by Mullineux *et al.* (1993). They found that a positive shock to the money supply led to an increase in output. Unlike Turner, they found that only a very small percentage (2.2%) of the variance of forecast error in GDP 24 months in advance was due to shocks to the money supply.

7.3.6   Credit models

7.3.6.1   *The credit shock hypothesis*

So far, as a simplification, we have assumed changes in credit outstanding result in changes in the money supply, which may or may not affect output. Indeed, since credit outstanding and deposits (part of the money supply) are on opposite sides of banks' balance sheets, and since an increase in debt would be spent and redeposited as deposits, so increasing the money supply, both changes in debt and the money supply would be expected to be very closely correlated (Jaffee and Stiglitz, 1990). However, it is possible that changes in the money supply and in debt outstanding may affect output separately. Several authors have considered the effect of *credit* rather than the *money supply* on output. Notice however, that very few have considered the effect of consumer credit alone; usually they consider the effect of total or business credit.

In general terms, if banks offer a greater volume of debt outstanding to consumers, and it is taken, then planned consumption will rise. Aggregate demand ($C + I + G$ from Equation (7.1)) will increase and firms will increase their output to avoid continuous unplanned reductions in inventories.

Bernanke (1986) and Blinder and Stiglitz (1983) have argued that variations in output can be caused by shocks to business and consumer loans. They argue that if financial intermediaries are unable or unwilling to grant new loans, some borrowers would be unable to gain new loans at all. This is because potential borrowers (especially consumers) cannot issue securities because of the high risk assessment costs of potential lenders and since they have been rejected by some intermediaries they will be regarded by others as high risk. If a shock occurs to the cost or availability of funds to intermediaries or in the creditworthiness of borrowers, intermediaries will make fewer loans, and investment expenditure by firms (and also, presumably, consumption expenditure by households) will decrease, reducing aggregate demand and output. Blinder and Stiglitz argue that credit shocks also affect aggregate supply. As an example of a shock, Bernanke considers a rise in interest rates, in the US, above the maximum rate which thrift institutions can pay. This reduces deposits into these organisations, which reduces mortgage lending. Assuming that no new alternative lenders increase mortgage lending to fill the deficiency and that the decrease in housing construction is not made up by increases in other sectors of aggregate demand, output would decrease. Other examples of such shocks include changes in financial regulation such as credit controls; financial innovations such as new lending instruments or computerisation; changes in financial intermediary liabilities and changes in the credit worthiness of borrowers. Blinder and Stiglitz argue that prices would not adjust with the change in aggregate demand to retain the natural output rate because of long term contracts (see above). Blinder and Stiglitz point out that if there is credit rationing (see Section 7.4.3) in equilibrium, then the many shocks may cause changes in the volume of debt with only very small changes in interest rates.

One of the earliest tests of the credit shock hypothesis was carried out by King (1985). He presented two sets of results. First, he added four lagged growth rates of demand deposits (part of the money supply), commercial and industrial loans, other loans (real estate loans and consumer credit) and total loans, separately, to a constant and four lagged growth rate of GNP variables to predict the growth rate of GNP for 1950–79. He found the addition of demand deposits to be statistically significant, but of the credit variables only 'other loans' were significant and only in a subset of the period. Secondly, he estimated vector autoregression (VAR) equations of the form:

$$Z_t = \sum_{i=1}^{T} \beta_i Z_{t-i} + \eta_t$$

where Z_t is an $n \times 1$ column vector of variables observed at time t, β is a matrix of constants to be estimated and significance tested and η_t is an $n \times 1$ column vector of random errors. The variables were as above. Concentrating on the equation where output is the dependent variable he found that demand deposits explained substantial proportions of the variance in future output but that loans only explained a substantial proportion when the correlation between current values of output and loans was assumed to represent causation from the latter to the former.

But Bernanke (1986) criticised King's method on two accounts. First, Bernanke argued that King's measure of credit was too narrow. Bernanke uses a broader measure of credit, the sum of loans made by savings and loans and commercial and mutual banks, and estimates standard VAR models. The results still support King's original findings.

Secondly, Bernanke argues that the VAR methodology used by King requires assumed restrictions to be applied to identify the structural shocks and so the impulse response functions and variance decompositions. These assumptions are not normally based on economic theory. But when they are based on economic theory, amended VAR models lend support to the credit shock hypothesis. To understand the basic idea, consider the following VAR:

$$\begin{bmatrix} 1 & a_{11} \\ a_{21} & 1 \end{bmatrix}\begin{bmatrix} w_t \\ x_t \end{bmatrix} = \begin{bmatrix} a_{10} \\ a_{20} \end{bmatrix} + \begin{bmatrix} a_{12} & a_{13} \\ a_{22} & a_{23} \end{bmatrix}\begin{bmatrix} w_{t-1} \\ x_{t-1} \end{bmatrix} + \begin{bmatrix} \eta_{wt} \\ \eta_{xt} \end{bmatrix}$$

or

$$AZ_t = B_0 + B_1 Z_{t-1} + \eta_t$$

where

$$A = \begin{bmatrix} 1 & a_{11} \\ a_{21} & 1 \end{bmatrix} \quad B_0 = \begin{bmatrix} a_{10} \\ a_{20} \end{bmatrix} \quad B_1 = \begin{bmatrix} a_{12} & a_{13} \\ a_{22} & a_{23} \end{bmatrix} \quad \eta_t = \begin{bmatrix} \eta_{wt} \\ \eta_{xt} \end{bmatrix}$$

$$Z_t = \begin{bmatrix} w_t \\ x_t \end{bmatrix}$$

and η_{wt} and η_{xt} are white noise sequences. This implies that

$$Z_t = C_0 + C_1 Z_{t-1} + e_t \tag{7.16}$$

where $C_0 = A^{-1}B_0$, $C_1 = A^{-1}B_1$, $e_t = A^{-1}\eta_t$. Equation (7.16) may be written as:

$$w_t = c_{10} + c_{11}w_{t-1} + c_{12}x_{t-1} + e_{1t} \tag{7.17}$$

$$x_t = c_{20} + c_{21}w_{t-1} + c_{22}x_{t-1} + e_{2t} \tag{7.18}$$

where the c_{ij} values are corresponding elements in the C_0 and C_1 matrices. Notice that the e_{1t}, e_{2t} values can be expressed in terms of the η_{wt} and η_{xt} values. But the e_{1t} and e_{2t} values have no economic interpretation. If we wish to consider the effects of shocks on w_t and on x_t we need to consider the values of η_{wt} and η_{xt}. This means that, since it is Equations (7.17) and (7.18) that are estimated, we need to be able to calculate the η_{wt} and η_{xt} values from the e_{1t} and e_{2t} values. The relationship between the two sets of values is given by $e_t = A^{-1}\eta_t$ from Equation (7.16). This implies that $\eta_t = Ae_t$.

In conventional VAR analysis, a restriction is imposed which makes the system recursive, for example the restriction $a_{21} = 0$ is imposed. This implies that x is not contemporaneous with w, but varies only with a lagged value of w. This results in the well known Cholesky decomposition of the residuals.

This restriction implies particular relationships between the η_{wt} and η_{xt} values and the e_{1t} and e_{2t} values. If these relationships are not economically meaningful then the values of η_{wt} and η_{xt} will not be properly identified and the effects of shocks on w_t and x_t, and the variance decompositions, will be misleading.

Bernanke argues that, rather than applying the Choleski decomposition, economically meaningful restrictions on the elements of the A matrix should be imposed. He estimates the e_t values from estimates of the reduced form models, Equations (7.17) and (7.18), and then estimates the relationships between the e_t values and the η_t values by using relationships between the e_t values suggested by economic theory. The variables used (playing the role of x_t and w_t above) were: real defence spending, the monetary base, money supply M1, credit, a price index and real GNP. He concluded that money and credit are equally important sources of changes in output.

7.3.6.2 *Introducing credit into a macroeconomic model*

Bernanke and Blinder (1988) have developed an amended version of the ISLM model of aggregate demand (see Fig. 7.5) which allows people to store their wealth not just as 'bonds' and money, but as bonds, money and bank loans. Using this trichotomy they amend the ISLM model to allow for changes in the supply of credit as opposed to that of money. Note that they do not distinguish between consumer credit and business credit in their model (but they do test it using data on both types of credit). We now summarise their model.

First consider the market for bank loans. The balance sheets of banks are simplified to consist, on the Assets side, of loans, reserves and bonds, and on the Liabilities side, to consist of deposits only. The demand for bank loans, C_d, is assumed to depend negatively on their interest rate, r_l, and positively on GNP, Y, the latter because of the desire for credit to pay for goods and services. Bank loan demand is also assumed to be positively related to the interest rate on other assets, r_b (credit can be used to buy such assets). Thus, $C_d = f_1(r_l, r_b, Y)$. The supply of loans, C_s, is assumed to depend positively on the interest rate banks receive when making them, r_l; negatively on the interest rate banks would receive if they held bonds, r_b, instead of loans; and positively on the deposits above the minimum they need to give people money on demand, $D(1 - \phi)$ where D is total deposits and ϕ is the proportion needed to be able to cover people's demand for their deposits. Thus, $C_s = f_2(r_b, r_l)D(1 - \phi)$. In equilibrium, demand equals supply, so:

$$f_1(r_l, r_b, Y) = f_2(r_b, r_l)D(1 - \phi) \tag{7.19}$$

Now, consider the demand and supply for deposits, the money placed in banks by the consumers, firms etc, and which are liabilities for banks. This is just the LM curve of Fig. 7.5. The supply of deposits, D_s to banks equals the banks' reserves, R, multiplied by the money multiplier, θ. This product, θR, represents the increase in deposits which results when cash is deposited into banks and lent and re-lent, increasing the deposits and so the money supply at each round. The multiplier depends positively on r_b because the greater the return on bonds, the lower the excess reserves banks will hold because they will prefer to hold a greater proportion of their assets as bonds. Thus, $D_s = \theta(r_b)R$. The demand for deposits, D_d, is due to the need for money for transactions. It therefore depends positively on output, Y, and is explained in the context of the Keynesian model. It also depends negatively on the interest rate on bonds because a person foregoes bonds, and so this rate, to hold money. Thus, $D_d = f_3(r_b, Y)$. In equilibrium, demand equals supply so:

$$D_d = D_s = D = f_3(r_b, Y) = \theta(r_b)R \qquad (7.20)$$

Finally, consider the market for goods and services. The only change needed to our analysis of the IS curve of Equation (7.4) is to argue that total demand depends negatively on the interest rate on loans, r_l and on bonds, r_b. On loans, because the lower is r_l the greater is the additional debt taken, and so the greater is consumption and investment demand at each level of output. On bonds, the greater the rate of return foregone by firms when they invest in capital equipment, the less equipment they will demand. Hence

$$Y = f_4(r_b, r_l) \qquad (7.21)$$

Bernanke and Blinder then substitute the right-hand side of Equation (7.20) into Equation (7.19) and after manipulation derive

$$r_l = f_5(r_b, Y, R) \qquad (7.22)$$

They then substitute Equation (7.22) into Equation (7.21) to derive

$$Y = f(r_b, f_5(r_b, Y, R)) \qquad (7.23)$$

Bernanke and Blinder call Equation (7.22) the 'commodities and credit'

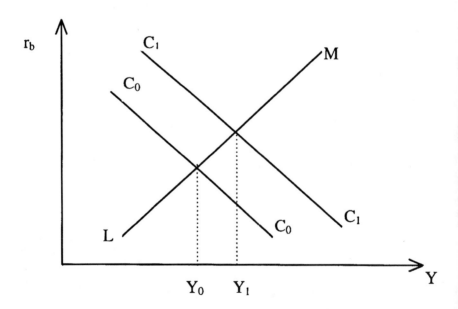

Figure 7.10   Equilibrium in the bonds, money and bank loans markets

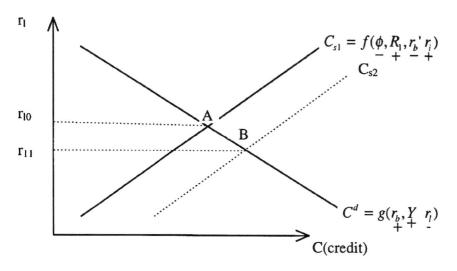

Figure 7.11   The bank loan market

curve. For a given value of R, Y and r_b are negatively related. To see this, suppose r_b rises. The opportunity cost of investing in equipment increases and firms' investment demand falls. This curve together with the LM curve of Equation (7.20) is shown in Fig. 7.10. Equilibrium occurs where the two curves intersect.

Bernanke and Blinder have computed the changes in the equilibrium values of output, money, credit and the interest rate on bonds when there is a separate rise in each of a number of variables such as credit supply, credit demand and the demand for money. To explain these intuitively we shall construct some diagrams. Figure 7.11 shows the demand and supply for credit, each as a function of the loan rate, r_l. The demand line is simply the left-hand side of Equation (7.19) after the right-hand side of Equation (7.20) has been substituted into it. Figure 7.12 shows Equation (7.21). The commodities and credit line in Fig. 7.10 is the line in Fig. 7.12.

Now suppose there is an increase in the supply of loans (they do not distinguish between consumer and business loans) – perhaps, as Bernanke and Blinder suggest, because banks perceive less risk. In Fig. 7.11 the C_s line moves to the right from C_{s1} to C_{s2}, representing the increased credit which is supplied at each loan rate. The lending rate is reduced from r_{10} to r_{11}. This shifts the Y line in Fig. 7.12 to the right from Y_0 to Y_1. At each value of r_b, more output is produced; Y is greater. So the CC curve in Fig. 7.10 shifts to the right, from C_0 to C_1. Given that the level of reserves, R, has not changed, the LM curve has not changed. So the equilibrium value of r_b has increased and so has the output, from Y_0 to Y_1. Interestingly, the model predicts that an *increase* in credit demand, a shift to the right in the C_d line in Fig. 7.11, results in a *decrease* in output.

Bernanke and Blinder offer empirical evidence on their model in the form of estimating the relative magnitudes of shocks to the demand for credit and to

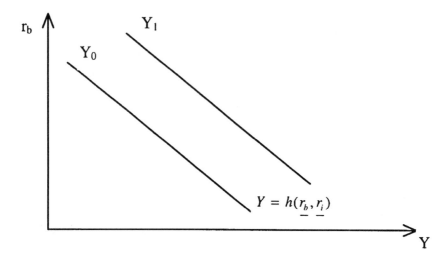

Figure 7.12   Aggregate demand and the interest rate on bonds

the demand for money. They do this by estimating the residuals from regression equations to explain, separately, the total demand for credit (household and business credit) and the demand for the money stock. Explanatory variables were output, a lagged dependent variable and interest rates. The residuals were interpreted as shocks to each demand function separately. The explanatory variables were found to have the signs predicted by the model but were, in many cases, insignificant. Their evidence suggested that shocks to the demand for money became greater relative to shocks to the demand for credit in the US economy during the 1980s. Note that the existence of any effects of these shocks on output were not empirically tested.

However, a weakness of this model is its assumption that the credit market clears; that is, that at the market interest rate demand equals supply and there is no rationing. We shall discuss this issue more fully in the next section of this chapter, although it is worth noting that Blinder (1987) has proposed a model which incorporates credit rationing, but does not include a substitute for bank credit. Blinder's model is explicitly in terms of business credit and both for this reason and because of its complexity is beyond the scope of this chapter.

7.4   Microeconomic models of consumer credit

In this section we consider microeconomic models of the consumer credit market. These will help us to consider the reverse direction of causation of Section 7.3: that the level of aggregate output determines different measures of consumer credit. The volume of net credit extended (new credit extended less repayments) in any period, is determined by factors affecting the *ex ante* demand and supply for it. By *ex ante* demand we mean the amount people wish to take at each possible value of income and other determining variables, and by *ex ante* supply we mean the amount which lenders wish to lend at each possible value of each determining variable. The *ex ante* amounts will only equal the actual amounts taken and lent when the market is in equilibrium. However, there are theoretical reasons and empirical evidence to suggest that credit is rationed (see Crook, 1996; Jaffee and Stiglitz, 1990). In this section we first consider demand, then supply, then we consider some econometric issues and finally empirical estimates of the income elasticity of demand for credit.

7.4.1   Demand for consumer credit

The most popular theory of the intertemporal utility maximising consumer is the life cycle/permanent income hypothesis (LCH/PIH) (Modigliani, 1986). Assume initially that an individual can borrow (or lend) any amount at the same interest rate. The intuitive idea is that over his/her lifetime the volume of consumption will rise and then fall, but not by the same amount as income. Early and late in life, an individual's income is less than his/her consumption. The individual will borrow early in life, repay these debts and save in the middle stages, and dissave in the later stages of his/her life. This is shown in Fig. 7.13.

The individual is assumed to maximise his/her utility over his/her lifetime subject to the constraint that the present value of his/her labour income equals that of his/her consumption. Given that future levels of consumption are

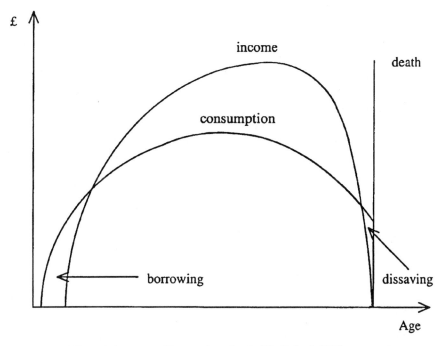

Figure 7.13   Life cycle/permanent income hypothesis (Modigliani, 1986)

uncertain and that delaying benefits reduces the present value of utility by the rate of time preference, we can more formally express the model as follows:

$$\max E_0 \sum_{t=1}^{T} \phi_i^t U(C_{it}) \qquad \text{where } \phi_i = 1/1 + \rho_i \tag{7.24}$$

such that

$$A_{it} = A_{it-1}(1 + r_{it}) + Y_{it} - C_{it} \qquad t = 1 \ldots T - 1 \qquad C_{it} \geqslant 0, \quad A_{iT} \geqslant 0 \tag{7.25}$$

where E is the expectation operator, ρ_i is individual i's rate of time preference, U is utility, C_{it} is the individual's consumption during period t, A_{it} is the individual's net wealth at the end of period t, r_{it} is the interest rate for individual i period t, and Y_{it} is individual i's labour income period t, and T is the length of the individual's life. If a particular functional form for the utility function is assumed it is possible to derive an expression for optimal C_{it}, C_{it}^*, in terms of Y_{it} and r_{it} (Japelli and Pagano, 1989), say

$$C_{it}^* = f(Y_{it}, r_{it}) \tag{7.26}$$

In any particular period, the individual's desired change in net assets is

$$A_{it} - A_{it-1}(1 + r_{it}) = b_{it} = Y_{it} - C_{it}^* \tag{7.27}$$

where C_{it}^* is the individual's utility maximising volume of consumption in period t (Hartropp, 1992). By substituting Equation (7.26) into Equation (7.27) we can derive an expression for the individual's desired volume of net change in assets in period t:

$$b_{it} = \phi(Y_{it}, r_{it}) = Y_{it} - f(Y_{it}, r_{it}) \tag{7.28}$$

Hartropp (1992) assumes income in period t is unknown in period 0 but that the expectations process can be proxied by a distributed lag function of previous income.

If the value of b_{it} is summed over only those for whom it is negative, the desired aggregate volume of net lending, B_t is derived: $B_t = \sum_{i\in\phi} b_{it}$ where ϕ is the set of individuals for whom $b_i < 0$. Following Jappelli and Pagano, if in period τ the value of b_{it} for an individual is cumulated (with accrued interest) from $t = 1$, we gain the stock of assets in period t for that individual

$$A_{it}^* = \sum_{t=1}^{\tau} b_{it}(1 + r_{it})^{\tau-t} \tag{7.29}$$

If this is summed over all such individuals for which A_t^* is negative, we gain the aggregate stock of desired debt outstanding in period t, D_t:

$$D_t = \sum_{i\in Q} A_{it}^* \tag{7.30}$$

where Q is the set of individuals for whom $A_{it}^* < 0$.

Hartropp (1992) argues that the nature of the effect of an increase in income is ambiguous. To explain why this is so, consider a two-period version of the LCH/PIH. This is shown in Fig. 7.14.

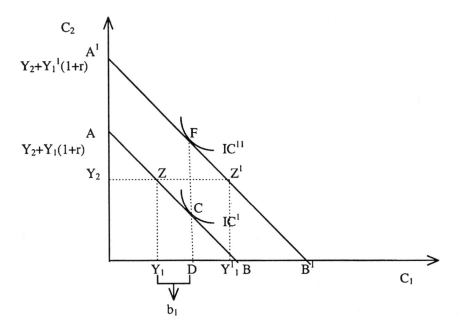

Figure 7.14   An individual's demand for credit

Suppose an individual receives income of Y_1 and Y_2 in periods 1 and 2 respectively, point Z. If he decides to save some of his income in period 1, and so reduce his consumption in period 1 by ΔC_1, he can increase his consumption in period 2 by $\Delta C_2 = -\Delta C_1(1+r)$. Therefore, $\Delta C_2/\Delta C_1 = -(1+r)$, which is the slope of the inter-temporal budget constraint AB. If he invests Y_1 at r, his total available funds for consumption in period 2 is $Y_2 + Y_1(1+r)$, which is represented by point A. If he borrows to consume more than Y_1 in period 1 he can increase consumption in period 1 by $Y_1 + Y_2/(1+r)$ (i.e. Y_1 plus Y_2, less the interest at rate r), which is point B.

The utility function is represented by an infinite series of lines like IC, each having the property of representing a constant level of ordinal utility. A utility maximising individual will locate where the budget line is just tangential to the indifference curve furthest from the origin (point C) and so will borrow b_1. The ambiguity of the effect of a change in income in period 1 from Y_1 to Y_1', say, can be observed directly from Fig. 7.14. If Y_1 does so increase, the budget line will move outwards parallel to itself to $A'B'$. The effect on borrowing depends on the utility function. If the utility function includes indifference curve IC'', consumption in period 1 remains constant at D and borrowing is reduced – in the diagram the individual becomes a net saver and saves $Y_1'D$. If the highest indifference curve is a tangent to $A'B'$ between Z' and B', the individual is still a borrower, but the desired amount of borrowing relative to the original amount, b_1, depends on exactly where the highest indifference curve is along $Z'B'$.

7.4.2 Supply of consumer credit

One model of the supply of credit by banks was described in Section 7.3.5. Most empirical studies which estimate the demand function for consumer credit assume that the volume of consumer credit supplied is infinitely elastic at an exogenously determined borrowing rate, i.e. the aggregate supply curve is horizontal (Crook, 1989; Hartropp, 1992; Garganas, 1975).

7.4.3 Credit constraints

7.4.3.1 *Theory*

It is well known that an individual may apply for a loan and have his/her application turned down by the credit supplier. If (s)he is unable to obtain the full amount of credit requested from any source to which (s)he is willing to apply at the given interest rates, the individual is said to be credit constrained. This situation for an individual may be represented as in Fig. 7.15. Line D represents the *ex ante* demand for an increase in net credit extended and line S, the supply. Suppose the interest rate is exogenously determined at r_0. If the individual is unable to gain as much net credit extended as is indicated by point X, he or she is credit constrained.

In terms of the LCH/PIH model of Section 7.4.1 an additional constraint would be added such as $A_t \geqslant 0$ (Zeldes, 1989). In the two-period diagrammatic version of Fig. 7.14 the budget line of the individual would become vertical

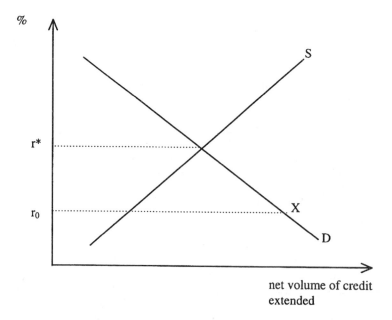

Figure 7.15 The demand and supply for credit

before it reached point B and the highest indifference curve tangent to the new budget line would touch it at the cusp (see Clayton *et al.*, 1979 for amendments in an imperfect market).

It may be optimal for a credit supplier to reject a particular credit application rather than charge a higher interest rate for a number of reasons. First, a supplier may be willing to offer more credit at higher interest rates where the higher interest rates cover the greater risk of granting a larger loan. However, beyond a certain loan size, the interest rate is so high that default is certain and so, at even higher rates, only smaller loans will be given so as not to increase even further the chance of default. It is then possible that at every conceivable interest rate the customer desires more credit than the supplier is willing to offer. Second, Jaffee and Modigliani (1969) argue that banks classify firms seeking loans into groups according to their characteristics, like size. All members of a group are charged the same interest rate despite differences in risk. Then, a member of a group could be offered less credit than it wishes, at the group rate. The rate is not raised for the firm because this would contravene the aim of the classification, which is to simplify the setting of rates. The same could, in theory, apply to consumers. In practice, lenders decide whether to offer a consumer a loan or not at the same rate as every other accepted applicant. The risk of each applicant is assessed, but any one supplier would not offer credit at a higher interest rate to one applicant, if the assessed risk was higher, than to another applicant. To gain the loan the rejected applicant would have to apply to higher rate suppliers.

Recent research has concentrated on a third reason for rationing, which is based on information theory. Stiglitz and Weiss (1981) argue that if the quoted rate of interest is increased, the return that a bank expects to receive rises, reaches a maximum and may eventually fall. Therefore, the bank will not raise the quoted rate above the level giving the maximum expected rate of return. At this quoted rate, demand for loans may exceed supply and non-price methods may be used to allocate credit. The explanation for the relationship between the expected and quoted interest rate is that if the quoted rate is increased, those who were likely to default at the lower rate will continue to apply and those who would be likely to repay reduce their demand for loans. So the overall risk of the supplier's loan portfolio increases.

The existence of credit constraints, called 'liquidity constraints' in the microeconomic theory of a consumer, enables us to describe a weakness of the earlier analyses in Section 7.3.2. In that section we said that when banks wish to increase their debt outstanding to persons in the economy, this credit would be taken, i.e. demanded. This implies that either the credit was formerly rationed at the current loan rate or that the rate which banks charge to borrowers is reduced (as in Bernanke and Blinder's model in Section 7.3.4).

To explain this, suppose the supply of loans curve shifts to the right, from S_0 to S_1 in Fig. 7.16. In the case of rationing the interest rate is r_0 and before the shift the volume of unsatisfied demand was $d_3 - d_1$. After the shift, $d_2 - d_1$ additional debt is borrowed. In the case of a decrease in the interest rate and no rationing, the initial equilibrium interest rate is r_1 and with the higher supply, the new rate is r_0.

The observed existence of rationing (Crook, 1996; Jappelli, 1990) suggests that the effect of an increase in credit supply does not reduce the interest rate

Interest rate

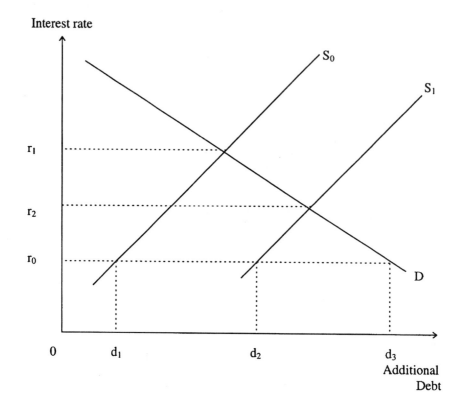

Figure 7.16   The effects of a liquidity constraint

on loans. However, the additional credit taken would increase aggregate demand by consumers and firms in the Keynesian and Monetarist models. Hence output would be increased.

A further implication for macroeconomic models is that a credit constrained consumer will be unable to consume as much as (s)he wishes and so her/his consumption expenditure will be more highly correlated with her/his current disposable income than a non-constrained consumer. This is often expressed by saying that consumption is 'excessively sensitive' to current income. Several authors (Hayashi, 1985; Flavin, 1985; Hall and Mishkin, 1982) have found that current income is correlated with current disposable income.

The prediction that changes in output would occur with very little, if any, change in interest rates has been empirically observed by a number of researchers (Wojnilower, 1980; Jaffee and Stiglitz, 1990).

7.4.3.2   *Empirical issues*

Econometrically there are several ways of dealing with a disequilibrium market when examining that market over time (Jaffee, 1971; Fair and Jaffee,

1972; Quandt, 1978; Ito and Ueda, 1981). These methods include, first, estimating cross-sectional credit supply functions over periods when rationing is expected to vary; secondly, gaining survey evidence of the magnitude of credit rationing; and thirdly, when building time series models of demand and supply of net credit, other techniques may be used. Consider the model:

$$Q_t^D = \alpha + \beta_1 r_t + \beta_2 Z_t + \varepsilon_t \qquad t = 1 \dots n \qquad (7.31)$$

$$Q_t^s = \gamma + \delta_1 r_t + \delta_2 X_t + \eta_t \qquad t = 1 \dots n \qquad (7.32)$$

where Q_t^D and Q_t^s are *ex ante* amounts of credit demanded and supplied respectively, r_t is the interest rate, Z and X are other exogenous variables in demand and supply respectively, and ε_t and η_t are error terms. In a market where there is no rationing $Q_t^D = Q_t^s = Q_t$, where Q is the volume of the good or service actually transacted. When there is rationing

$$Q_t = \min(Q_t^D, Q_t^s) \qquad (7.33)$$

If all observations are related to periods of rationing, they all relate to *ex ante* supply, not demand.

Jaffee (1971) and Fair and Jaffee (1972) explained several methods which might be applied in these cases. It is important at this point to distinguish between a market which is a rationed *equilibrium* and one in a rationed *disequilibrium*. In the former there is excess demand but the interest rate is not adjusting, albeit even slowly, to remove the excess and bring the market into equilibrium. In the second case the interest rate is slowly adjusting to bring the market into equilibrium.

If the market is of the second type, a rational disequilibrium, observations may be ascribed to either demand or supply, according to whether the interest rate has increased or decreased over the last period. Alternatively, the change in price may be made a function of excess demand (see Fair and Jaffee, 1972; Sealey, 1979). A further alternative (Ito and Ueda, 1981) is to model the current interest rate as a weighted average of the previous period's rate and the equilibrium rate, with the weights differing according to whether the observed rate increases or decreases over the last period. By substitution, an expression for the equilibrium rate in terms of the X and Z variables (see above) and the demand and supply functions may be identified.

A further technique is to include a proxy for the size of rationing. Thus, $Q_t^D = Q_t^s + J_t$, where J_t is rejected demand. Let $J_t = f(M_t)$ where M_t proxies rejected demand. Then $Q^D = \gamma + \delta_1 r_t + \delta_2 X_t + f(M_t) + e_t$ from Equation (7.32).

Of the relatively few econometric estimations of demand functions for consumer credit, most have not modelled rejections explicitly, for example, Garganas (1975) and Hartropp (1992) in the UK and Barth *et al.* (1983) for the US personal loan market. Methods which have been used to include rejections include: gaining estimates of rejection rates by survey and empirically modelling them, for example Greer (1973) for the US personal loan market, and using a proxy, for example Crook (1989), for credit extended by British retailers.

Table 7.2 Elasticities of demand for consumer credit in the UK

Variable	Crook (1989)		Elasticities		
	Long run	Short run	Garganas (1975)	Hartropp (1992)	Allard (1979)
			Short run	Long run	
Personal disposable income	+1.73	+1.02	+0.48 to +1.41	+0.46	+0.61
Real interest rate	—	—	−0.20 to −0.38	−0.75	—
Terms control	−0.22	−0.13	−0.36 to −0.49	−2.3	—
Vacancies	+0.07	+0.04	—	—	—

7.4.4 Results

The results of estimates of demand functions for credit extended for sectors within the UK are shown in Table 7.2 in the form of elasticities.

It should be noted that unlike the LCH/PIH theory above, the dependent variable of each study, except for Hartropp, relates to gross credit extended. The range of income elasticities varies from +1.73 (Crook for retailers) to +0.46 (Hartropp). But all studies show credit demand as being procyclical.

However, many of these studies may be criticised. None of the studies test for the existence of a cointegrating relationship between the variables. The economic theory of the optimal volume of the desired volume of net credit extended is an equilibrium model: each individual is assumed to be in equilibrium when (s)he is in a utility maximising position. Econometrically, in equilibrium there is a meaningful relationship between the current dated variables. Yule (1926) showed that it is possible to gain excellent t-statistics by regressing one time trended variable on another, yet all that is being correlated is the time trend. A careful examination of the residuals may show serious autocorrelation of various orders. For a meaningful identification of a long run equilibrium relationship, variables must be cointegrated.

A series is said to be integrated order d, denoted $I(d)$ if, after differencing d times, it has a stationary invertible non-deterministic ARMA representation. Suppose a vector X_t consists of k variables. Engle and Granger (1987) defined variables to be cointegrated of order d, b denoted $CI(d, b)$, if all of the variables are $I(d)$ and there exists a linear combination of them $Z_t = \alpha' X_t$ which is $I(d - b)$ where $b > 0$. The vector of constants, α', is known as the cointegrating vector. If $d = b = 1$ and y_t and x_t are cointegrated then there is a stable relationship between the levels of y_t and x_t over time and the cointegrating linear combination represents a long run equilibrium relationship. If a long run relationship in this sense exists then the variables must be cointegrated (Muscatelli and Hurn, 1992). Furthermore, Engle and Granger (1987) proved that if two variables, y_t and x_t, are $CI(1, 1)$ there exists an Error Correction Mechanism (ECM) of the form:

$$\Delta Y_t = \alpha + \sum_{i=0}^{n} \beta_{1i} \Delta x_{t-i} + \sum_{i=1}^{n} \beta_{2i} \Delta Y_{t-i} + \beta_3 z_{t-1} + \varepsilon_t$$

where $z_t = y_t - \alpha - \beta x_t$ and $\varepsilon_t \sim iid(0, \sigma^2)$.

The ECM representation allows one to separate explanations of short run movements in y_t from the long run relationship, z_t.

Garganas (1975) and Crook (1989) assumed that a stock adjustment model represents changes in gross credit extended. In both cases one or more lagged variables were included in the empirical model, so they cannot have estimated an equilibrium relationship. Most of Garganas' models indicate severe first-order autocorrelation and the model that does not, has a second-order autoregressive process to model the residuals. Crook, however, takes fourth differences to remove seasonal variation and did not find evidence of first-order autocorrelation. A difficulty with Hartropp's (1992) work is that he finds that the stock of debt (and income and wealth) has a unit root in levels; that is, the levels are $I(1)$. Therefore, there could be a cointegrating relationship between the levels of debt and the other variables. However, the theoretical LCH/PIH model (above) specifies an equilibrium relationship between the *change* in debt outstanding and *levels* of explanatory variables. Therefore, if the first difference in debt is stationary, but the levels of the explanatory variables are $I(1)$, these variables cannot cointegrate. Hartropp regresses the first difference in debt on the first differences in the explanatory variables. Whilst they are all stationary, the result is not a long run relationship nor is it implied from the theory.

Whilst we have concentrated on studies relating to UK data, no papers relating to US data are known to test for cointegration.

7.4.5 Conclusions

The volume of net consumer credit extended in any period depends on the amounts demanded and supplied at various values of the endogenous variables. Rationing makes the demand function difficult to estimate. Various models have been estimated and they give a range of income elasticities, but all suggest that credit extended is procyclical. Only Hartropp has begun to test the variables for a cointegrating relationship and his results can be interpreted as suggesting that the net change in credit extended is not cointegrated with income. Rationing also means that the transmission mechanism of many standard macroeconomic models as we described them may require amendment.

7.5 Summary

This chapter has considered alternative theories of the relationship between credit and output in an economy. The role of statistical analysis has been to provide evidence to support, or otherwise, hypothesised relationships between these two variables.

Data suggest that cycles in the volume of net credit extended lead cycles in GDP for the UK economy by approximately three quarters. As a simplification we assumed that changes in the volume of debt outstanding, which has been extended by banks, result in changes in the money supply. Much of these changes in debt will be in the form of loans to consumers. The standard models of the economy differ as to the effect of a change in the money supply on output. A traditional Keynesian model would predict a positive relationship whereas Monetary Cycle Theorists believe that only unexpected changes in the money supply lead to deviations of output away from the natural rate. Real business cycle theorists do not regard changes in the money supply as a determinant of business cycles, only an effect of such cycles. New Keynesians see changes in the money supply as causing cycles. However, if we allow the possibility that changes in the volume of credit may act in a separate way compared with changes in the money supply, Bernanke and Blinder's theoretical model shows that increases in credit supply still result in increased output, but an increase in credit demand results in an output fall.

The statistical evidence has taken various forms, which we have briefly summarised in this chapter. One approach has been to test output for difference, as opposed to trend, stationarity using, for example, the augmented Dickey–Fuller test. Another method has been to estimate the residuals from a regression of the demand for money on explanatory variables, to regard these as unexpected changes in the money supply, and then to regress output on these and other variables. A third method has been to estimate vector autoregressive models of an economy and examine the impulse response functions of a shock in the money supply (or demand) on output and to examine the proportion of the variance of forecast errors in output due to monetary shocks. The VAR method has also been used to distinguish between the effects of shocks in credit on output and shocks to money supply M1. Finally, the relative magnitude of residuals from regression estimates of the demand for credit and the demand for money functions have been compared.

The results suggest that changes in real factors such as changes in technology have important effects on output. Whilst contrary evidence exists, there is support for the hypothesis that unexpected changes in the money supply also cause changes in output, and some evidence has been found that even expected changes in the money supply also affect output. However. recent VAR models have revealed widely differing magnitudes for the effects of shocks in the demand and supply of money. The small amount of evidence which separates credit from the money supply supports the idea that they work in parallel. Clearly more research on the effects of changes in the demand and supply of credit on output is needed.

On the other hand, variations in output – business cycles – determine variations in net credit extended. The amount of net credit extended depends on the *ex ante* demand and *ex ante* supply of it. However, credit is rationed; this implies that the macroeconomic models of Section 7.3 may need revision.

The LCH/PIH is the standard model of the optimum volume of credit which an individual will demand. A statistical difficulty which has to be addressed is how to model rationing. Some studies have attempted this by using a proxy, ascribing observations to periods of excess demand or of excess supply, modelling the adjustment process of an interest rate and other methods. Econometric models of the demand for credit have found it to be procyclical with an elasticity of between 1.73 and 0.46. However, many of these studies are flawed and further research in this area is also sorely needed.

Acknowledgements

I would like to thank Stuart Sayer, Phil Bowers and Nicholas Terry for very helpful discussions and an anonymous referee for helpful comments. None are in any way responsible for any errors or omissions from this chapter.

Appendix 7.1

Real business cycle models

Real business cycle models are rather complex and we give here only a very simple summary, following Attfield *et al.* (1991), of that by Kydland and Prescott. Assume that there is a large number of identical individuals who each wish to maximise their expected utility (U), which in turn is a function of their consumption C_{t+i}, and leisure, l_{t+i}, say:

$$U = E_t\left[\sum_{i=0}^{\infty}\gamma^i(\phi \log(C_{t+i}) + (1 - \phi)\log l_{t+i})\right]$$

where γ is the discount factor which individuals apply to a future utility to compensate for the delay in its receipt.

Output, Y_t, is assumed to be a function of inputs of labour, N_t, capital K_t, and the state of technology, Z_t, say: $Y_t = Z_t N_t^{\delta} K_t^{1-\delta}$. Equilibrium requires aggregate output to equal aggregate demand. Then McCallum (1989) shows that if technological shocks, Z_t, follow a first-order autoregressive path, capital stock and consumption will follow a second-order autoregressive process.

The intuition behind this is as follows. 'Real aggregate demand', i.e. the demand for goods and services at each level of real interest rate, is determined independently of the level of the money supply – real aggregate demand is determined by the equality of planned expenditures and actual output, the Keynesian IS curve. This is shown in Fig. 7.17. It has a negative gradient because investment and consumption demands are negatively related to the real interest rate. If the interest rate rises, consumption and investment demands decrease, aggregate demand falls and so does its equality with expenditure.

On the other hand real aggregate supply is positively related to the real interest rate because at a higher real interest rate, the real wage in the current period is greater relative to the real wage in future periods. Therefore, workers supply more services in the current period, relative to future periods, than at a lower interest rate. In real business cycle theory, short run fluctuations in output are caused by shifts in the 'real aggregate supply' and/or the 'real aggregate demand' functions. Two examples of such causes are changes in fiscal policy and changes in technology. Thus, a change in technology may increase the marginal product of labour. At each possible real wage, more labour is demanded, the demand for labour schedule shifts to the right and more labour, and so more output, is produced, all at each possible interest rate. Thus, the real aggregate supply function in Fig. 7.17 shifts to the right, to RAS_1. The new technology may also lead to greater demand at each interest

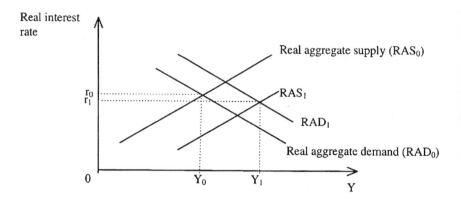

Figure 7.17   The effect of a technological shock on output

rate so the real aggregate demand shifts to the right also, to RAD_1 and output rises.

References

Allard, R. (1979) Credit restrictions and lag distributions on consumer durable equations. *Applied Economics*, **11**(2), 171–84.

Attfield, C. L. F., Dremery, D. and Duck, N. W. (1981) Unanticipated monetary growth, output and the price level: UK 1946–1977. *European Economic Review*, **16**, 367–385.

Attfield, C. L. F., Demery, D. and Duck, N. W. (1991) *Rational Expectations in Macroeconomics*, 2nd edition. Oxford: Blackwell.

Barro, R. J. (1977) Unanticipated money growth and unemployment in the United States. *American Economic Review*, **67**, 101–115.

Barro, R. J. (1978) Unanticipated money, output and the price level in the United States. *Journal of Political Economy*, **86**, 549–581.

Barro, R. J. (1979) Unanticipated money growth and unemployment in the United states: reply. *American Economic Review*, **69**, 1004–1009.

Barro, R. L. and Rush, M. (1980) Unanticipated money and economic activity. In S. Fischer (Ed), *Rational Expectations and Economic Policy*, Chicago: Chicago University Press.

Barth, J. R., Gotur, P., Manage, N. and Yezer, A. J. M. (1983) The effect of government regulations on personal loan markets: a tobit estimation of a microeconomic model. *The Journal of Finance*, **38**, 1233–1251.

Bernanke, B. S. (1986) *Alternative Explanations of Money-income Correlation*, Carnegie-Rochester series on Public Policy, **25**, 49–100.

Bernanke, B. S. and Blinder, A. S. (1988) Is it money or credit or both, or neither? *American Economic Review Papers and Proceedings*, **78**, 435–439.

Blanchard, O. (1989) A traditional interpretation of macroeconomic fluctuations. *American Economic Review*, **79**, 1146–1164.

Blinder, A. S. (1987) Credit rationing and effective supply failures. *Economic Journal*, **97**, 327–352.

Blinder, A. and Stiglitz, J. (1983) Money, credit constraints and economic activity. *American Economic Review*, **73**, 257–276.

Campbell, J. Y. and Mankiw, N. G. (1987) Are output fluctuations transitory? *Quarterly Journal of Economics*, **102**, 857–880.

Campbell, J. Y. and Mankiw, N. G. (1989) International evidence on the persistence of economic fluctuations. *Journal of Monetary Economics*, **23**, 319–333.

Clayton, G., Crook, J. N. and Sedgewick, R. (1979) An analysis of the determinants of the demand for consumer credit in the UK; the model. Paper submitted to Office of Fair Trading.

Cooper, R. and John, A. (1988) Co-ordinating co-ordination failures in Keynesian models, *Quarterly Journal of Economics*, **103**, 441–463.

Crook, J. N. (1989) The demand for retailer financed instalment credit: an econometric analysis. *Managerial and Decision Economics*, **10**, 311–319.

Crook, J. N. (1996) Credit constraints and US households. *Applied Financial Economics*, **6**, 477–485.

Dickey, D. A. and Fuller, W. A. (1979) Distribution of the estimators for autoregressive time series with a unit root. *Journal of the American Statistical Association*, **74**, 427–431.

Dickey, D. A. and Fuller, W. W. (1981) Likelihood ratio statistics for autoregressive time series with a unit root. *Econometrica*, **49**, 1057–1071.

Engle, R. F. and Granger, C. W. J. (1987) Cointegration and error correction: representation, estimation and testing. *Econometrica*, **55**, 251–276.

Fair, R. C. and Jaffee, D. M. (1972) Methods of estimation for markets in disequilibrium. *Econometrica*, **40**, 497–514.

Fischer, S. (1977) Long term contracts, rational expectations and the optimal money supply rule. *Journal of Political Economy*, **89**, 974–1009.

Flavin, M. (1985) Excess sensitivity of consumption to current income: liquidity constraints or myopia? *Canadian Journal of Economics*, **18**, 117–136.

Friedman, M. (1968) The role of monetary policy. *American Economic Review*, **58**, 1–17.

Friedman, B. M. and Kuttner, K. N. (1993) Economic activity and the short-term credit markets: an analysis of prices and quantities. *Brookings Papers on Economic Activity*, **1993(2)**, 193–266.

Fuller, W. A. (1976) *Introduction to Statistical Time Series*. New York: Wiley.

Garganas, N. C. (1975) An analysis of consumer credit and its effects on purchases of consumer durables. In G. Renton (Ed), *Modelling the UK Economy*, London: Heinemann.

Gordon, R. J. (1987) *Macroeconomics*, Boston: Little Brown.

Granger, C. W. J. (1969) Investigating causal relations by econometric models and cross spectral methods. *Econometrica*, **37**, 424–438.

Greer, D. (1973) An econometric analysis of the personal loan market, Technical Study No. IV, National Commission on Consumer Finance.

Hall, R. E. and Mishkin, F. S. (1982) The sensitivity of consumption to transitory income: estimates from panel data on households. *Econometrica*, **50**, 461–481.

Hartropp, A (1992) Demand for consumer borrowing in the UK, 1969–90. *Applied Financial Economics*, **2,** 11–20.

Hayashi, R. F. (1985) The effect of liquidity constraints on consumption: a cross-sectional analysis, *Quarterly Journal of Economics*, **100,** 83–206.

Hoover, K. D. (1988) *On The New Classical Macroeconomics*, Oxford: Basil Blackwell.

Ito, T. and Ueda, K. (1981) Tests of the equilibrium hypothesis in disequilibrium econometrics: an international comparison of credit rationing. *International Economic Review*, **22,** 691–708.

Jaffee, D. M. (1971) *Credit Rationing and the Commercial Loan Market*, New York: Wiley.

Jaffee, D. and Modigliani, F. (1969) A theory and test of credit rationing. *American Economic Review*, **59,** 850–872.

Jappelli, T. and Pagano, M. (1989) Consumption and capital market imperfections: an international comparison. *American Economic Review*, **79,** 1088–1105.

Jaffee, D. and Stiglitz, J. (1990) Credit rationing. In B. Friedman and F. Hahn (Eds), *Handbook of Monetary Economics*, Vol. 2, North Holland.

Jappelli, T. (1990) Who is credit constrained in the US? *Quarterly Journal of Economics*, **105,** 219–234.

King, S. (1985) Monetary transmission: through bank loans, or bank liabilities. Manuscript, Stanford University.

King, R. G. and Plosser, C. I. (1984) Money, credit and prices in a real business cycle. *American Economic Review*, **74,** 363–380.

Kormendi, R. C. and Meguire, P. G. (1984) The real output of monetary shocks: cross country tests of rational expectations propositions. *Journal of Political Economy*, **92,** 875–908.

Kydland, F. E. and Prescott, E. C. (1982) Time to build and aggregate fluctuations. *Econometrica*, **50,** 1345–1370.

Long, J. B. and Plosser, C. I. (1983) Real Business cycles. *Journal of Political Economy*, **91,** 39–69.

Lucas, R. E. Jr. (1973) Some international evidence on output-inflation trade offs. *American Economic Review*, **63,** 326–334.

Lucas, R.R. Jr. (1975) An equilibrium model of the business cycle. *Journal of Political Economy*, **83,** 1113–1144.

MacKinnon, J. G. (1990a) Critical values for cointegration tests. In R. F. Engle and C. W. J. Granger (Eds), *Long Run Economic Relationships*, Oxford: Oxford University Press.

MacKinnon, J. G. (1990b) Critical values for cointegration tests. UC San Diego Discussion paper, 90–4.

McCallum, R. T. (1986) On 'real' and 'sticky price' theories of the business cycle. *Journal of Money, Credit and Banking*,**18,** 397–414.

McCallum; B. T. (1989) Real business cycle models. In R. J. Barro (Ed), *Modern Business Cycle Theory*, Oxford: Basil Blackwell.

Mankiw, N. G. (1985) Small menu costs and large business cycles a macroeconomic model of monopoly. *Quarterly Journal of Economics*, **100,** 529–537.

Makiw, G. N. (1994) *Macroeconomics*, New York: Worth.

Mishkin, F. (1982a) Does anticipated monetary policy matter? An econometric investigation. *Journal of Political Economy*, **980**, 22–50.

Mishkin, F. (1982b) Does anticipated aggregate demand policy matter? *American Economic Review*, **72**, 788–802.

Modigliani, F. (1977) The monetarist controversy, or should we forsake stabilisation policies?, *American Economic Review*, **67**, 1–19.

Modigliani, F. (1986) Life cycle, individual thrift and the wealth of nations. *American Economic Review*, **76**, 297–313.

Mullineux, A., Dickinson, D. G. and Peng, W. (1993) *Business Cycles*, Oxford: Blackwell.

Muscatelli, V. A. and Hurn, S. (1992) Cointegration and dynamic time series models. *Journal of Economic Surveys*, **16**, 1–43.

Nelson, C. R. and Plosser, C. I. (1982) Trends and random walks in macroeconomic time series. *Journal of Monetary Economics*, **10**, 139–162.

Osborn, D. R. (1990) A survey of seasonality in macroecoomic variables. *International Journal of Forecasting*, **6**, 327–336.

Quandt, R. E. (1978) Tests of the equilibrium versus the disequilibrium hypothesis. *International Economic Review*, **19**, 435–452.

Sargent, T. J. and Wallace, N. (1975) Rational expectations, the optimal monetary instrument and the optimal money supply rule. *Journal of Political Economy*, **83**, 241–254.

Sealey, C. W. (1979) Credit rationing in the commercial loan market: estimates of a structural modal under conditions of disequilibrium. *The Journal of Finance*, **34**, 689–702.

Sherman, H. J. (1991) *The Business Cycle: Growth and Crisis under Capitalism*. Princeton: Princeton University Press.

Sims, C. A. (1972) Money, income and causality. *American Economic Review*, **62**, 540–552.

Sims, C. A. (1980) Comparisons of inter-war and post-war business cycles: monetarism reconsidered. *American Economic Review*, **70**, 250–257.

Stiglitz, J. E. and Weiss, A. (1981) Credit rationing in markets with imperfect information. *American Economic Review*, **71**, 393–410.

Taylor, J. (1979) Staggered price setting in a macromodel. *American Economic Review*, **69**, 108–113.

Turner, O. M. (1993) A structural vector autoregression model of the UK business cycle. *Scottish Journal of Political Economy*, **40**, 143–164.

Wojnilower, A. (1980) The central role of credit crunches in recent financial history. *Brookings Papers on Economic Activity*, **1980(2)**, 277–340.

Yule, G. U. (1926) Why do we sometimes get nonsense correlations between time series. A study in sampling and the nature of time series. *Journal of the Royal Statistical Society*, **89**, 1–64.

Zeldes, S. P. (1989) Consumption and liquidity constraints: an empirical investigation. *Journal of Political Economy*, **97**, 305–346.

Part III

FINANCIAL MARKETS

8

Probability in Finance: An Introduction

Saul D. Jacka

8.1 A brief survey

An introductory chapter of this sort is no place to attempt a historically accurate review of the use of probability in finance, nevertheless a very brief sketch might be in order.

The use of probabilities to measure uncertainty in a market of traded assets dates back at least to Bachelier (1900), but a rigorous justification based on an axiomatic treatment of the theory of choice and utility representation of preferences may be found in Kreps (1988) (see also Debreu, 1972).

Arrow's (1953) invention of a general equilibrium model of security markets based on agents trading in an idealised market and the search for 'market clearing' provides the foundation for the key idea of 'state prices' – positive discount factors, one for each date and 'state of the world' – such that any security price is the state-price weighted sum of its future pay-offs. The existence of state prices is equivalent to the existence of 'risk-neutral' probabilities for states (Arrow, 1970).

All of this work was done in a discrete-time setting, Merton (1969, 1971) initiated the study of continuous-time financial modelling – introducing his general equilibrium model in 1973 (Merton, 1973), in the same year as the publication of the Black–Scholes option pricing formula (Black and Scholes, 1973). Since 1970 there has been an explosion of interest in the theoretical aspects of finance; we shall only mention the key explanatory works on option pricing: Harrison and Kreps (1979) and Harrison and Pliska (1981). For a comprehensive survey of these ideas we refer the reader to the following books: Jarrow (1988), Merton (1990) and Duffie (1988), and to Chapter 9 in this volume by Hodges.

8.2 Some reminders on stochastic calculus

In continuous-time finance a key tool is stochastic calculus. General semi-martingale calculus is a powerful tool, largely produced by French probabilists in the 1960s and 1970s. An excellent reference text is Jacod's seminal work (Jacod, 1979), but for a more pleasant introduction it would be better to read Øksendal (1985) followed, perhaps, by Protter (1990). In this section we shall simply recall a few key facts: the reader is referred to Protter (1990) for any unexplained terms.

First recall that if A is a predictable process and M is a semi-martingale then the stochastic integral $(A \cdot M)_t = \int_0^t A_s \, dM_s$ is well-defined and is a semi-martingale; moreover, if M is a local martingale then so is $(A \cdot M)$.

Ito's generalised formula gives the basis of a full stochastic calculus: the version for continuous semi-martingales states that if f is $C^{2,1}$ then

$$f(X_t, t) = f(X_0, 0) + \int_0^t f_x(X_s, s)\, dX_s + \frac{1}{2}\int_0^t f_{xx}(X_s, s)d\langle X\rangle_s + \int_0^t f_s(X_s, s)\, ds$$

For a general semi-martingale S, the square bracket process $[S]$ may be defined as

$$[S]_t = \mathbb{P} - \lim \Sigma (S_{t_{i+1}} - S_{t_i})^2$$

the $\{t_i\}$ forming a dissection of $[0, t]$ and the limit in probability being taken as the mesh size of the dissection tends to zero. It follows that $[S]$ may also be defined by the formula $S_t^2 = S_0^2 + 2\int_0^t S_{s-}dS_s + [S]_t$.

Under a local square integrability condition, the angle-bracket process $\langle S\rangle_t$ is defined as the dual pre-visible projection of $[S]$, that is $\langle S\rangle$ is the unique predictable, increasing process such that $[S] - \langle S\rangle$ is a local martingale. It follows that the square- and angle-bracket processes agree for continuous semi-martingales, and for a general semi-martingale:

$$[S]_t = \langle S^c\rangle_t + \sum_{s \leqslant t}(\Delta S_s)^2$$

where S^c is the continuous martingale part of S.

For semi-martingales M and N, $\langle M, N\rangle$ and $[M, N]$ are defined by polarisation: $[M, N] = \frac{1}{2}([M+N]-[M-N])$ and $\langle M, N\rangle = \frac{1}{2}(\langle M+N\rangle - \langle M-N\rangle)$.

We recall, also, the Doob–Meyer decomposition for supermartingales: under suitable integrability conditions, if S is a supermartingale then there is a unique decomposition

$$S = M_t - A_t$$

where M is a martingale and A is a predictable, increasing, process with $A_0 = 0$.

Finally, recall that any predictable martingale of bounded variation is constant.

These, together with the ideas of change of measure, optimal stopping and martingale representation (all of which are more specialised and will be covered later in this introduction) are the elements of stochastic calculus used (so far) in finance.

8.3   The Black–Scholes framework

8.3.1   The Black–Scholes formula

Before we proceed with the financial theory it is worth stating the assumptions made throughout the rest of this chapter about securities markets. These are usually summarised as follows.

(1)   Agents are price takers.
(2)   The market is frictionless.
(3)   Short sales are allowed.

Assumption (1) indicates that any trade is assumed to be marginal so that the volume of the trade is insufficient to cause a shift of prices (the market is already in equilibrium); assumption (2) is that there are no trading costs whatsoever, whilst assumption (3) is that an agent or trader may sell what they do not possess – thus acquiring the responsibility to provide monetary payments that the buyer would be entitled to if they had actually received the securities.

Following a well-established tradition I shall now introduce the Black–Scholes framework for pricing an option on a stock. The ideas used have far more general relevance than is generally appreciated and as many times as the key concepts are explained they are confused, so it is worth attempting the explanation again as it will serve as a guide and exemplar to the more general theory that will follow.

Black and Scholes' key idea (Black and Scholes, 1973) is as follows: suppose you possess an option on a stock or share S (we follow tradition in conflating the name and monetary value of the stock): that is, you are the possessor of a contract which entitles you to some function $f(S_T)$ at time T. The stock is freely traded and the only other medium of investment is a riskless savings account which pays interest at rate r. We denote the accumulated value of an initial investment of £1 in this account by D_t (so D_t is just e^{rt}). The question is 'what is a fair price for the option'. Black and Scholes observed that if you can invest an initial sum V_0 in the market, and then trade it in such a way as to guarantee to have $f(S_T)$ at time T then V_0 is the only fair price for the option. This follows since if the option trades at any other price there is a risk-free way of making money (either by selling the option and trading or by buying the option and selling short in the market).

The model Black and Scholes chose for the stock-price process was as follows:

$$S_t = S_0 \exp\left(\sigma W_t + \int_0^t (\mu_s - \tfrac{1}{2}\sigma^2)\,ds\right) \tag{8.1}$$

or $dS_t = \sigma S_t\,dW_t + \mu_t S_t\,dt$, where W is a Brownian motion.

What they showed was that under these conditions there is a self-financing hedging portfolio for the option: that is, a pair of (predictable) processes (ϕ, ψ) satisfying the following two conditions: if $V_t \stackrel{def}{=} \phi_t S_t + \psi_t D_t$ then

(1) $dV_t = \phi_t\,dS_t + \psi_t\,dD_t$ (the self-financing condition)

and

(2) $V_T = f(S_T)$ (the hedging condition),

and that

$$V_0 = \mathbb{E}e^{-rT}f(\bar{S}_T) \tag{8.2}$$

where $\bar{S}_0 = S_0$ and $d\bar{S}_t = \sigma\bar{S}_t\,dW_t + r\bar{S}_t\,dt$, provided that the expression for V_0 was finite.

In other words they showed that there was a hedging portfolio and that its initial (set-up) cost was the expected discounted pay-off from the option under revised dynamics for S. Notice that the revised dynamics for S (those of $\bar{S}$) are such that $\bar{S}_t/D_t$ is a martingale – this is no coincidence.

The proof is both short and informative so we include a sketch.

The first thing to do is to denominate everything in terms of $D: \tilde{S} \equiv S/D$, $\tilde{V} = V/D$; then the self-financing constraint just becomes

$$d\tilde{V}_t = \phi_t\,d\tilde{S}_t \tag{8.3}$$

Now change probability measure to $\mathbb{Q}$ so that the dynamics of $\tilde{S}$ are those of $\bar{\tilde{S}}$ (this can be done, as we shall see later). As we have remarked, under these dynamics $d\tilde{S}_t = \sigma\tilde{S}_t\,dW_t$, so that Equation (8.3) becomes

$$d\tilde{V}_t = \phi_t\sigma\tilde{S}_t\,dW_t \tag{8.4}$$

Now if we define $\hat{V}_t = \mathbb{E}_\mathbb{Q}[f(S_T)/D_t \mid \mathcal{F}_t]$ then it is automatic that $\hat{V}$ is a martingale, so, to satisfy Equation (8.4), it is only necessary that it can be written as a stochastic integral with respect to N.

Now, by the Markov property (of S under $\mathbb{Q}$) we see that

$$\hat{V}_t = e^{-rT}f(S_t, T - t) = e^{-rT}f(e^{rt}\tilde{S}_t, T - t) \tag{8.5}$$

for a suitable f with $f(x, 0) \equiv f(x)$.

Itô's formula then gives us that

$$d\hat{V}_t = e^{-rT}\{f_x(S_t, T - t)e^{rt}\,d\tilde{S}_t + (\tfrac{1}{2}\sigma^2 f_{xx}S_t^2 + rS_t f_x - f_t)\,dt\}$$

and the fact that $\hat{V}$ is a martingale (under $\mathbb{Q}$) tells us that the second term in the brace must be a martingale differential under $\mathbb{Q}$ (since the first is); but it corresponds to a predictable process of bounded variation so must be null!

Hence $\hat{V}$ satisfies Equation (8.4) under $\mathbb{Q}$ and, because stochastic integrals are invariant under changes of measure, it also satisfies Equation (8.3) under any measure $\mathbb{P} \sim \mathbb{Q}$, including our original measure. Thus, we have created our self-financing hedging portfolio and given a fair price for the option (ϕ and ψ may be extracted from Equation (8.5) and the definition of V).

Of course, since $\log S_T$ is normally distributed under $\mathbb{Q}$ we can give an explicit formula for V_0.

8.3.2 Issues arising from the formula

The issues which arise from the argument above can be summarised in the following questions.

(1)   Might not there be another, cheaper, hedging portfolio?
(2)   The argument used is fairly general, might it work for more realistic securities markets?

(3)   What was special about using D as the unit of account?

The answers to these questions are respectively 'No', 'Yes' and 'Nothing', but each issue will be covered in more detail later.

8.4   Arbitrage and option pricing in a general securities market

We suppose now that we have a more general securities market than that in the Black–Scholes model. To be precise we have securities $S^0, \ldots, S^n$ (all assumed to have non-negative prices) and our beliefs about the world are given by $\mathbb{P}$.

We assume that, under $\mathbb{P}$, S^0 is actually strictly positive and we shall use it as our unit of account so that $\tilde{S}_i \equiv S^i/S^0$ gives the value of the ith security in our unit of account. We also give a horizon, events beyond which we will not concern ourselves with, called T.

8.4.1   Arbitrage

Informally, an arbitrage opportunity is the opportunity to make a risk free profit within the market. More formally, it is defined as follows: $\phi = (\phi^0, \ldots, \phi^n)$ is an arbitrage opportunity if ϕ is a self-financing portfolio satisfying

$$V(\phi)_0 \equiv \sum_{i=0}^{n} \phi_0^i S_0^i = 0 \tag{8.6}$$

$$V(\phi)_t \equiv \sum_{i=0}^{n} \phi_t^i S_t^i \geq 0, \quad \forall t \in [0, T] \tag{8.7}$$

and

$$\mathbb{P}(V(\phi)_T > 0) > 0 \tag{8.8}$$

Thus, ϕ represents a chance of making a profit with no countervailing chance of making a loss.

Let us suppose that there is a measure $\mathbb{Q}$ satisfying

$$\left. \begin{array}{l} \mathbb{Q} \sim \mathbb{P} \\[2em] \text{under } \mathbb{Q}, \text{ each } \tilde{S}^i \text{ is a martingale} \end{array} \right\} \tag{8.9}$$

and

(a measure satisfying Equation (8.9) is termed an equivalent martingale measure, abbreviated to EMM).

If we now consider a possible arbitrage opportunity, and denominate in our unit of account, then Equations (8.6) to (8.8) are unchanged except for substituting $\tilde{S}$ for S and $\tilde{V}$ for V (with $\tilde{V} = V/S^0$ and $\tilde{S}^0 \equiv 1$) whilst the self-financing condition becomes

$$d\tilde{V}(\phi)_t = \sum_{i=1}^{n} \phi^i \, d\tilde{S}_t^i \tag{8.10}$$

Now, under an EMM $\mathbb{Q}$, Equation (8.10) tells us that $\tilde{V}$ is a local martingale, whilst Equation (8.7) tells us that $\tilde{V}$ is non-negative so that $\tilde{V}(\phi)$ is supermartingale (under $\mathbb{Q}$). It follows immediately that $0 = \tilde{V}(\phi)_0 \geqslant \mathbb{E}_\mathbb{Q} \tilde{V}(\phi)_T$. But $\tilde{V}(\phi)_T \geqslant 0$ so we must have $\mathbb{Q}(\tilde{V}(\phi)_T > 0) = 0 \Rightarrow \mathbb{Q}(V(\phi)_T > 0) = 0$, but $\mathbb{Q}$ is equivalent, to $\mathbb{P}$, so Equation (8.8) cannot hold. Thus, the existence of an EMM guarantees the *absence* of arbitrage opportunities. The amazing thing, with profound modelling implications, is that the reverse implication also holds so that a model for a securities market is arbitrage-free if and only if there is at least one EMM. The full result is due to Stricker (1984), but see also Harrison and Pliska (1981).

The reason why the result is significant lies in our economic assumptions; if there is an arbitrage opportunity then traders will profit from it, and the volume of trade will be such that prices will be affected until the arbitrage opportunity disappears, conflicting with our 'price taker' assumption.

It follows that our market model must possess an EMM and this implies that we are restricted to assuming that all discounted security prices are semi-martingales under $\mathbb{P}$ (see the next section).

8.4.2   General option pricing and martingale representation

The general theory of pricing (European) options is due essentially to Harrison and Pliska (1981) building on the work in discrete-time by Harrison and Kreps (1979).

The framework is a general securities market specified above. A general option (European) is simply an entitlement to be paid a (random) amount X at time T (with $X \geqslant 0$). Harrison and Pliska refined the arguments of Black and Scholes essentially as follows: let $\tilde{X} = X/S_T^0$, then define $\tilde{X}_t = \mathbb{E}_\mathbb{Q}[\tilde{X} \mid \mathcal{F}_t]$ (so that $\tilde{X}_T = \tilde{X}$), where $\mathbb{Q}$ is an EMM.

Suppose $\tilde{X}_t$ can be written as a stochastic integral with respect to the (vector) process $\tilde{S}$:

$$\tilde{X}_t = \tilde{X}_0 + \int_0^t \phi_t \, d\tilde{S}_t$$

then ϕ gives a self-financing portfolio which hedges the claim X, and its initial value is $X_0 \equiv \tilde{X}_0 S_0^0$ so that, since there is no arbitrage opportunity, X_0 is the unique fair price for the claim.

The relevance of the EMM $\mathbb{Q}$ seems dubious, at first sight, but Harrison and Pliska give the following martingale representation result (due originally to Jacod, 1979): every $\mathbb{Q}$ martingale can be written as a stochastic integral with respect to $\tilde{S}$ if and only if $\mathbb{Q}$ is the unique EMM. Thus, in the option pricing context we see that every contingent claim (with $\tilde{X}$ integrable) can be hedged, and thus priced uniquely, if and only if there is a unique EMM.

A market with a unique EMM is termed complete, whilst one with more than one EMM is incomplete. Clearly, in an incomplete market some claims can be hedged (and thus priced uniquely) – a diagnostic for such claims is provided in Jacka (1992): essentially a claim X is hedgeable if $\mathbb{E}_\mathbb{Q} \tilde{X}$ is constant as $\mathbb{Q}$ runs through all the EMMs.

Note that if $(\mathcal{F}_t)$ is the filtration of a d-dimensional process W,

$$\tilde{S}_t = \tilde{S}_0 + \int_0^t \sigma_s \, dW_s$$

for some $n \times d$ matrix process σ of full rank with $n \geqslant d$, and under some $\mathbb{Q} \sim \mathbb{P}$, W is a Brownian motion then we have the martingale representation property for $\tilde{S}$ and the market is complete – this applies in the Black–Scholes market with $n = d = 1$.

8.5   Changes of measure

8.5.1   The Cameron–Martin–Girsanov formula

We have used the idea of changes of measure throughout this chapter without defining how it is done. The basic idea is just the Radon–Nikodym theorem specialised to probability measures (thereafter abbreviated to p.ms.):

Suppose $\mathbb{P}$ is a p.m. on $(\Omega, \mathcal{F})$ and Λ is a non-negative random variable on $(\Omega, \mathcal{F})$ with $\mathbb{E}_{\mathbb{P}}\Lambda = 1$, then, defining $\mathbb{Q} : \mathcal{F} \to \mathbb{R}$ by

$$\mathbb{Q}(A) = \int_A \Lambda \, d\mathbb{P} = \mathbb{E}_{\mathbb{P}}\Lambda 1_A \qquad (8.11)$$

$\mathbb{Q}$ is a p.m. on $(\Omega, \mathcal{F})$ with $\mathbb{Q} \ll \mathbb{P}$ and, for any random variable X, $\mathbb{E}_{\mathbb{Q}}X = \mathbb{E}_{\mathbb{P}}\Lambda X$; conversely, if $\mathbb{Q} \ll \mathbb{P}$, then there exists such a Λ for which Equation (8.11) holds (this *is* the Radon–Nikodym theorem for p.ms.).

Let us now determine the effect of a change of measure on a semimartingale: to do this we need the 'Bayes formula for change of measure'. Suppose we work with a full filtered probability space; $\mathbb{Q} \ll \mathbb{P}$ and $d\mathbb{Q}/d\mathbb{P} = \Lambda$. If we look at the restrictions of $\mathbb{Q}$ and $\mathbb{P}$ to $\mathcal{F}_t$ then Λ will not do as a density because Λ is not necessarily $\mathcal{F}_t$-measurable; however $\Lambda_t \stackrel{\text{def}}{=} \mathbb{E}_{\mathbb{P}}[\Lambda \mid \mathcal{F}_t]$ will because, for $X \in \mathcal{F}_t$

$$\mathbb{E}_{\mathbb{P}}\Lambda X = \mathbb{E}_{\mathbb{P}}[\mathbb{E}_{\mathbb{P}}[\Lambda X \mid \mathcal{F}_t]] = \mathbb{E}_{\mathbb{P}}[X\Lambda_t]$$

From this observation we may check (using the definition of conditional expectation) that, for any $X \in \mathcal{F}$

$$\mathbb{E}_{\mathbb{Q}}[X \mid \mathcal{F}_t] = \mathbb{E}_{\mathbb{P}}[\Lambda X \mid \mathcal{F}_t]/\Lambda_t$$

so that if $X \in \mathcal{F}_s$

$$\mathbb{E}_{\mathbb{Q}}[X \mid \mathcal{F}_t] = \mathbb{E}_{\mathbb{P}}[\Lambda_s X \mid \mathcal{F}_t]/\Lambda_t$$

Now suppose S is a $\mathbb{P}$-semi-martingale with decomposition

$$S = M + A$$

where M is a $\mathbb{P}$-martingale, and A is a process of bounded variation, and denote by λ the solution to $\lambda_t = \int_0^t d\Lambda_t/\Lambda_t$. The Cameron–Martin–Girsanov change of measure formula states that if $M\Lambda$ is locally integrable then

$$S_t = \tilde{M}_t + \tilde{A}_t$$

where $\tilde{M}_t$ is a $\mathbb{Q}$ local martingale (given by $\tilde{M}_t = M_t - \langle M, \lambda \rangle_t$) and $\tilde{A}_t$ is a process of bounded variation (given by $\tilde{A}_t = A_t + \langle M, \lambda \rangle_t$) (there is a formula for the general case: simply replace $\langle M, \lambda \rangle_t$ by $\int_0^t d[M, \lambda]_t/(1 + \Delta\lambda_t)$), and it follows immediately from it that semi-martingales remain semi-martingales under changes of measure.

To give an example, consider the case (in the Black–Scholes model) where

$$dS_t = \sigma S_t \, dW_t + \mu_t S_t \, dt$$

and under $\mathbb{P}$, W is a BM. Define

$$\Lambda_t = \exp\left(\int_0^t \frac{(\mu_s - r)}{\sigma} \, dW_s - \frac{1}{2} \int_0^t \frac{(\mu_s - r)^2}{\sigma_2} \, ds \right)$$

and assume that, under $\mathbb{P}$, Λ_T has expectation 1. It is easy to check from Itô's formula that $d\Lambda_t = \Lambda_t - (\mu_s - r)/\sigma \, dW_t$, so that $d\lambda_t = (\mu_s - r)/\sigma \, dW_t$, so that the Cameron–Martin–Girsanov formula implies that under $\mathbb{Q}$ (with $d\mathbb{Q}/d\mathbb{P} = \Lambda_T$), $\tilde{W}_t = W_t + \int_0^t (\mu_s - r)/\sigma \, ds$ is a martingale on $[0, T]$ (and since the square brackets process is unchanged by changes of measure, $\tilde{W}$ is a BM) and it follows that

$$dS_t = \sigma S_t \, dW_t + \mu_t S_t \, dt$$
$$= \sigma S_t \, d\tilde{W}_t + r S_t \, dt$$

where $\tilde{W}$ is a $\mathbb{Q}$-BM.

8.5.2   Change of numeraire

We now address the third question at the end of Section 8.3.

Suppose we wish to change our unit of account (or discounting process) from S^0 to $S^k (1 \leqslant k \leqslant n)$? We shall (have to) assume that S^k is a strictly positive process. What we shall show is that, denoting by $\tilde{R}$ the vector of security prices discounted by S^k, there is a bijection between EMMs for $\tilde{S}$ and EMMs for $\tilde{R}$ and that all pricing formulae remain unchanged so that there is no unique role for S^0. The process of changing the unit of account (and associated changes in EMMs) is normally termed 'change of numeraire' and is, at times, a powerful technique for obtaining explicit pricing formulae (see, for example, Geman *et al.*, 1991).

Given $\mathbb{Q}$, an EMM for $\tilde{S}$, we define $\Lambda_t = \tilde{S}_t^k / \tilde{S}_0^k$. Since $\mathbb{Q}$ is an EMM and $\tilde{S}_0^k$ is positive, Λ is a positive martingale with expectation 1, so $\hat{\mathbb{Q}}$, defined by $d\hat{\mathbb{Q}}/d\mathbb{Q} = \Lambda_T$ is a p.m. and, since $\Lambda > 0$, $\hat{\mathbb{Q}} \sim \mathbb{Q} \sim \mathbb{P}$. We define $\tilde{R}_t^i = S_t^i / S_t^k = \tilde{S}_t^i / \tilde{S}_t^k$.

Now, using Bayes' formula:

$$\mathbb{E}_{\hat{\mathbb{Q}}}[\tilde{R}_{t+s}^i \mid \mathcal{F}_t] = \mathbb{E}_{\mathbb{Q}}[\Lambda_{t+s} \tilde{R}_{t+s}^i \mid \mathcal{F}_t]/\Lambda_t$$

$$= \mathbb{E}_{\mathbb{Q}}\left[\frac{\tilde{S}_{t+s}^k}{\tilde{S}_0^k} \cdot \frac{\tilde{S}_{t+s}^i}{\tilde{S}_{t+s}^k} \,\middle|\, \mathcal{F}_t \right] \bigg/ \frac{\tilde{S}_t^k}{\tilde{S}_0^k}$$

$$= \mathbb{E}_{\mathbb{Q}}[\tilde{S}_{t+s}^i \mid \mathcal{F}_t]/\tilde{S}_t^k = \tilde{S}_t^i / \tilde{S}_t^k = \tilde{R}_t^i \quad \text{(since } \mathbb{Q} \text{ is an EMM for } \tilde{S}\text{)}.$$

Thus, since i was arbitrary, $\hat{\mathbb{Q}}$ is an EMM for $\tilde{R}$. The argument is symmetric in the indices k and 0 so we have established the required bijection.

Moreover, $S_0^k \mathbb{E}_{\hat{\mathbb{Q}}}[X/S_T^k] = S_0^k \mathbb{E}_{\mathbb{Q}}[\tilde{S}_T^k X/\tilde{S}_0^k S_T^k] = S_0^0 \mathbb{E}_{\mathbb{Q}}[X/S_T^k]$ so pricing formulae (in currency) are unaffected by the choice of numeraire.

8.6 American options

An American option contract differs from a European one in the following way: the option may be exercised at any time up to T, the expiry date of the contract. It follows that a general model for such contracts needs to specify a claims process $(X_t : 0 \leqslant t \leqslant T)$. If the option is exercised at t then the amount receivable is X_t. Clearly the holder of such a contract may exercise it at any stopping time $\tau \leqslant T$. Assuming the usual model for the securities market and the existence of an EMM, the natural question is 'can we hedge such an option'? The answer to this question is intimately connected with the technique of optimal stopping, an excellent summary of which, together with its application to pricing, is contained in Chapter 10, 'American options' by Lamberton, in this book (see also Karatzas, 1988). Here we shall content ourselves with establishing the connection (for a much fuller discussion see Jacka, 1997).

Notice first that, since holders of option contracts may choose different exercise times, a hedging portfolio will not necessarily exactly meet the claim. We search therefore for an American hedging portfolio which we define as follows.

ϕ is a hedging portfolio with value V if ϕ is predictable, with $V_t(\phi) = \phi_t \cdot S_t$,

$$dV_t(\phi) = \phi_t \, dS_t \quad \text{(the self-financing condition)} \tag{8.12}$$

$$V_t(\phi) \geqslant X_t \text{ for all } t \in [0, T]$$
$$\text{(the portfolio is always sufficient to meet the claim)} \tag{8.13}$$

and

there exists a stopping time $\tau \leqslant T$ such that

$$V_\tau(\phi) = X_\tau \text{ a.s. } \quad \text{(the portfolio is minimal)} \tag{8.14}$$

To see the connection with optimal stoppings let us, for the time being, assume the market is complete with EMM $\mathbb{Q}$. Let the discounted claims process be denoted $\tilde{X}$.

Define

$$V_t = \operatorname*{ess\,sup}_{T \geqslant \tau \geqslant t} \mathbb{E}_\mathbb{Q}[\tilde{X}_\tau \mid \mathcal{F}_t]$$

(see Chapter 10, by Lamberton), and assume that V is regular. Then V may be written as

$$V_t = M_t - A_t \tag{8.15}$$

where M is a $\mathbb{Q}$-martingale and A is a predictable increasing process with $A_0 = 0$ (Theorem 10.2.2 of Chapter 10).

Now, since M is a $\mathbb{Q}$-martingale and we have a complete market we see that M_t is hedgeable and we can write

$$M_t = V_0 + \int_0^t \phi_s \cdot d\tilde{S}_s$$

moreover, since A is increasing, $M_t \geqslant V_t \geqslant \tilde{X}_t$.

We see that with this choice of ϕ, $V(\phi)_t = V_0 + \int_0^t \phi_s \, dS_s$. To show that ϕ also satisfies conditions (8.13) and (8.14) we need only observe that $V(\phi)_t = S_t^0 M_t \geq X_t$ and that if $\tau = \inf\{t \geq 0 : V_t = \tilde{X}_t\}$ then (Theorem 10.2.9 in Chapter 10), $V_\tau = \tilde{X}_\tau$ a.s. and (Lemma 10.2.10) $A_\tau = A_0 \equiv 0$ so $V_\tau = M_\tau$ and thus $V(\phi)_\tau \equiv S_\tau^0 M_\tau = S_\tau^0 V_\tau = S_\tau^0 \tilde{X}_\tau = X_\tau$. So ϕ is an (American) hedge. Moreover, if ϕ' is another hedging portfolio then we must have $\tilde{V}(\phi')_0 \geq \mathbb{E}_{\mathbb{Q}}[\tilde{X}_\tau] = \tilde{V}(\phi)_0$, so $V(\phi')_0 \geq V(\phi)_0$, and conversely, since ϕ' is a hedge there is a σ such that $\tilde{V}(\phi')_\sigma = \tilde{X}_\sigma$, and so

$$\tilde{V}(\phi')_0 = \mathbb{E}_{\mathbb{Q}} \tilde{X}_\sigma \leq \mathbb{E}_{\mathbb{Q}} \tilde{V}(\phi)_\sigma \ \text{(from (8.13))}$$
$$= \tilde{V}(\phi)_0$$

so $V(\phi')_0 \leq V(\phi)_0$, implying that any two (American) hedges have the same set-up cost and this must therefore be the unique fair price for the option.

In the case of an incomplete market, life is more difficult, but the following result has been obtained in Jacka (1997):

$$\text{define } V_t^{\mathbb{Q}} = \text{ess sup}_{T \geq \tau \geq t} \mathbb{E}_{\mathbb{Q}}[\tilde{X}_\tau \mid \mathcal{F}_t]$$

then, provided $V^{\mathbb{Q}}$ is regular for each $\mathbb{Q}$; if $\mathbb{E}_{\mathbb{Q}} V_t^{\mathbb{Q}}$ is independent of $\mathbb{Q}$ (as $\mathbb{Q}$ runs through the collection of EMMs) for all $t \in [0, T]$ then in fact $V^{\mathbb{Q}}$ is independent of $\mathbb{Q}$ and the martingale in Equation (8.15) is representable, so that there is a hedging portfolio; moreover any hedging portfolio has set-up cost V_0, so this is the unique fair price for the option.

8.7   Some final remarks

This chapter has concentrated on considerations of no-arbitrage and option pricing to the exclusion of much theoretical work in finance; in addition we have neglected the issue of dividends (without which any self-respecting economist would view a paper security as worthless), and we have not touched upon the specialist areas of fixed-interest securities and term-structure models, and forward and futures contracts.

To redress the balance a little we mention some of the recent work on portfolio and consumption optimisation: an excellent list of papers is given on pp. 168 and 169 of Duffle (1992). We also single out the series of papers by Foldes (1979, 1991a, b and c) and the paper by Karatzas (1989).

Touching on the subject of dividends, no great changes to the framework enunciated earlier are required: provided dividends are assumed to be invested in a denominated security (S^0 say), we may reproduce our previous analysis almost without change; this applies whether dividends are paid just on traded securities or, additionally, on the option requiring valuation.

Finally, we refer the reader to Chapter 11 for a very brief introduction to term-structure models.

References

Arrow, K. (1953) Le rôle des valeurs boursières pour la repartition la meillures des risques. *Econometrie* Colloq. Internat. Centre Nationale de la Recherche Scientifique, **40**.

Arrow, K. (1970) *Essays in The Theory of Risk Bearing*. London: North-Holland.

Bachelier, L. (1900) Théorie de la speculation. *Annales scientifiques de l'école normale supérieure*, 3rd ser., **17**, 21–88. Translated in P. Cootner (Ed), *The Random Character of Stock Market Prices*, Cambridge: MIT Press, 1964.

Black, F. and Scholes, M. (1973) The pricing of options and corporate liabilities. *Journal of Political Economies*, **81**, 637–654.

Debreu, G. (1972) Smooth preferences. *Econometrica*, **40**, 603–615.

Duffie, D. (1988) *Security Markets: Stochastic Models*. Boston: Academic Press.

Duffie, D. (1992) *Dynamic Asset Pricing Theory*. Princeton, NJ: Princeton.

Foldes, L. (1979) Optimal saving and risk in continuous time. *Review of Economics Studies*, **46**, 39–65.

Foldes, L. (1991a) Certainty equivalence in the continuous-time portfolio-cum-saving model. In M. H. A. Davis and R. J. Elliott (Eds), *Applied Stochastic Analysis*, London: Gordon and Breach.

Foldes, L. (1991b) Existence and uniqueness of an optimum in the infinite-horizon portfolio-cum-saving model with semi-martingale investments. London School of Economics Financial Markets Group, discussion paper No. 109, London School of Economics.

Foldes, L. (1991c) Optimal sure portfolio plans. *Mathematical Finance*, **1**, 15–55.

Geman, H., El Karoui, N. and Rochet, J.-C. (1991) Probability changes and option pricing. Probability Laboratory, Université of Paris VI, research report.

Harrison, J. M. and Kreps, D. M. (1979) Martingales and arbitrage in multiperiod securities markets. *Journal of Economic Theory*, **20**, 381–408.

Harrison, J. M. and Pliska, S. (1981) Martingales and stochastic integrals in the theory of continuous trading. *Stochastic Processes and Applications*, **11**, 215–260.

Harrison, J. M. and Pliska, S. (1983) A stochastic calculus model of continuous trading: complete markets. *Stochastic Processes and Applications*, **15**, 313–316.

Jacka, S. D. (1992) A martingale representation result and an application to incomplete financial markets. *Mathematical Finance*, **2**, 23–34.

Jacka, S. D. (1997) Pricing American options in incomplete markets. Department of Statistics, University of Warwick, research report.

Jacod, J. (1979) *Calcul Stochastique et Problémes de Martingales*. Lecture Notes in Maths 714, Berlin-Heidelberg-New York: Springer.

Jarrow, R. A. (1988) *Finance Theory*. Englewood Cliffs, NJ: Prentice-Hall.

Karatzas, I. (1988) On the pricing of American options. *Applied Mathematical Optimization*, **17**, 37–60.

Karatzas, I. (1989) Optimization problems in the theory of continuous trading. *SIAM. Journal of Control Optimization*, **27**, 1221–1259.

Kreps, D. (1988) *Notes on the Theory of Choice.* Boulder, CO and London: Westview Press.

Merton, R. C. (1969) Lifetime portfolio selection under uncertainty: the continuous time case. *Review of Economics and Statistics*, **51**, 247–257.

Merton, R. C. (1971) Optimum consumption and portfolio rules in a continuous time model. *Journal of Economic Theory*, **3**, 373–413.

Merton, R. C. (1973) An intertemporal capital asset pricing model. *Econometrica*, **41**, 867–888.

Merton, R. C. (1990) *Continuous-Time Finance.* Oxford: Basil Blackwell.

Øksendal, B. (1985) *Stochastic Differential Equations with Applications.* Berlin: Springer.

Protter, P. (1990) *Stochastic Integration and Differential Equations.* New York: Springer.

Stricker, C. (1984) Integral representation in the theory of continuous trading. *Stochastics*, **13**, 249–257.

9

Introduction to Financial Economics

Stewart D. Hodges

9.1   Introduction

This chapter provides an introduction to some of the key ideas in financial economics. In particular, we describe three aspects of capital markets. These are:

(1)   the various roles which capital markets play in the economy;
(2)   the state-preference framework for representing securities in economies with a finite number of states and securities. This also leads to the idea of riskless arbitrage, and the fundamental pricing relationship which holds provided no such arbitrage opportunities exist;
(3)   the paradigm of expected utility maximisation, including its axiomatic basis, and some of its weaknesses. We also describe the ideas of risk aversion and stochastic dominance.

These ideas provide the basic principles for the role and valuation of derivatives in both complete and incomplete markets.

9.2   The role of the capital market

In the 1930s the economist Irving Fisher described a number of important aspects of financial markets, at least ignoring issues to do with uncertainty or risk. Later work by Arrow, Debreu and others, in the mid-1960s, extended his insights to treat future outcomes as random variables. We will describe the case under certainty first, and then develop a framework that allows us to analyse conditions where risk is important.

9.2.1   Efficient Consumption and Investment

The first idea is that the capital market enables individuals' consumption through time to be made in an efficient way. The introduction of a capital market enables individuals either to borrow or lend money. Without such a market, an individual with excess cash today but small prospective income for a later date might be prepared to lend money at a very low interest rate. Conversely, another individual with small current wealth but good prospective future income, might be willing to pay a high interest rate to borrow money. By trading with each other they can both become better off. The capital market provides a single interest rate and, after it has been introduced, all individuals adjust their borrowing so that it is optimised relative to the market rate. At this point consumption plans are efficient in the sense that no individual can be made better off without making another worse off. Economists refer to this type of efficiency as Pareto efficiency. The key result is that the financial market leads to a Pareto efficient allocation of consumption across individuals, provided that lending and borrowing is available without frictions (i.e. at the same interest rate)[1] for any maturity date.

The same idea applies to investment opportunities. Without a capital market, our cash rich individual might be willing to undertake new real investments offering a very low return in the absence of any better opportunities. Conversely, our cash poor individual might have to forego exploiting very profitable opportunities because of an absence of funds. With a proper capital market any real investment with a return greater than the interest rate is worth exploiting. The cash poor individual can borrow the necessary funds and make money from undertaking new investments which return more than the interest rate. Similarly, the rich investor will now lend at the interest rate instead of making real investments which yield less than the market rate. Thus, the capital market leads to investments being undertaken efficiently, as well as enabling individuals to consume efficiently.

9.2.2   Separation of ownership and management

The existence of the market also makes it possible for the choice and management of investments to be delegated from the owners to professional managers. We have described a rule for choosing investments: accept investments which provide a return greater than the interest rate. Shareholders can subscribe capital to companies, and employ managers to run the companies, secure in the knowledge that the managers can be given clear instructions of how to act in the interests of the owners. If every potential owner of shares had a different required rate of return, as they would without the capital market, the delegation of management would be much more problematic.

9.2.3   Information for decisions

Another benefit which stems from the existence of the market is also that it provides information to investors and managers to enable them to make the appropriate decisions. Interest rates (and share prices) are published con-

---

[1] The form of how interest rates depend on maturity is termed the term structure of interest rates, and the issues involved have given rise to a considerable academic literature, both before and since the first continuous-time formulations of Merton, 1973, and Vasicek, 1977.

tinuously and give essential information for making decisions concerning individual consumption and investment, and the profitability of new real investments.

Without this flow of information, it would be much more difficult for management and ownership to be separated.

9.2.4 Central market place

The provision of a central market place is also important. It reduces the search costs of those wishing to trade. Instead of having to look among all the individuals in the economy for someone willing to trade, it is much more effective for everyone wishing to trade to go to a central market place. With n individuals in an economy, each individual just communicates with a single market, instead to having to search among the remaining $(n-1)$ individuals for a potential trading partner.

9.2.5 Liquidity

When a company sells shares in order to make a real investment, such as building a factory, the share issue is said to be a *primary* market transaction. New shares are created and the proceeds are used by the company for new and typically irreversible investments. However, most transactions on the stock exchange are not of this kind. They simply consist of one shareholder selling to another investor, with no new money going to the company. These transactions are called *secondary* market transactions. Occasionally we hear criticisms of this from commentators who do not really understand these markets. They ask: 'Isn't it terrible that such a small proportion of transactions occurs as primary issues? The city is just a gambling den, and it starves industry of the capital it needs!' This type of comment is rather silly for a number of reasons. The factories of ICI (or whoever) are built to last a long time. Few investors (either private or institutional) would be prepared to commit permanent capital to such an enterprise, if they could not subsequently sell their claim to someone else. Just as banks manufacture liquidity by taking demand deposits from investors, and lending them on to corporations as term loans, so too does the share market provide liquidity by enabling claims on the equity of companies to be sold from one investor to another.

9.2.6 Risk

The final function of the capital market is that it also provides a market for risk, and enables efficient sharing of the risk inherent in the economy among the various individuals. We will revisit this idea after developing the state-preference framework as the basis of analysis.

9.3   The state preference framework

9.3.1   Definitions and assumptions

The driving assumption of this framework of analysis is that we can adequately model risk as not knowing which *state* of nature, $s = 1, 2, \ldots, S$, will occur at any future date, t. For simplicity, we shall assume S is finite, but there are also important extensions to infinite-dimensional state spaces.

We shall assume that all investors can agree on the definitions of these states and on the cash paid, D_{sj}, by each security $j = 1, 2, \ldots, J$ in each state s. (We consider a single future date t and suppress the t subscript to keep the notation under control!) Investors may disagree on the probability of each state. This very general framework turns out to have enormous power. It was pioneered by Debreu (1959), Arrow (1964), Hirshleifer (1964, 1965).

In this section we will first look at an individual's portfolio choice problem. We next use a pair of numerical examples to describe how markets may differ depending on the number of securities available. This introduces the idea of a complete market versus an incomplete one. This leads to a key result in finance, often called the fundamental theorem of finance: the equivalence between the absence of riskless arbitrage opportunities and the existence of a linear pricing operator (and hence the possibility of pricing securities as expectations under an appropriately chosen probability measure). Finally, we revisit the ideas (already described) of Irving Fisher, about the efficiency of consumption and investments, to see how they generalise to this framework.

9.3.2   Portfolio selection and equilibrium

We consider the problem faced by a single investor, who wishes to invest an amount b_0 in the J securities, whose prices are given by the vector p. The investor is assumed to choose the portfolio which maximises a function $U(c)$ of the vector of consumption in each state. Note that this is a very general and unrestrictive criterion. For example, the individual may favour some states more than others. We simply assume that $U(c)$ is differentiable with respect to the elements of c and that the partial derivatives are non-negative, a condition known as non-satiation, meaning that in any state more consumption is preferred to less. The choice problem is formulated as:

Maximise $U(c)$ subject to
$$p'x = b_0 \quad \text{(budget constraint)}$$
$$c = Dx \quad \text{(defines consumption in each state)}$$

where c is the vector $c_1, \ldots, c_S$ of consumption in each state, p = security prices, and x = amounts purchased of each security.

The first-order conditions are:

$$\frac{\partial L}{\partial x_j} = \sum_{s=1}^{S} \frac{\partial U}{\partial c_s} D_{sj} - \lambda p_j = 0$$

If we now define

$$q_s = \frac{\partial U}{\partial c_s} / \lambda \quad (\text{=ratio of marginal utilities})$$

we have $p_j = \Sigma q_s D_{sj}$, or

$$p' = q'D \tag{9.1}$$

By non-satiation $q \geqslant 0$. Even though aggregation may be difficult, Equation (9.1) always holds as long as investors can costlessly sell short. If, instead, we have the constraints $x \geqslant 0$ then we get

$$p' \geqslant q'D \tag{9.2}$$

The Equations (9.1) and (9.2) are important characterisations of market prices. We will comment further on them later.

9.3.3   Some examples

We consider next two numerical examples which illustrate the importance of the number of securities in the market, and whether or not they enable any conceivable pattern of future consumption to be obtained.

Example 1

In our first example we have as many (linearly independent) securities as we have possible future states. This is called a *complete market*, and any vector of required cash flows can be obtained.

Security	Price	Pay-off in	
		State 1	State 2
A	13	20	10
B	9	10	10

We can find q by solving:

$$20q_1 + 10q_2 = 13$$
$$10q_1 + 10q_2 = 9$$

which gives $q_1 = 0.4$, $q_2 = 0.5$. q_1 and q_2 are the prices of *pure* securities (or Arrow–Debreu securities), which pay:

Pure security	Price	Pay-off in	
		State 1	State 2
1	0.4	1	0
2	0.5	0	1

These pure securities, which pay one unit if and only if a particular state occurs, form a natural basis for the entire security market.

Note we can construct them as portfolios of A and B.

To do this, for example to manufacture pure security 1, we can find the portfolio holdings x by solving:

$$20x_1 + 10x_2 = 1$$
$$10x_1 + 10x_2 = 0$$

which gives $x_1 = 0.1$, so $x_2 = -0.1$, with a cost of $0.1 \times 13 - 0.2 \times 9 = 0.4$ as previously.

Example 2

In our second example we have more states than securities. The example is constructed showing two securities and three states. The risk free bond, R_f, gives a 5% return regardless of which state occurs. The second asset M represents the market portfolio of equities and offers a 10% return in state 2, but it may go down, or it may go up even more.

Security	Price	Pay-off in		
		State 1	State 2	State 3
R_f	100	105	105	105
M	100	90	110	130

If the probabilities of each state happened to be equal, then the market asset would have an expected return of 10%, offering a risk premium of 5% above the bond. Since we now have more states than securities, different investors may have different q vectors, and the prices of pure securities are not unique. This is called an *incomplete market*. We could make this complete if we added an extra security, for example a single call option on the market would do the trick. Contingent claims, such as options, and also dynamic trading through time both play a role in completing markets. A paper by Ross (1976) was one of the first contributions to draw attention to this.

9.3.4   The no-arbitrage condition

In this section we will establish the key relationship, often called the fundamental theorem of finance, that the absence of riskless arbitrage opportunities is equivalent to the existence of a linear pricing operator. A result from linear algebra called Farkas' Lemma enables us to establish this proposition immediately, for the case we are considering of finite numbers of states and securities. Farkas' lemma may be stated as follows:

$$Dx \geqslant 0 \Rightarrow p'x \geqslant 0 \quad \text{if and only if}$$
$$\text{there exists } q \geqslant 0 \quad \text{such that } q'D = p'$$

Let us examine the economic interpretation of this result. The first line is a statement that no arbitrage is possible. *Arbitrage* means (to academics) the simultaneous sale and purchase of equivalent securities for immediate (riskless) profit. (Practitioners use the term a little more loosely.) The mathematical statement of it is saying that any portfolio which will deliver non-negative cash flows in all states of the world ($Dx \geqslant 0$) must have a non-negative price ($p'x \geqslant 0$). The second line tells us that this is equivalent to the existence of a non-negative pricing vector ($q \geqslant 0$) which explains the price of each security according to its state-contingent cash flows ($q'D = p'$). Note that q is precisely the vector of state prices (or prices of Arrow–Debreu pure securities) that we introduced earlier. In a complete market q is unique, in an incomplete market it is not, but provided there is no risk-free arbitrage then such a q will always exist. Farkas' Lemma can be proved reasonably simply, making use of the separating hyperplane theorem (see, for example, Mas-Colel et al., 1995). However, we may perhaps gain more intuition into this result by instead considering a linear programming formulation of how we might look for arbitrage, and see how the dual linear program gives us the state prices.

To look for an arbitrage we could solve the following linear programming problem to find the minimum cost way to obtain a non-negative vector of future cash flows:

$$\text{Minimise} \quad p'x$$
$$\text{subject to} \quad Dx \geqslant 0$$
$$x \text{ unrestricted.}$$

$x = 0$ is a feasible solution and, in the case where there is no arbitrage, it is also optimal. In this case there is also a feasible solution to the dual. In the case where arbitrage exists it is possible to find a negative initial cost portfolio which delivers non-negative future cash flows. Since the right-hand side is zero, the positions and the profit can be scaled to an arbitrary degree and so the LP is unbounded. The dual LP is therefore infeasible if arbitrage is possible.

The dual LP is:

$$\text{Maximise} \quad 0'q$$
$$\text{subject to} \quad D'q = p$$
$$q \geqslant 0.$$

Thus, a pricing vector $q \geqslant 0$, which explains all the prices in p, exists if and only if there is no arbitrage.

9.3.5   Martingale pricing

Modern contingent claims valuation (particularly in a continuous-time framework) has made considerable use of the above linear pricing result. Most elegantly this is done by using it to justify representing the processes of relative prices as martingales. In our discrete framework the argument runs as follows.

The price of any security j can be represented as:

$$p_j = \sum_s q_s d_{sj}$$

We choose a particular security N to use as numeraire. In selecting this security we must make sure that its future value is strictly positive in every state at the future date (or dates) which concerns us (i.e. $d_{sN} > 0$). This numeraire security is priced in the same way as:

$$p_N = \sum_s q_s d_{sN}$$

Now in terms of prices relative to the numeraire security we have

$$\frac{p_j}{p_N} = \frac{\sum_s q_s d_{sj}}{\sum_s q_s d_{sN}} = \frac{\sum_s (q_s d_{sN}) \frac{d_{sj}}{d_{sN}}}{\sum_s q_s d_{sN}}$$

$$= \sum_s \pi_s^N \frac{d_{sj}}{d_{sN}} \quad \text{where } \pi_s^N = \frac{q_s d_{sN}}{\sum_s q_s d_{sN}}$$

This result provides us with an easy way to value options and other contingent claims. Provided the securities which trade in the market span the claim we are interested in, the claim can be valued simply as an expectation under a suitable probability measure. Thus:

$$p_j = p_N E^N \left[\frac{d_{sj}}{d_{sN}} \right]$$

where E^N denotes the expectation under the probability measure π^N associated with numeraire N. This approach is used extensively in modern contingent claims valuation. The most useful and commonly used numeraire is the value of a bank account in which interest is continuously reinvested. Indeed, the measure corresponding to this is often simply called the martingale measure. Under it, the prices of futures contracts are martingales. The price of a discount bond with maturity corresponding to the date at which the payment on the contingent claim is determined is also useful for some applications. Under the probability measure, forward contracts for that date are martingales. El Karoui *et al.* (1995) provide a useful description of these techniques.

9.3.6 Fisher separation revisited

Earlier we described how what is known as 'Fisher separation' applies in a capital market which is perfect (i.e. no frictions, perfect competition and informationally efficient) and where future outcomes are certain. The Fisher result gave us the efficiency of consumption and investment and made it possible for ownership and management to be separated in a conceptually simple way. In order for this efficiency result to hold it must be possible to create bonds for any required maturity.

A key contribution of the Arrow–Debreu state preference framework is that it shows us that the Fisher result can be generalised to the case where outcomes are risky. All we have to do is to take the labels which we used to denote different dates in the certain world, and use them instead to label different possible states of the world for a given date. In this way all the good things which arose out of the previous analysis continue to hold provided that we have sufficient securities to make the security market complete. There must be sufficient securities to span the state-space, i.e. Rank $(D) = S$. In principle, if the market is complete, managers can use state prices to value uncertain projects and no welfare losses result from separating management and ownership. In practice too, it may be possible to value particular investment projects as 'real options' making use of the information available from traded options and other financial contracts.

The state-preference framework has become extremely important for at least four reasons. First, it is very general: we are not required to make strong assumptions about either investors' preferences, or the kinds of probability distributions involved. This contrasts, for example, with what is necessary to justify the mean–variance framework of the capital asset pricing model. Second, the extension to many periods (with discrete time and finite states) is fairly straightforward, and we can also manage the somewhat more complicated extensions to continuous time and a continuum of states. Third, although results become difficult unless we assume the market is complete, we can often assume that they are 'dynamically complete'. This means that while the market would be incomplete if we were forced to hold static portfolios, by continuous trading we can replicate any pay-off, as in the famous Black and Scholes (1973) option valuation method. Finally, options may enable us to complete the market (or at least span wide classes of pay-offs) even when the options themselves are not redundant (i.e. cannot be replicated exactly through dynamic trading). Derivative instruments play a particular role in completing the market and enabling risks to be shared. These securities include futures, swaps and options. Their pay-offs are fixed by reference to the values of security prices, or rates (such as interest rates or exchange rates). Most derivatives arise as zero–sum bets among market participants and are not part of any company's primary finance. Legitimate concerns exist as to how much the disadvantages of incentives for manipulation or dysfunctional speculation stack up against the advantages of improved sharing of risks, especially in markets which are thin.

9.4 The expected utility paradigm

9.4.1 The general framework

Maximisation of expected utility is the conventional model adopted by economists to represent how individuals make decisions under conditions of risk. It was first introduced by von Neuman and Morgenstem in 1947 in the context of the theory of games. It forms the basis of much of our understanding of market equilibrium and approaches to valuing assets in incomplete markets.

The general approach is as follows. A function $U(w)$ is used to represent a single individual's utility of different levels of wealth w at some future date.[2] Decisions are taken on the basis of choosing whatever leads to a maximisation of the statistical expectation of utility, under the individual's particular

---

[2] It is sometimes convenient to assume a utility function for an entity such as a company. This can be rather problematic for two separate reasons. First, it is more satisfactory in finance theory for managers to maximise share value. Assuming a utility function instead amounts to a conflict of interest between shareholders and managers. Second, analysis by Arrow (his impossibility theorem) shows that decision making within a firm (his model is one where a voting system is used to arrive at decisions) may not lead to decisions that satisfy the axioms which characterise expected utility. He provides an example where the transitivity assumption is violated.

probability beliefs. This framework is a special, and more restrictive, case of the more general kind of preference function already introduced in the framework of state preference theory. By wealth we really have in mind real wealth, or consumption, and not just monetary wealth as measured in pounds or dollars. Usually, we assume that the function $U(w)$ is twice differentiable, and we also make the following further assumptions.

More wealth is preferred to less (non-satiation):

this implies that $U'(w) > 0$.

Individuals have diminishing marginal utility:

this implies that $U''(w) < 0$.

For reasons we shall see shortly, this is also the same as assuming that the individual is risk averse.

We might also ask the question: 'What units are the function $U(w)$ measured in?' It is clear that we have some freedom as to how we scale U, so the units are somewhat irrelevant. Because expectation is a linear operator, it is easy to see that any affine transformation of U (e.g. $V(w) = a + bU(w)$, with $b > 0$) will preserve the same rankings among alternatives.

9.4.2   An example

We will next develop a simple example to show how a risk averse investor would choose the optimal amount to bet to maximise expected utility.

My initial wealth is 20. How much should I bet (x) if I gain $2x$ or lose x with equal probability, when my utility function is $U(w) = \sqrt{w}$?

Analysis

The investor's maximisation problem is:

$$\max E(U) = \tfrac{1}{2}\{(100 + 2x)^{1/2} + (100 - x)^{1/2}\}$$

The first-order condition for the optimal x value is then:

$$4 \times \frac{\partial}{\partial x} = \frac{2}{(100 + 2x)^{1/2}} - \frac{1}{(100 - x)^{1/2}} = 0$$

$$\therefore 4(100 - x) = 100 + 2x$$

$$\therefore 6x = 300, \quad \underline{x = 50}$$

So I should bet 50, and my final wealth will be 200 or 50.

9.4.3   Axiomatic basis

The expected utility paradigm can be justified (or at least rationalised) from the following four reasonably plausible assumptions (see, for example, Herstein and Milnor, 1953, for a more extended treatment):

(1) Comparability: Given two risky prospects x and y, we can always decide that y is preferred to x, that x is preferred to y, or that we are indifferent between x and y. Symbolically we write:

$$x \prec y, x \succ y, \quad \text{or} \quad x \sim y$$

(2) Transitivity: If x is preferred to y, and y is preferred to z, then x is preferred to z. Algebraically we write:

$$x \succ y, y \succ z \Rightarrow x \succ z$$

(3) Strong independence (choice between lotteries): If we are indifferent between x and y, then we are indifferent between a lottery which gives x with probability α and any other safe or risky prospect z with probability $(1 - \alpha)$ and another lottery which gives y with probability α and z with probability $(1 - \alpha)$.

$$x \sim y \Rightarrow \begin{bmatrix} x : \Pr \alpha \\ z : \Pr(1 - \alpha) \end{bmatrix} \sim \begin{bmatrix} y : \Pr \alpha \\ z : \Pr(1 - \alpha) \end{bmatrix}$$

(4) Continuity: If x is preferred to y and y is preferred to z, then there is some unique probability α such that the individual is indifferent between y and a lottery which gives x with probability α, or z with probability $(1 - \alpha)$.

If $x \succ y \succ z$ then there is a unique α

$$\text{such that } y \sim \begin{bmatrix} x : \Pr \alpha \\ z : \Pr(1 - \alpha) \end{bmatrix}$$

9.4.4   Construction of a utility function

We now describe how we could construct the form of the utility function for an individual whose choices were always consistent with the above axioms. We begin by arbitrarily assigning utility values to two distinct wealth levels. We know that we are allowed to do this because decisions are preserved under any affine transformation of the utility function. We can, for example, choose $U(£0) = 0$ and $U(£100\,000) = 100$. The next step is to determine what probability α makes

$$£50\,000 \sim \begin{bmatrix} £100\,000 & : \Pr \alpha \\ 0 & : \Pr(1 - \alpha) \end{bmatrix}$$

Construct $U(£50, 000)$ as equal to $(1 - \alpha)U(£0) + \alpha\, U(£100\,000) = 100\alpha$.

This process is repeated, bisecting the wealth level until the desired precision is obtained. For example, the next step would be to find the probability β, for which the individual is indifferent between £25 000 for certain, or £50 000 with probability β or £0 with probability $1 - \beta$. Using the same principle as before, we construct $U(£25\,000) = (1 - \beta)U(£0) + \beta\, U(£50\,000) = 100\alpha\beta$.

9.4.5   Consistency of choices

How can we be sure that the utility function we have obtained by this process will consistently predict all subsequent choices under risk? The answer to this lies in the axioms. If the axioms we have assumed are true, then consistency is guaranteed. For example, suppose that we have obtained the utility of wealth levels of £0, £25 000, £50 000 and £100 000 by the construction described above. We now have a model which predicts choices among any lottery involving these particular wealth levels. For example, we might ask what probability γ would make the individual indifferent between £25 000 for certain, and a lottery of £100 000 with probability γ or £0 with probability $(1 - \gamma)$. Provided that the strong independence axiom is true, then the individual is not only indifferent between £25 000 and the lottery £50 000 with probability β and £0 with probability $(1 - \beta)$, but also between these and the case where the £50 000 payment is replaced by a lottery of £100 000 with probability α and £0 with probability $(1 - \alpha)$. Thus, £25 000 for certain gives the same expected utility as £100 000 with probability $\alpha\beta$ and £0 with probability $(1 - \alpha\beta)$, which confirms the consistency and the previously obtained value of $U(£25\,000) = 100\alpha\beta$.

9.4.6   Risk aversion

Some investors are more risk averse than others. They will demand a greater expectation of profit in order to accept a given risk or, conversely, will pay a greater insurance premium in order to avoid it. We examine next how this idea of risk aversion is related to properties of the utility function.

Consider the following choice. What (insurance) premium p would we be willing to pay which would make us indifferent between paying the premium for certain, or facing a 50:50 chance of gaining or losing an amount h? We assume that the individual's initial wealth is an amount a, and that the outcome will occur immediately. The question then is what does p have to be to make:

$$(a - p) \sim \begin{bmatrix} a - h \colon \Pr\frac{1}{2} \\ a + h \colon \Pr\frac{1}{2} \end{bmatrix}$$

We have:

$$U(a - p) = \tfrac{1}{2}U(a - h) + \tfrac{1}{2}U(a + h)$$

We assume h is sufficiently small that we can use the Taylor expansion to give:

$$U(a) - pU'(a) + \ldots = \tfrac{1}{2}\left\{ \begin{array}{l} U(a) - hU'(a) + \tfrac{1}{2}h^2 U''(a) + \ldots \\ U(a) + hU'(a) + \tfrac{1}{2}h^2 U''(a) + \ldots \end{array} \right\}$$

giving

$$-pU'(a) \approx \tfrac{1}{2}h^2 U''(a)$$

so

$$p \approx -\tfrac{1}{2}h^2 \frac{U''(a)}{U'(a)}$$

Thus, the premium p, is the product of two terms: the variance of the risk, h^2, times $-\tfrac{1}{2}U''/U'$ which measures the (local) *absolute risk aversion* $(-\tfrac{1}{2}U''/U' = \text{ARA})$. Its reciprocal is sometimes referred to as *risk tolerance* $(-2U'/U'' = \text{RT})$. It is worth noting that, from the condition that $U'(w) > 0$, p is positive provided that the other condition

$U''(w) < 0$ holds. Our measure of risk aversion is also invariant with respect to affine transformations of the utility function.

In applications, the easiest utility functions to work with are ones where the risk tolerance is linear in wealth. This family is referred to either as *linear risk tolerance* (LRT) utility functions, or more commonly as *hyperbolic absolute risk aversion* (HARA) ones. Choosing the case of constant absolute risk aversion gives the utility function $U(w) = -e^{-\lambda w}$. This function has the particular simplifying property that choices are unaffected by the individual's current wealth level. The remaining functions are all of the form

$$U(w) = \frac{1}{1-b}(a + w)^{1-b}$$

for $b > 0$, and giving $U(w) = \ln(a + w)$ corresponding to the limit as b tends to one. These functions also lead to simple forms of portfolio decisions.

9.4.7   Stochastic dominance

As a final theme, we consider conditions under which all investors with particular properties of their utility functions (for example, non-satiation, or non-satiation and risk aversion) will agree that one probability distribution of wealth is preferred to another. This principle is known as stochastic dominance. A distribution F stochastically dominates distribution G if all investors (of a particular type) prefer F to G.

First-order stochastic dominance

First-order stochastic dominance provides a partial ordering relationship under the assumption that individuals prefer more wealth to less (non-satiation). The notation $\genfrac{}{}{0pt}{}{F > G}{FSD}$ is used to denote that distribution F is preferred to distribution G by all such individuals. In other words, $\genfrac{}{}{0pt}{}{F > G}{FSD}$ if and only if $E[U]$ is greater under F for all U with $U' > 0$.

First-order stochastic dominance is equivalent to the cumulative distribution function of F always lying below (i.e. to the right of) the distribution function of G.

Second-order stochastic dominance

Second-order stochastic dominance provides a partial ordering relationship under the assumption that individuals are risk averse as well as preferring more wealth to less. The notation $\genfrac{}{}{0pt}{}{F > G}{SSD}$ is used to denote that distribution F is preferred to distribution G by all such individuals. In other words, $\genfrac{}{}{0pt}{}{F > G}{SSD}$ if and only if $E[U]$ is greater under F for all U with $U' > 0$ and $U'' < 0$.

Second-order stochastic dominance occurs when the cumulative area between the two distribution functions remains positive $\left(\int_{-\infty}^{W}(G(x) - F(x))\,dx > 0\right)$.

These relationships provide convenient ways of obtaining partial orderings of probability distributions of future wealth based on only the weakest assumptions about preferences.

9.4.8 Applications

Applications of the expected utility paradigm abound and it is worth describing a few of them here. They are often less well known than they deserve to be and, in particular, by people working in the derivatives area with its emphasis on valuation in a complete markets/exact replication framework. As soon as we depart from problems of exact replication we need some criterion like expected utility to characterise optimal decisions. Such situations arise in investment management or where jump processes or market frictions, such as transactions costs, make spanning prohibitively costly. Davis and Norman (1990) have shown how investment under transactions costs can be analysed within a dynamic portfolio optimisation, and Hodges and Neuberger (1989) first applied this kind of approach to hedging options under transactions costs. Dybvig's (1988), 'How to throw away a million dollars on the stock market' makes use of expected utility and stochastic dominance concepts to show the inefficiency of particular investment management rules including stop–loss strategies and rolling over portfolio insurance. Expected utility is often used to characterise market equilibrium. Early literature often assumed utility of the HARA family in order to do this. For example, Bick (1987), gives an expected utility equilibrium model which supports Black and Scholes option pricing. More recently, papers by Hodges and Carverhill (1993), and He and Leland (1993), describe the evolution of the equity risk premium in economies characterised by an expected utility maximising representative investor. A further burgeoning recent literature explores, the 'equity premium paradox' where time additive expected utility models seem incapable of reconciling the apparently low level of real interest rates with the high risk premium on equities. Campbell *et al.* (1997, Chapter 8) provides a useful introduction to this literature, and indeed to the general problems of developing satisfactory intertemporal models of preferences.

9.4.9 Weaknesses of the expected utility paradigm

This brings us to other weaknesses of the expected utility approach. Do individuals really behave as if they maximise $E[U]$, even in a single period context? The most dubious assumption is the strong independence axiom. Consider the following gambles:

$$A: \begin{bmatrix} 0: & \Pr\frac{1}{2} \\ £30: & \Pr\frac{1}{2} \end{bmatrix}, \qquad B: \begin{bmatrix} 0: & \Pr 0.9 \\ £200: & \Pr 0.1 \end{bmatrix}, \qquad C: \begin{bmatrix} 0: & \Pr 0.7 \\ £30: & \Pr 0.25 \\ £200: & \Pr 0.05 \end{bmatrix}$$

It is easy to see that $C = \begin{bmatrix} A: & \Pr\frac{1}{2} \\ B: & \Pr\frac{1}{2} \end{bmatrix}$, so under the axiom we should have $A \succ C \succ B$ or $A \prec C \prec B$. Are your rankings consistent with this? Allais and others have shown that many individuals' preferences actually violate these rankings, and so we should treat the verity of the expected utility paradigm with a certain amount of scepticism despite its undoubted elegance. Recent and continuing research goes on[3] to try to find alternative ways of modelling investors' preferences which are more realistic and yet also sufficiently tractable. For the time being, the expected utility paradigm still has a lot to offer us, despite its well understood shortcomings.

9.5 Summary

This chapter has described the microeconomic foundations of finance.[4] We described first the various roles of the capital market. It enables both consumption and production (i.e. real investment) decisions to be made efficiently. It provides information to companies and to individuals facilitating the separation of ownership from management. It creates liquidity: even though a company's capital is permanent, it does not have to look to the same investors to provide that capital at all times. Investors are happy to buy shares precisely because they know that they will be able to sell them at a reasonably fair price at any time in the future. The central market place of the stock exchange also minimises the search costs of such transactions; it is unnecessary to go out looking for individuals who want to trade, instead each individual goes to the market. The capital mark*et al*so enables risk to be shared in an efficient way.

Next, we described state-preference theory, which gives us a way of describing security markets where the outcomes on securities are uncertain. Individuals may disagree as to the probabilities of different outcomes occurring, but we assume that they can agree as to how we label the outcomes which might occur. Not only does this framework provide a very general setting for work in securities markets, but it also enables us to develop a pricing theory based on the absence of arbitrage. This identity is the key to understanding modern continuous-time analysis. The ideas of no-arbitrage pricing, spanning and completeness are fundamental in this regard, and so too is the distinction between whether a market is complete in a static sense, or whether it is only made complete by dynamic trading or by the addition of new securities. In continuous time analysis our finite dimensional vector spaces of portfolios and their possible outcomes are replaced by more general function spaces. This creates many technical difficulties, but always the key intuitions come from the finite case. Indeed the continuous models are best regarded as convenient approximations to a discrete reality, and whenever they provide different conclusions our assumptions must be wrong.

Finally, there are many situations where it is necessary for us to model investors' decision making under risk in an explicit fashion; for example, in the management of investment portfolios, for understanding market

---

[3] Machina (1982) seems to have been the first to relax the assumption of the independence axiom. See also, Ingersoll (1987, p. 34) for an example.

[4] For more extended texts on finance see, for example, Brealey and Myers (1996), Copeland and Weston (1988) or Ingersoll (1987). These are listed in order of increasing technical sophistication and, conversely, decreasing amounts of detail concerning the capital markets themselves. For extended texts on microeconomics see for example Mas-Colel *et al.* (1995), or Varian (1994).

equilibrium, and for managing risk in incomplete markets. This chapter reviews the expected utility paradigm, including its axiomatic basis and some of its weaknesses.

References

Arrow, K. (1964) The role of securities in the optimal allocation of risk bearing. *Review of Economic Studies*, **31**, 91–96.

Bick, A. (1987) On the consistency of the Black–Scholes model with a general equilibrium framework. *Journal of Financial and Quantitative Analysis*, **23**, 153–161.

Black, F. and Scholes, M. (1973) The pricing of options and corporate liabilities, *Journal of Political Economy*, **81**, 637–654.

Brealey, R. A. and Myers, S. C. (1996) *Principles of Corporate Finance*, 5th Edition. New York: McGraw-Hill.

Campbell, J. Y., Lo, A. W. and MacKinlay, A. C. (1997) *The Econometrics of Financial Markets*. Princeton, New Jersey: Princeton University Press.

Copeland, T. E. and Weston, J. F. (1988) *Financial Theory and Corporate Policy*, 3rd Edition. New York: Addison-Wesley.

Davis, M. H. A. and Norman, A. R. (1990) Portfolio selection with transactions costs. *Mathematics of Operations Research*, **15**, 676–713.

Debreu, G. (1959) *Theory of Value*. New York: Wiley.

Dybvig, P. H. (1988) Inefficient dynamic portfolio strategies, or how to throw away a million dollars in the stock market. *The Review of Financial Studies*, **1**, 67–88.

El Karoui, N., Geman, H. and Rochet, J. C. (1995) Changes of numeraire, changes of probability measure and option pricing. *Journal of Applied Probability*, **32**, 443–458.

He, H. and Leland, H. (1993) On equilibrium asset price processes. *Review of Financial Studies*, **6**, 593–617.

Herstein, I. N. and Milnor, J. (1953) An axiomatic approach to expected utility. *Econometrica*, **21**, 291–297.

Hirshleifer, J. (1964) Efficient allocation of capital in an uncertain world. *American Economic Review*, **54**, 77–85.

Hirshleifer, J. (1965) Investment decisions under uncertainty: choice-theoretic approaches. *Quarterly Journal of Economics*, **79**, 509–536.

Hodges, S. D. and Carverhill, A. P. (1993) Quasi mean reversion in an efficient stock market: the characterisation of economic equilibria which support Black–Scholes option pricing. *Economic Journal*, **103**, 395–405.

Hodges, S. D. and Neuberger, A. (1989) Optimal replication of contingent claims under transactions costs. *The Review of Futures Markets*, **8**, 223–239.

Ingersoll, J. E. (1987) *Theory of Financial Decision Making*. New Jersey: Rowman and Littlefield.

Machina, M. (1982) Expected utility analysis without the independence axiom. *Econometrica*, **50**, 277–323.

Mas-Colel, A., Whinston, M. D. and Green, J. R. (1995) *Microeconomic Theory*. New York: Oxford University Press.

Merton, R. C. (1973) Theory of rational option pricing. *Bell Journal of Economics*, **4**, 171–183.

Ross, S. A. (1976) Options and efficiency. *Quarterly Journal of Economics*, **90**, 75–89.

Varian, H. R. (1994) *Microeconomic Analysis*, third edition. New York: W. W. Norton and Co.

Vasicek, O. A. (1977) An equilibrium characterization of the term structure. *Journal of Financial Economics*, **5**, 177–188.

von Neuman, J. and Morgenstern, O. (1947) *Theory of games and economic behaviour*. Princeton: Princeton University Press.

10

American Options

Damien Lamberton

The modern theory of American options originated in papers by Bensoussan (1984) and Karatzas (1988), who highlighted the connection between American option hedging/pricing and the mathematical theory of optimal stopping. Recent survey papers such as Karatzas (1989) and Myneni (1992) provide good introductions to American option pricing in a rather general setting. See also the introductory part of this book. The aim of this chapter is to provide an introduction to the theory of optimal stopping and to derive the basic properties of American option prices in classical models.

The first two sections are devoted to optimal stopping (with discrete time in the first section and continuous time in the second section). In the third section, we introduce the value function of an American option and discuss its major features. The fourth section is a brief survey of numerical methods. The fifth section is an appendix devoted to the proofs of some of the results of Section 10.3.

Throughout this chapter, (in)equalities between random variables are to be understood in the almost sure sense.

10.1 Optimal stopping: discrete time

In this section, the underlying probability space $(\Omega, \mathcal{F}, \mathbf{P})$ is equipped with a discrete filtration $(\mathcal{F}_n)_{n=0,1,\dots,N}$, with a finite horizon $N \in \mathbf{N}$. We will denote by $T_{n,N}$ the set of all stopping times with values in $\{n, n+1, \dots, N\}$. Our basic reference for optimal stopping in discrete time is Neveu (1975, ch. 6). Applications to option pricing in discrete time can be found in Lamberton and Lapeyre (1996, ch. 2).

10.1.1 The Snell envelope

Let $Z = (Z_n)_{0 \leqslant n \leqslant N}$ be an adapted (finite) sequence of real-valued integrable random variables. The optimal stopping problem for Z consists of maximising

$E(Z_v)$ over all stopping times v (note that Z_v is integrable because v takes on finitely many values). The main tool for solving this problem is the so-called *Snell envelope*, of Z. In order to introduce this notion, we have to recall the definition of the essential upper bound.

Definition 10.1.1:   Let $(X_i)_{i \in I}$ be a (possibly uncountable) family of real valued random variables. The essential upper bound of $(X_i)_{i \in I}$ is the (unique up to null events) random variable $\bar{X}$, with values in $\bar{R} = [-\infty, +\infty]$, satisfying the following two properties:

(1)   for every $i \in I$, $X_i \leqslant \bar{X}$ a.s.;
(2)   If X is a random variable with values in $\bar{R}$ such that $X_i \leqslant X$ a.s., for every $i \in I$, then $\bar{X} \leqslant X$ a.s.

In this definition, we emphasised that the inequalities are up to null events. However, as mentioned earlier, we will skip the label 'a.s.' in subsequent inequalities between random variables. The usual notation for $\bar{X}$ is ess-sup$_{i \in I} X_i$. We refer the reader to Neveu (1975, ch. 6) or Dacunha-Castelle and Duflo (1986, vol. 2, ch. 5), for the existence and the basic properties of the essential upper bound. In particular, recall that the family $(X_i)_{i \in I}$ is said to be a *lattice* (or to have the lattice property) if for any two indices $i, j \in I$, there exists an index $k \in I$ such that $X_k \geqslant \max(X_i, X_j)$. If $(X_i)_{i \in I}$ is a lattice, there exists a sequence $(i_n)_{n \in \mathbb{N}}$ in I such that $(X_{i_n})_{n \in \mathbb{N}}$ is a non-decreasing sequence of random variables satisfying $\lim_{n \to +\infty} X_{i_n} = $ ess-sup$_{i \in I} X_i$ (cf. Neveu, 1975, ch. 6).
     The Snell envelope can now be defined as follows.

Definition 10.1.2:   Let $Z = (Z_n)_{0 \leqslant n \leqslant N}$ be an adapted sequence of real-valued integrable random variables. The *Snell envelope* of Z is the sequence $U = (U_n)_{0 \leqslant n \leqslant N}$ defined by

$$U_n = \text{ess-}\sup_{v \in T_{n,N}} E(Z_v \mid \mathcal{F}_n), \quad 0 \leqslant n \leqslant N$$

In the following, we consider a fixed sequence Z as above and denote by U its Snell envelope. We first state the well known *principle of dynamic programming* for optimal stopping, which provides a useful algorithm for computing the Snell envelope.

Theorem 10.1.3:   *The Snell envelope U of Z satisfies the following properties:*

(1) $U_N = Z_N$
(2) $U_n = \max(Z_n, E(U_{n+1} \mid \mathcal{F}_n))$, *for* $0 \leqslant n \leqslant N - 1$.

For the proof of Theorem 10.1.3, we will need the following lemma.

Lemma 10.1.4:   *Let* $n \in \{0, 1, \ldots, N\}$. *The family* $(E(Z_v \mid \mathcal{F}_n), v \in T_{n,N})$ *has the lattice property.*

Proof: Let $v_1, v_2 \in T_{n,N}$ and $X_i = E(Z_{v_i} \mid \mathcal{F}_n)$ $(i = 1, 2)$. Define a stopping time v by setting:

$$v = v_1 \mathbf{1}_{\{X_1 \geqslant X_2\}} + v_2 \mathbf{1}_{\{X_1 < X_2\}}$$

Clearly, $v \in \mathcal{T}_{n,N}$ and $E(Z_v \mid \mathcal{F}_n) \geq E(Z_{v_i} \mid \mathcal{F}_n)$, for $i = 1, 2$. ☐

Proof of Theorem 10.1.3: The equality $U_N = Z_N$ is obvious from the definition. Now, let $0 \leq n \leq N - 1$. It is clear that $U_n \geq Z_n$. Moreover, the lattice property of the family $(E(Z_v \mid \mathcal{F}_{n+1}), v \in \mathcal{T}_{n+1,N})$ enables us to write

$$E(U_{n+1} \mid \mathcal{F}_n) = E\left(\text{ess-} \sup_{v \in \mathcal{T}_{n+1,N}} E(Z_v \mid \mathcal{F}_{n+1}) \bigg| \mathcal{F}_n \right) = \text{ess-} \sup_{v \in \mathcal{T}_{n+1,N}} E(E(Z_v \mid \mathcal{F}_{n+1}) \mid \mathcal{F}_n)$$

as can be seen by approximating the essential upper bound by a non-decreasing sequence. But, for $v \in \mathcal{T}_{n+1,N}$

$$E(E(Z_v \mid \mathcal{F}_{n+1}) \mid \mathcal{F}_n) = E(Z_v \mid \mathcal{F}_n) \leq U_n$$

where the last inequality follows from the inclusion $\mathcal{T}_{n+1,N} \subset \mathcal{T}_{n,N}$. Hence $U_n \geq \max(Z_n, E(U_{n+1} \mid \mathcal{F}_n))$.

Now, if $v \in \mathcal{T}_{n,N}$, we have $v \vee (n+1) \in \mathcal{T}_{n+1,N}$ (where $a \vee b$ denotes the maximum of the numbers a and b) and

$$\begin{aligned}
E(Z_v \mid \mathcal{F}_n) &= Z_n \mathbf{1}_{\{v=n\}} + E(Z_v \mathbf{1}_{\{v>n\}} \mid \mathcal{F}_n) \\
&= Z_n \mathbf{1}_{\{v=n\}} + \mathbf{1}_{\{v>n\}} E(Z_{v \vee (n+1)} \mid \mathcal{F}_n) \\
&\leq Z_n \mathbf{1}_{\{v=n\}} + \mathbf{1}_{\{v>n\}} E(U_{n+1} \mid \mathcal{F}_n) \\
&\leq \max(Z_n, E(U_{n+1} \mid \mathcal{F}_n))
\end{aligned}$$

Therefore, $U_n \leq \max(Z_n, E(U_{n+1} \mid \mathcal{F}_n))$. ☐

Corollary 10.1.5: *The Snell envelope U is the minimal supermartingale which dominates Z.*

Proof: The inequalities $U_n \geq E(U_{n+1} \mid \mathcal{F}_n)$ and $U_n \geq Z_n$ (both implied by property 2 in Theorem 10.1.3) prove that U is a supermartingale which dominates Z. On the other hand, if V is a supermartingale majorant of Z, it is easy to prove that $V_n \geq U_n$ by backward induction. ☐

Remark 10.1.6: The efficiency of the principle of dynamic programming is quite apparent in a Markovian setting. Indeed, assume that $Z_n = f(n, X_n)$, where $X = (X_n)_{0 \leq n \leq N}$ is a Markov chain with values in some measurable state space $(E, \mathcal{E})$ and $f(n, \cdot)$ a measurable function with non-negative values. More precisely, assume that, for $n = 0, 1, \ldots, N - 1$, the conditional distribution of X_{n+1} given $\mathcal{F}_n$ is given by $P_n(X_n, \cdot)$, where P_n is a transition probability on $E \times \mathcal{E}$ (cf. Neveu, 1965, ch. 3). Define U(n,x), for $n = 0, 1, \ldots N$ and $x \in E$, from the following algorithm:

$$\begin{cases}
U(N, x) = f(N, x) \\
U(n, x) = \max\{f(n, x), P_n[U(n+1, \cdot)](x)\} \quad \text{for } 0 \leq n \leq N - 1.
\end{cases}$$

Using Theorem 10.1.3, it is easy to check that the process $(U(n, X_n))_{0 \leq n \leq N}$ is the Snell envelope of Z. ☐

10.1.2 Optimal stopping times

A stopping time $\bar{v} \in T_{0,N}$ is said to be *optimal* if it satisfies $\mathbf{E}(Z_{\bar{v}}) = \sup_{v \in T_{0,N}} \mathbf{E}(Z_v)$. Optimal stopping times can be characterised with the help of the Snell envelope, as the following theorem shows.

Theorem 10.1.7:
(1)   *A stopping time $\bar{v} \in T_{0,N}$ is optimal if and only if the following conditions are both satisfied:*
   (a) $U_{\bar{v}} = Z_{\bar{v}}$
   (b) *the stopped process $U^{\bar{v}}$, defined by $U_n^{\bar{v}} = U_{\bar{v} \wedge n}, 0 \leqslant n \leqslant N$, is a martingale.*
(2)   *The stopping time $v_0 = \min \{n \in \mathbb{N} \mid U_n = Z_n\}$ is optimal.*

Proof: Assume that conditions $(1a)$ and $(1b)$ are fulfilled. Then

$$\mathbf{E}(Z_{\bar{v}} \mid \mathcal{F}_0) = \mathbf{E}(U_{\bar{v}} \mid \mathcal{F}_0) = U_0$$

and, since $U_0 \geqslant \mathbf{E}(Z_v \mid \mathcal{F}_0)$ for all $v \in T_{0,N} \bar{v}$ is optimal.

Before proving the converse, we prove that v_0 is optimal. Condition $(1a)$ is obviously satisfied by v_0 and it suffices to prove that U^{v_0} is a martingale. Observe that

$$U_{v_0 \wedge (n+1)} - U_{v_0 \wedge n} = (U_{n+1} - U_n)\mathbf{1}_{\{n < v_0\}}$$

Now, on the set $\{v_0 > n\}$, $U_n > Z_n$ and, consequently, $U_n = \mathbf{E}(U_{n+1} \mid \mathcal{F}_n)$. Hence

$$
\begin{aligned}
\mathbf{E}(U_{n+1}^{v_0} - U_n^{v_0} \mid \mathcal{F}_n) &= \mathbf{E}\big((U_{n+1} - U_n)\mathbf{1}_{\{n < v_0\}} \mid \mathcal{F}_n\big) \\
&= \mathbf{E}\big((U_{n+1} - \mathbf{E}(U_{n+1} \mid \mathcal{F}_n))\mathbf{1}_{\{n < v_0\}} \mid \mathcal{F}_n\big) \\
&= \mathbf{1}_{\{n < v_0\}}\mathbf{E}(U_{n+1} - \mathbf{E}(U_{n+1} \mid \mathcal{F}_n) \mid \mathcal{F}_n) \\
&= 0
\end{aligned}
$$

It follows that U^{v_0} is indeed a martingale. Also, note that $\mathbf{E}(U_0) = \mathbf{E}(Z_{v_0})$.

Finally we prove that an optimal stopping time satisfies $(1a)$ and $(1b)$. Let $\bar{v}$ be an optimal stopping time. Then $\mathbf{E}(Z_{\bar{v}}) = \mathbf{E}(Z_{v_0}) = \mathbf{E}(U_0)$. On the other hand, since U is a supermartingale majorant of Z, we have $\mathbf{E}(Z_{\bar{v}} \mid \mathcal{F}_0) \leqslant \mathbf{E}(U_{\bar{v}} \mid \mathcal{F}_0) \leqslant U_0$. Hence, the last two inequalities must be equalities. Therefore, $Z_{\bar{v}} = U_{\bar{v}}$ and $U_0 = \mathbf{E}(U_{\bar{v}} \mid \mathcal{F}_0)$ which, together with the supermartingale property, implies that $U^{\bar{v}}$ is a martingale.   □

It follows from Theorem 10.1.7 that v_0 is the smallest optimal stopping time. The largest one can be characterised through the Doob decomposition of U. Recall that there exists a unique martingale $M = (M_n)_{0 \leqslant n \leqslant N}$ and a unique non-decreasing predictable process $A = (A_n)_{0 \leqslant n \leqslant N}$ with $A_0 = 0$ such that

$$U_n = M_n - A_n, \qquad 0 \leqslant n \leqslant N$$

This is the so-called Doob decomposition of the supermartingale U (cf. Neveu, 1975, ch. 7).

Theorem 10.1.8:   *The stopping time v_1 defined by $v_1 = \inf \{n \mid A_{n+1} > 0\}$, with the convention $\inf \emptyset = N$, is optimal and is the largest optimal stopping time.*

Proof: Note that the fact that v_1 is a stopping time follows from the predictability of A. Now, it follows from the definition of v_1 that $U^{v_1} = M^{v_1}$. Therefore U^{v_1} is a martingale. Also, note that if v is an optimal stopping time, the martingale property of U^v yields that $\mathbf{E}(U_v) = \mathbf{E}(U_0) = \mathbf{E}(M_v - A_v) = \mathbf{E}(M_0) - \mathbf{E}(A_v)$, hence, since $M_0 = U_0$, $\mathbf{E}(A_v) = 0$ and $v \leqslant v_1$.

In order to prove that v_1 is optimal, it remains to check that $Z_{v_1} = U_{v_1}$. Now

$$U_{v_1} = \sum_{j=0}^{N} U_j \mathbf{1}_{\{v_1 = j\}}$$

$$= Z_N \mathbf{1}_{\{v_1 = N\}} + \sum_{j=0}^{N-1} U_j \mathbf{1}_{\{v_1 = j\}}$$

Let $j \leqslant N - 1$. On $\{v_1 = j\}$, we have $U_j = M_j$ and $\mathbf{E}(U_{j+1} \mid \mathcal{F}_j) = M_j - A_{j+1} < U_j$, therefore $U_j = \max(Z_j, \mathbf{E}(U_{j+1} \mid \mathcal{F}_j)) = Z_j$. $\qquad\square$

10.2 Optimal stopping: continuous time

The theory of optimal stopping in continuous time can be developed in a very general setting. Our purpose, in this section, is to derive sufficiently general results for financial applications, without facing too many technical difficulties. The proofs are inspired by El Karoui (1981), to which we refer the reader for more complete results.

We consider a probability space $(\Omega, \mathcal{F}, \mathbf{P})$ with a continuous time filtration $(\mathcal{F}_t)_{t\in[0,T]}$, where T is a finite horizon. The filtration is assumed to satisfy the so-called *usual conditions* (cf. Karatzas and Shreve, 1988, ch. 1). We will denote by $\mathcal{T}_{t,T}$ the set of all stopping times with values in the interval $[t, T]$. The following terminology will be used in the following.

Definition 10.2.1: A right-continuous adapted process $(X_t)_{0 \leqslant t \leqslant T}$ is said to be

(i) regular if X_τ is integrable for all $\tau \in \mathcal{T}_{0,T}$ and, for every non-decreasing sequence of stopping times $(\tau_n)_{n\in\mathbb{N}}$ with $\tau = \lim_{n\to\infty} \tau_n$, we have $\lim_{n\to\infty} \mathbf{E}(X_{\tau_n}) = \mathbf{E}(X_\tau)$;

(ii) of class D if the family $(X_\tau)_{\tau\in\mathcal{T}_{0,T}}$ is uniformly integrable. $\qquad\square$

Note that a regular process may have discontinuous paths.

10.2.1 The Snell envelope in continuous time

Throughout this section, $Z = (Z_t)_{0 \leqslant t \leqslant T}$ will denote a right-continuous, adapted, regular process satisfying:

$$\forall t \in [0, T], \quad Z_t \geqslant 0 \quad \text{and} \quad \mathbf{E}\left(\sup_{0 \leqslant t \leqslant T} Z_t\right) < \infty$$

Such a process is obviously of class D. In this setting, the Snell envelope of Z can be defined as follows.

Theorem 10.2.2:   *Let $U = (U_t)_{0 \leqslant t \leqslant T}$ be the process defined by*

$$U_t = \text{ess-} \sup_{\tau \in T_{t,T}} \mathbf{E}(Z_\tau \mid \mathcal{F}_t), \quad 0 \leqslant t \leqslant T$$

(1)   *U is a supermartingale.*
(2)   *For all $t \in [0, T]$, $\mathbf{E}(U_t) = \sup_{\tau \in T_{t,T}} \mathbf{E}(Z_\tau)$.*
(3)   *U admits a right-continuous modification.*

The right-continuous modification of U is called the Snell envelope of Z. We will still denote it by U. For the proof of Theorem 10.2.2, we will need the following lemma, which can be proved in exactly the same way as Lemma 10.1.4.

Lemma 10.2.3:   *Let $t \in [0, T]$. The family $(\mathbf{E}(Z_\tau \mid \mathcal{F}_t), \tau \in T_{t,T})$ has the lattice property.*

Proof of Theorem 10.2.2:
(1)   Let $0 \leqslant s \leqslant t \leqslant T$. It follows from Lemma 10.2.3 that

$$\mathbf{E}(U_t \mid \mathcal{F}_s) = \text{ess-} \sup_{\tau \in T_{t,T}} \mathbf{E}(\mathbf{E}(Z_\tau \mid \mathcal{F}_t) \mid \mathcal{F}_s)$$

$$= \text{ess-} \sup_{\tau \in T_{t,T}} \mathbf{E}(Z_\tau \mid \mathcal{F}_s) \leqslant U_s$$

which proves that U is a supermartingale. For the last inequality, we have used the inclusion $T_{t,T} \subset T_{s,T}$.
(2)   The second statement is an easy consequence of the lattice property.
(3)   Since the filtration satisfies the usual conditions, the existence of a right-continuous modification will follow from the right-continuity of $t \mapsto \mathbf{E}(U_t)$ (cf. Karatzas and Shreve, 1988, ch. 1, section 3). Let $(t_n)_{n \in \mathbb{N}}$ be a sequence in $[t, T]$, with t_n decreasing to t. We have $\mathbf{E}(U_{t_n}) = \sup_{\tau \in T_{t_n,T}} \mathbf{E}(Z_\tau) \leqslant \mathbf{E}(U_t)$, for every $n \in \mathbb{N}$. Now let $\tau \in T_{t,T}$ and $\tau_n = (\tau + t_n - t) \wedge T$. Then, $\tau_n \in T_{t_n,T}$ and $\lim_{n \to \infty} Z_{\tau_n} = Z_\tau$, as follows from the right-continuity of Z. Hence, $\mathbf{E}(Z_\tau) \leqslant \liminf_{n \to \infty} \mathbf{E}(Z_{\tau_n}) \leqslant \liminf_{n \to \infty} \mathbf{E}(U_{t_n})$. Therefore, since $\tau \in T_{t,T}$ is arbitrary, $\mathbf{E}(U_t) \leqslant \liminf_{n \to \infty} \mathbf{E}(U_{t_n})$ and hence $\lim_{n \to \infty} \mathbf{E}(U_{t_n}) = \mathbf{E}(U_t)$.   □

Remark 10.2.4:   Note that U is of class D. This follows from the fact that $0 \leqslant U \leqslant M$, where M is the martingale defined by $M_t = \mathbf{E}(\sup_{s \in [0,T]} Z_s \mid \mathcal{F}_t)$

Corollary 10.2.5:   *The Snell envelope U is the smallest right-continuous supermartingale which dominates Z.*

Proof: Let V be a right-continuous supermartingale majorant of Z. It follows from the optional sampling theorem that for every $\tau \in T_{t,T}$

$$\mathbf{E}(Z_\tau \mid \mathcal{F}_t) \leqslant \mathbf{E}(V_\tau \mid \mathcal{F}_t) \leqslant V_t$$

Therefore, $U_t \leqslant V_t$.   □

10.2.2 Optimal and suboptimal stopping times

A stopping time τ^* will be called *optimal* if $\mathbf{E}(z_{\tau^*}) = \sup_{\tau \in \mathcal{T}_{0,T}} \mathbf{E}(Z_\tau)$. We are now in a position to characterise optimal stopping times, in analogy with the discrete time case.

Theorem 10.2.6: *A stopping time $\tau^* \in \mathcal{T}_{0,T}$ is optimal if and only if the following conditions are both satisfied:*
(1) $U_{\tau^*} = Z_{\tau^*}$
(2)   *the stopped process U^{τ^*}, defined by $U_t^{\tau^*} = U_{\tau^* \wedge t}, 0 \leqslant t \leqslant T$, is a martingale.*

Proof: First assume that both conditions are satisfied. Then, by the optional sampling theorem, $\mathbf{E}(Z_{\tau^*}) = \mathbf{E}(U_{\tau^*}) = \mathbf{E}(U_0)$, which proves that τ^* is optimal.
  Conversely, if τ^* is optimal, we have $\mathbf{E}(Z_{\tau^*}) = \mathbf{E}(U_0)$. But, since U is a supermartingale majorant, $U_0 \geqslant \mathbf{E}(U_{\tau^*} \mid \mathcal{F}_0) \geqslant \mathbf{E}(Z_{\tau^*} \mid \mathcal{F}_0)$. Hence, $Z_{\tau^*} = U_{\tau^*}$ and $U_0 = \mathbf{E}(U_{\tau^*} \mid \mathcal{F}_0)$, which yields that U^{τ^*} is a martingale.   □

So far, we have not proved the existence of an optimal stopping time. Indeed the optimality of the natural candidate $\tau_0 = \inf\{t \in [0, T] \mid U_t = Z_t\}$ is harder to prove than in the discrete time case. For that purpose, we need to define *suboptimal* stopping times in the following way. For $0 < \lambda < 1$ and $0 \leqslant t < T$, let

$$D_t^{(\lambda)} = \inf \{s > t \mid Z_s \geqslant \lambda U_s\}$$

and $D_T^{(\lambda)} = T$. Since the filtration satisfies the usual conditions, $D_t^{(\lambda)}$ is a stopping time (cf. Revuz and Yor, 1994, ch. 1, section 4). We obviously have $D_t^{(\lambda)} \in \mathcal{T}_{t,T}$ and $s \leqslant t \Rightarrow D_s^{(\lambda)} \leqslant D_t^{(\lambda)}$.

Theorem 10.2.7:
(1)   *For every $\lambda \in (0, 1)$, $(D_t^{(\lambda)})_{0 \leqslant t \leqslant T}$ is a (non-adapted) right-continuous process.*
(2)   *For every $\lambda \in (0, 1)$ and every $t \in [0, T]$, we have $U_t = \mathbf{E}(U_{D_t^{(\lambda)}} \mid \mathcal{F}_t)$.*

Note that the second statement implies that $\mathbf{E}(Z_{D_t^{(\lambda)}} \mid \mathcal{F}_t) \geqslant \lambda U_t$. In particular $\mathbf{E}(Z_{D_0^{(\lambda)}}) \geqslant \lambda \mathbf{E}(U_0)$, which is close to optimality if λ is close to 1.

Proof of Theorem 10.2.7:
(1)   If $D_t^{(\lambda)} > t$, we have $D_u^{(\lambda)} = D_t^{(\lambda)}$ for every $u \in (t, D_t^{(\lambda)})$, hence $D_{t+}^{(\lambda)} = D_t^{(\lambda)}$. If $D_t^{(\lambda)} = t$, one can find a (strictly) decreasing sequence $(t_n)_{n \in \mathbb{N}}$ with $\lim_{n \to \infty} t_n = t$ and $Z_{t_n} \geqslant \lambda U_{t_n}$. Now, for every $n \in \mathbb{N}$, $D_{t_{n+1}}^{(\lambda)} \leqslant t_n$. Hence $D_{t+}^{(\lambda)} \leqslant \lim_{n \to \infty} t_n = t$. This proves the right-continuity of $D^{(\lambda)}$.
(2)   Let $U_t^{(\lambda)} = \mathbf{E}(U_{D_t^{(\lambda)}} \mid \mathcal{F}_t)$, for $0 \leqslant t \leqslant T$. We want to prove that $U^{(\lambda)} = U$. The inequality $U^{(\lambda)} \leqslant U$ follows from the supermartingale property of U. Also, if $s \geqslant t$, we have $U_{D_t^{(\lambda)}} \geqslant \mathbf{E}(U_{D_s^{(\lambda)}} \mid \mathcal{F}_{D_t^{(\lambda)}})$, which, after conditioning with respect to $\mathcal{F}_t$, yields that $U^{(\lambda)}$ is a supermartingale. Moreover, it follows from the right-continuity of $D^{(\lambda)}$ and the fact that U is of class D that $t \mapsto \mathbf{E}(U_t^{(\lambda)}) = \mathbf{E}(U_{D_t^{(\lambda)}})$ is right-continuous. Hence, $U^{(\lambda)}$ admits a right-continuous modification, which we still denote by $U^{(\lambda)}$. Consider the supermartingale $\bar{U}$ defined by: $\bar{U}_t = \lambda U_t + (1 - \lambda) U_t^{(\lambda)}$. We have:

$$\bar{U}_t = \lambda U_t + (1 - \lambda)\mathbf{E}(U_{D_t^{(\lambda)}} \mid \mathcal{F}_t)$$
$$= \lambda U_t + (1 - \lambda)\mathbf{E}(1_{\{Z_t > \lambda U_t\}} U_{D_t^{(\lambda)}} + 1_{\{Z_t \leq \lambda U_t\}} U_{D_t^{(\lambda)}} \mid \mathcal{F}_t)$$
$$\geq \lambda U_t + (1 - \lambda)\mathbf{E}(1_{\{Z_t > \lambda U_t\}} U_t \mid \mathcal{F}_t)$$
$$= \lambda U_t + (1 - \lambda)U_t 1_{\{Z_t > \lambda U_t\}}$$
$$= U_t 1_{\{Z_t > \lambda U_t\}} + \lambda U_t 1_{\{Z_t \leq \lambda U_t\}}$$
$$\geq Z_t$$

where the first inequality follows from the fact that U is non-negative (because Z is) and the equality $D_t^{(\lambda)} = t$ on the set $\{Z_t > \lambda U_t\}$. Hence $\bar{U} \geq U$ and $U^{(\lambda)} \geq U$. Therefore, $U^{(\lambda)} = U$.   □

Corollary 10.2.8:   *For every $\lambda \in (0, 1)$ and every stopping time τ, we have $U_\tau = \mathbf{E}(U_{D_\tau^{(\lambda)}} \mid \mathcal{F}_\tau)$.*

Proof: If τ is a 'discrete' stopping time (i.e. with values in a finite subset of $[0, T]$), the statement follows easily from property 2 in Theorem 10.2.7. Now, for any stopping time τ, one can find a non-increasing sequence $(\tau_n)_{n \in \mathbf{N}}$ of discrete stopping times with $\lim_{n \to \infty} \tau_n = \tau$. For every $n \in \mathbf{N}$, we have

$$U_{\tau_n} = \mathbf{E}(U_{D_{\tau_n}^{(\lambda)}} \mid \mathcal{F}_{\tau_n}) \tag{10.1}$$

Since U is right-continuous and of class D, as $n \to \infty$, $U_{\tau_n} \to U_\tau$ and $U_{D_{\tau_n}^{(\lambda)}} \to U_{D_\tau^{(\lambda)}}$ almost surely and in L^1. Moreover

$$\mathbf{E}\left| \mathbf{E}(U_{D_{\tau_n}^{(\lambda)}} \mid \mathcal{F}_{\tau_n}) - \mathbf{E}(U_{D_\tau^{(\lambda)}} \mid \mathcal{F}_\tau) \right| \leq \mathbf{E}\left| \mathbf{E}(U_{D_{\tau_n}^{(\lambda)}} - U_{D_\tau^{(\lambda)}} \mid \mathcal{F}_{\tau_n}) \right|$$
$$+ \mathbf{E}\left| \mathbf{E}(U_{D_\tau^{(\lambda)}} \mid \mathcal{F}_{\tau_n}) - \mathbf{E}(U_{D_\tau^{(\lambda)}} \mid \mathcal{F}_\tau) \right|$$
$$\leq \mathbf{E}\left| U_{D_{\tau_n}^{(\lambda)}} - U_{D_\tau^{(\lambda)}} \right| + \mathbf{E}\left| \mathbf{E}(U_{D_\tau^{(\lambda)}} \mid \mathcal{F}_{\tau_n}) - \mathbf{E}(U_{D_\tau^{(\lambda)}} \mid \mathcal{F}_\tau) \right|$$

Hence, using the right-continuity of the filtration,

$$\lim_{n \to \infty} \mathbf{E}(U_{D_{\tau_n}^{(\lambda)}} \mid \mathcal{F}_{\tau_n}) = \mathbf{E}(U_{D_\tau^{(\lambda)}} \mid \mathcal{F}_\tau)$$

in L^1, and the corollary follows by passing to the limit in Equation (10.1).   □

Theorem 10.2.9:   *The Snell envelope is a regular process and the stopping time $\tau_0 = \inf\{t \geq 0 \mid U_t = Z_t\}$ is the smallest optimal stopping time.*

Proof: Let $(\tau_n)_{n \in \mathbf{N}}$ be a non-decreasing sequence of stopping times. Let $\tau = \lim_{n \to \infty} \tau_n$. The sequence $(\mathbf{E}(U_{\tau_n}))_{n \in \mathbf{N}}$ is non-increasing and $\mathbf{E}(U_{\tau_n}) \geq \mathbf{E}(U_\tau)$ for every n. On the other hand, if $\lambda \in (0, 1)$, $(D_{\tau_n}^{(\lambda)})_{n \in \mathbf{N}}$ is a non-decreasing sequence with limit $\bar{\tau} \geq \tau$. We have, since Z is regular

$$\mathbf{E}(U_\tau) \geq \mathbf{E}(U_{\bar{\tau}}) \geq \mathbf{E}(Z_{\bar{\tau}}) = \lim_{n \to \infty} \mathbf{E}(Z_{D_{\tau_n}^{(\lambda)}})$$

But $Z_{D_{\tau_n}^{(\lambda)}} \geq \lambda U_{D_{\tau_n}^{(\lambda)}}$. Hence, using Corollary 10.2.8,

$$\mathbf{E}(U_\tau) \geq \lambda \lim_{n \to \infty} \mathbf{E}(U_{\tau_n})$$

for every $\lambda \in (0, 1)$. Therefore, U is regular.

Now, let $(\lambda_n)_{n \in \mathbb{N}}$ be an increasing sequence such that $\lim_{n \to \infty} \lambda_n = 1$ and $\tau_n^* = D_0^{(\lambda_n)}$. The sequence $(\tau_n^*)_{n \in \mathbb{N}}$ is non-decreasing. Moreover

$$\mathbf{E}(U_0) = \mathbf{E}(U_{\tau_n^*}) \leqslant \frac{1}{\lambda_n} \mathbf{E}(Z_{\tau_n^*})$$

Now let n tend to infinity and use the regularity of Z to see that $\tau^* = \lim_{n \to \infty} \tau_n^*$ is optimal. But this implies (by Theorem 10.2.6) that $Z_{\tau^*} = U_{\tau^*}$ and U^{τ^*} is a martingale, hence $\tau^* \geqslant \tau_0$, and U^{τ_0} is also a martingale, so that τ_0 is optimal. It follows from Theorem 10.2.6 that any optimal stopping time must dominate τ_0. $\qquad\square$

10.2.3 The Doob–Meyer decomposition and optimal stopping

The characterisation of the largest optimal stopping time can also be achieved in this setting. Indeed, since U is a regular supermartingale of class D, it admits a Doob–Meyer decomposition (cf. Karatzas and Shreve, 1988, ch. 1, section 4, especially Theorem 4.14). Namely, there exists a unique right-continuous martingale M and a unique non-decreasing, continuous, adapted process A with $A_0 = 0$ such that

$$U_t = M_t - A_t, \quad t \in [0, T]$$

The following lemma shows that A does not increase as long as $U > Z$.

Lemma 10.2.10: *Let $t \in [0, T]$ and $\tau_t = \inf \{s \geqslant t \mid U_s = Z_s\}$. Then, $A_t = A_{\tau_t}$.*

Proof: Let $(\lambda_n)_{n \in \mathbb{N}}$ be an increasing sequence of positive numbers such that $\lim_{n \to \infty} \lambda_n = 1$. The equality $U_t = \mathbf{E}(U_{D_t^{(\lambda_n)}} \mid \mathcal{F}_t)$ yields $A_t = A_{D_t^{(\lambda_n)}}$. Since A is continuous, $A_t = A_{\tau_t^*}$, where $\tau_t^* = \lim_{n \to \infty} D_t^{(\lambda_n)}$. We also have, from the regularity of U

$$\mathbf{E}(U_{\tau_t^*}) = \lim_{n \to \infty} \mathbf{E}(U_{D_t^{(\lambda_n)}})$$

and $Z_{D_t^{(\lambda_n)}} \geqslant \lambda_n U_{D_t^{(\lambda_n)}}$. Hence, using the regularity of Z, $\mathbf{E}(U_{\tau_t^*}) \leqslant \mathbf{E}(Z_{\tau_t^*})$. Therefore, $U_{\tau_t^*} = Z_{\tau_t^*}$ and $\tau_t \leqslant \tau_t^*$. Since A is non-decreasing, $A_t = A_{\tau_t}$. $\qquad\square$

Theorem 10.2.11: *The stopping time $\tau_1 = \inf \{t \geqslant 0 \mid A_t > 0\}$ (with the convention $\inf \emptyset = T$) is optimal and is the largest optimal stopping time.*

Proof: Let τ^* be an optimal stopping time. Then U^{τ^*} is a martingale. Therefore, $\mathbf{E}(U_0) = \mathbf{E}(U_{\tau^*})$. But $\mathbf{E}(U_{\tau^*}) = \mathbf{E}(M_{\tau^*}) - \mathbf{E}(A_{\tau^*})$ and $\mathbf{E}(M_{\tau^*}) = \mathbf{E}(U_0)$. Hence $\mathbf{E}(A_{\tau^*}) = 0$ and $\tau^* \leqslant \tau_1$.

We now prove that τ_1 is optimal. We know from Lemma 10.2.10 that $A_{\tau_1} = A_{\bar{\tau}_1}$, where $\bar{\tau}_1 = \inf \{s \geqslant \tau_1 \mid U_s = Z_s\}$. The definition of τ_1 yields that $\bar{\tau}_1 = \tau_1$. Hence, $U_{\tau_1} = Z_{\tau_1}$ and τ_1 is optimal. $\qquad\square$

Remark 10.2.12: If Z is a continuous semimartingale, the process A can be related to the local time at 0 of the semimartingale $U - Z$ (see Jacka, 1993, for details and applications. See also El Karoui, 1982). $\qquad\square$

The following theorem is a characterisation of the Snell envelope in terms of its Doob–Meyer decomposition. It will enable us to relate the optimal stopping problem to variational inequalities (see Remark 10.3.9 below).

Theorem 10.2.13:   *Let $\hat{U}$ be a right-continuous regular supermartingale of class D, with Doob–Meyer decomposition $\hat{U} = \hat{M} - \hat{A}$. $\hat{U}$ is the Snell envelope of Z if and only if the following conditions are satisfied:*

(1) $\hat{U} \geqslant Z$;
(2) $\hat{U}_T = Z_T$;
(3)   *for every $t \in [0, T]$, $\hat{A}_t = \hat{A}_{\hat{\tau}_t}$, where $\hat{\tau}_t = \inf \{s \geqslant t \mid \hat{U}_s = Z_s\}$.*

Proof: The 'only if' part follows from Lemma 10.2.10. Conversely, if $\hat{U}$ satisfies the above conditions, the minimality of the Snell envelope implies $\hat{U} \geqslant U$. Moreover

$$\hat{A}_t = \hat{A}_{\hat{\tau}_t} \Rightarrow \mathbf{E}(\hat{U}_t) = \mathbf{E}(\hat{U}_{\hat{\tau}_t}) = \mathbf{E}(Z_{\hat{\tau}_t})$$

Hence $\mathbf{E}(\hat{U}_t) \leqslant \mathbf{E}(U_t)$ and $\hat{U} = U$.   □

10.2.4   Optimal stopping and Brownian motion

In this subsection, we will assume that $(\mathcal{F}_t)_{0 \leqslant t \leqslant T}$ is the augmented filtration of some standard one-dimensional Brownian motion $B = (B_t)_{0 \leqslant t \leqslant T}$ (cf. Karatzas and Shreve, 1988, section 2.7). We will assume that

$$Z_t = f(t, B_t)$$

where f is a bounded continuous function on $[0, T] \times \mathbf{R}$, with values in $[0, +\infty)$. It is clear that Z satisfies all the assumptions made at the beginning of section 10.2.1. The following theorem shows that the Snell envelope can then be expressed as a function of t and B_t. Similar results can be proved in a much more general setting (see El Karoui, 1988).

Theorem 10.2.14:   *Let F be the function defined on $[0, T] \times \mathbf{R}$ by*

$$F(t, x) = \sup_{\tau \in \mathcal{T}_{0, T-t}} \mathbf{E}(f(t + \tau, x + B_\tau))$$

F is a continuous function and the process $(F(t, B_t))_{0 \leqslant t \leqslant T}$ is the Snell envelope of $Z = (f(t, B_t))_{0 \leqslant t \leqslant T}$.

Proof: First, observe that if $\tau \in \mathcal{T}_{0, T-t}$ and $x, y \in \mathbf{R}$

$$\left| \mathbf{E}f(t + \tau, x + B_\tau) - \mathbf{E}f(t + \tau, y + B_\tau) \right| \leqslant \mathbf{E} \sup_{0 \leqslant s \leqslant T - t} \left| f(t + s, x + B_s) - f(t + s, y + B_s) \right|$$

$$\leqslant \mathbf{E} \sup_{0 \leqslant s, s' \leqslant T} \left| f(s', x + B_s) - f(s', y + B_s) \right|$$

This implies that $x \mapsto F(t, x)$ is continuous (uniformly with respect to t).
   Now, let $0 \leqslant t \leqslant t' \leqslant T$. If $\tau \in \mathcal{T}_{0, T-t'}$, we have $\tau \in \mathcal{T}_{0, T-t}$ and

$$\mathbf{E}f(t' + \tau, x + B_\tau) = \mathbf{E}f(t + \tau, x + B_\tau) + \mathbf{E}(f(t' + \tau, x + B_\tau) - f(t + \tau, x + B_\tau))$$
$$\leqslant F(t, x) + \mathbf{E} \sup_{0 \leqslant s \leqslant T - t'} |f(t' + s, x + B_s) - f(t + s, x + B_s)|$$

On the other hand, if $\tau \in \mathcal{T}_{0,T-t}$, then $\tau' = (\tau + t - t')^+ \in \mathcal{T}_{0,T-t'}, t' + \tau' = t' \vee (t + \tau)$ and

$$\mathbf{E}f(t + \tau, x + B_\tau) = \mathbf{E}f(t' + \tau', x + B_{\tau'}) + \mathbf{E}(f(t + \tau, x + B_\tau) - f(t' + \tau', x + B_{\tau'}))$$
$$\leqslant F(t', x) + \mathbf{E} \sup_{t \leqslant u \leqslant T} |f(u, x + B_{u-t}) - f(u \vee t', x + B_{(u-t')^+})|$$

The continuity of F follows easily from the preceding estimates.

Now, let U be the Snell envelope of $Z = (f(t, B_t))_{0 \leqslant t \leqslant T}$. We want to prove that $U_t = F(t, B_t)$. We have

$$U_t = \text{ess-} \sup_{\tau \in \mathcal{T}_{t,T}} \mathbf{E}(f(\tau, B_\tau) \mid \mathcal{F}_t)$$

We will proceed by discrete approximation. For $n \in \mathbf{N}$, let

$$U_t^n = \text{ess-} \sup_{\tau \in \mathcal{T}_{t,T}^n} \mathbf{E}(f(\tau, B_\tau) \mid \mathcal{F}_t)$$

where $\mathcal{T}_{t,T}^n$ is the set of all stopping times with values in $\{t + j(T - t)/n, j = 0, 1, \ldots, n\}$. Using dynamic programming and the independence of the increments of Brownian motion, we see that $U_t^n = F^n(t, B_t)$, where $F^n(t, \cdot)$ is computed through the following algorithm:

$$\begin{cases} F^n(T, x) = f(T, x) \\ F^n(t_j^n, x) = \max\left(f(t_j^n, x), \mathbf{E}F^n(t_{j+1}^n, x + B_{t_{j+1}^n} - B_{t_j^n}) \right), & 0 \leqslant j \leqslant n - 1 \end{cases}$$

where $t_j^n = t + j(T - t)/n$. Inspection of this algorithm shows that

$$F^n(t, x) = \sup_{\tau \in \mathcal{T}_{0,T-t}'^n} \mathbf{E}f(t + \tau, x + B_\tau)$$

where $\mathcal{T}_{0,T-t}'^n$ is the set of all stopping times with values in $\{j(T - t)/n, j = 0, 1 \ldots n\}$. The desired equality follows by letting n go to infinity.   □

Remark 10.2.15:   Theorem 10.2.14 remains valid if f is no longer bounded but satisfies a growth condition such as $f(t, x) \leqslant C(1 + e^{\alpha x})$ for some constants C and α. To see this, it suffices to apply Theorem 10.2.14 with $f(t, x)$ replaced by $f(t, x) \wedge M$, where $M > 0$, and to take limits as $M \to \infty$.

10.3   The value function of an American option

10.3.1   The basic model

We will consider American options written on a single dividend-paying stock. Let S_t be the stock-price at time t. In the Black–Scholes setting, the stock-price process satisfies the stochastic differential equation

$$\frac{dS_t}{S_t} = (r - \delta) \, dt + \sigma \, dB_t \tag{10.2}$$

where, under the so-called risk neutral probability measure (which in the following will be the reference probability, denoted by **P**), B is a standard one-dimensional Brownian motion. The positive constant σ is the volatility of the stock, the non-negative constant r is the interest rate and the non-negative constant δ is the dividend rate. We will denote by $(\mathcal{F}_t)_{0 \leqslant t \leqslant T}$ the augmented filtration of B. Note that solving Equation (10.2) yields

$$S_t = S_0 \exp\left(\left(r - \delta - \frac{\sigma^2}{2}\right)t + \sigma B_t\right) \qquad (10.3)$$

An American option with maturity T is characterised by an adapted process $(Z_t)_{0 \leqslant t \leqslant T}$, where Z_t is the pay-off of the option if exercise occurs at time t. We will restrict our study to pay-off processes given by

$$Z_t = \psi(S_t)$$

where ψ is a continuous non-negative function with (sub-)linear growth (i.e. $\psi(x) \leqslant C(1 + x)$, for some constant C). The main two examples are calls (resp. puts) for which $\psi(x) = (x - K)^+$ (resp. $\psi(x) = (K - x)^+$), where K is the strike price.

The modern theory of American option pricing asserts that the discounted value of an American option is the Snell envelope of the discounted pay-off process under the risk neutral measure (cf. Karatzas, 1988, 1989; Myneni, 1992; Duffie, 1992, ch. 7).

Proposition 10.3.1:   *Let $\psi: [0, +\infty) \to [0, +\infty)$ be a continuous function with (sub-)linear growth. The value at time t of the American option with pay-off function ψ is given by $V(t, S_t)$, where*

$$V(t, x) = \sup_{\tau \in T_{0, T-t}} \mathbf{E} e^{-r\tau} \psi\left(x \exp\left[\left(r - \delta - \frac{\sigma^2}{2}\right)\tau + \sigma B_\tau\right]\right)$$

Proof: Due to Equation (10.3), it suffices to apply Theorem 10.2.14 with

$$f(t, x) = e^{-rt} \psi\left(S_0 \exp\left(\left(r - \delta - \frac{\sigma^2}{2}\right)t + \sigma x\right)\right)$$

Note that f satisfies $f(t, x) \leqslant C(1 + e^{\sigma x})$, so that Remark 10.2.15 applies.   □

The function V in Proposition 10.3.1 will be called the value function of the American option in the following. The following corollary states the well-known equivalence between American and European call options on a non-dividend-paying stock.

Corollary 10.3.2:   *Assume $\delta = 0$. The value function of an American call with strike price K is given by*

$$V(t, x) = \mathbf{E} e^{-r(T-t)} \left(x \exp\left[\left(r - \frac{\sigma^2}{2}\right)(T - t) + \sigma B_{T-t}\right] - K\right)^+$$

Proof: We may assume that $t = 0$. Let $\tau \in T_{0, T}$. We have

$$\mathbf{E}\left[e^{-rT}\left(x\exp\left[\left(r-\frac{\sigma^2}{2}\right)T+\sigma B_T\right]-K\right)^+ \Big| \mathcal{F}_\tau\right] \geqslant \mathbf{E}\left[x\exp\left[\sigma B_T-\frac{\sigma^2}{2}T-Ke^{-rT}\right]\Big|\mathcal{F}_\tau\right]$$

$$= x\exp\left[\sigma B_\tau-\frac{\sigma^2}{2}\tau\right]-Ke^{-rT}$$

$$\geqslant x\exp\left[\sigma B_\tau-\frac{\sigma^2}{2}\tau\right]-Ke^{-rt}$$

where we have used the martingale property of $\exp[\sigma B_t - (\sigma^2/2)t]$ and the fact that $r \geqslant 0$. Since the conditional expectation is non-negative, we may replace the minorant by its positive part. This yields

$$\mathbf{E}e^{-rT}\left(x\exp\left[\left(r-\frac{\sigma^2}{2}\right)T+\sigma B_T\right]-K\right)^+ \geqslant \mathbf{E}e^{-rt}\left(x\exp\left[\left(r-\frac{\sigma^2}{2}\right)\tau+\sigma B_\tau\right]-K\right)^+$$

Since $\tau \in \mathcal{T}_{0,T}$ is arbitrary, the corollary is proved.   □

Remark 10.3.3:   A similar argument shows that, if $r = 0$, the American put is equivalent to the European put.   □

10.3.2   Calls and puts

The following proposition establishes a relation between call and put prices. It was observed in the context of options on foreign exchange by Grabbe (1993) and subsequently by several people (see Charretour *et al.*, 1992; McDonald and Schroder, 1990). In order to clarify our statement we will mention the dependence of call and put prices on K, r, δ. Namely, let

$$C(t, x; K, r, \delta) = \sup_{\tau \in \mathcal{T}_{0,T-t}} \mathbf{E}e^{-r\tau}\left(x\exp\left[\left(r-\delta-\frac{\sigma^2}{2}\right)\tau+\sigma B_\tau\right]-K\right)^+$$

and

$$P(t, x; K, r, \delta) = \sup_{\tau \in \mathcal{T}_{0,T-t}} \mathbf{E}e^{-r\tau}\left(K - x\exp\left[\left(r-\delta-\frac{\sigma^2}{2}\right)\tau+\sigma B_\tau\right]\right)^+$$

Proposition 10.3.4:   *We have*

$$C(t, x; K, r, \delta) = P(t, K; x, \delta, r) = xP(t, K/x; 1, \delta, r)$$

Proof: For the proof see Section 10.5.   □

10.3.3   Analytic properties

We will now concentrate on the American put to derive some regularity properties of the value function. Note that, due to Proposition 10.3.4, this implies no loss of generality. We will use the notation $P(t, x)$ for $P(t, x; K, r, \delta)$. Therefore

$$P(t, x) = \sup_{\tau \in \mathcal{T}_{0,T-t}} \mathbf{E}e^{-r\tau}\psi\left(x\exp\left[\left(r-\delta-\frac{\sigma^2}{2}\right)\tau+\sigma B_\tau\right]\right) \qquad (10.4)$$

with

$$\psi(x) = (K - x)^+$$

We will also assume that r is positive (recall that if $r = 0$, American puts are equivalent to European puts).

The following properties are easily derived from Equation (10.4).

(1)   For every $x \in [0, +\infty)$, $t \mapsto P(t, x)$ is a non-increasing function.
(2)   For every $t \in [0, T]$, $x \mapsto P(t, x)$ is a non-increasing convex function.
(3)   For every $(t, x) \in [0, T] \times [0, +\infty)$, $P(t, x) \geqslant \psi(x) = P(T, x)$.

Note that the second property follows from the fact that ψ is a non-increasing convex function.

The continuity properties of P are stated in the following proposition.

Proposition 10.3.5:
(1)   *For every $(t, x) \in [0, T] \times [0, +\infty)$, we have*

$$P(t, x) = \sup_{\tau \in T_{0,1}} \mathbf{E} e^{-r\tau(T-t)} \psi\left(x \exp\left[\left(r - \delta - \frac{\sigma^2}{2} \right) \tau(T-t) + \sigma \sqrt{T - t} B_\tau \right] \right) \quad (10.5)$$

(2)   *For every $t \in [0, T]$, and for $x, y \geqslant 0$, $|P(t, x) - P(t, y)| \leqslant |x - y|$.*
(3)   *There exists a constant $C > 0$ such that, for every $x \in [0, +\infty)$, and for $t, s \in [0, T]$,*

$$|P(t, x) - P(s, x)| \leqslant C |(T - t)^{1/2} - (T - s)^{1/2}|$$

Proof: For the proof see Section 10.5.   □

Remark 10.3.6:   It follows from the Lipschitz properties of P, as stated in Proposition 10.3.5, that the first-order partial derivatives of P (in the sense of distributions) are locally bounded in the open set $(0, T) \times (0, +\infty)$. More precisely, $\|\partial P/\partial x\|_{L^\infty([0,T]\times[0,+\infty))} \leqslant 1$ and, for $t \in [0, T)$, $\|(\partial P/\partial t)(t, \cdot)\|_{L^\infty([0,+\infty))} \leqslant C/(T - t)^{1/2}$. We refer the reader to Rudin (1991) for the basics of distribution theory.   □

10.3.4   The variational inequality

The purpose of this subsection is to derive the variational inequality satisfied by the value function P of the American put. Actually, it is convenient to make the following logarithmic change of variable. If we set $X_t = \log(S_t)$ in Equations (10.2) and (10.3), we see that X satisfies the simple stochastic differential equation

$$dX_t = \mu \, dt + \sigma \, dB_t \quad (10.6)$$

where $\mu = r - \delta - \sigma^2/2$. We will denote by X^x the solution of Equation (10.6) which satisfies $X_0^x = x$, namely $X_t^x = x + \mu t + \sigma B_t$. The semi-group of the diffusion X can be defined as follows. For $t \geqslant 0$ and ϕ a non-negative Borel function on $\mathbf{R}$, set

$$Q_t \phi(x) = \mathbf{E}(\phi(X_t^x))$$

The infinitesimal generator of the semi-group $(Q_t)_{t \geqslant 0}$ is the operator $\frac{\sigma^2}{2}(\partial^2/\partial x^2) + \mu(\partial/\partial x)$. We refer the reader to Karatzas and Shreve (1988, chs 3 and 5) for the Markov properties of stochastic differential equations.

We now have $P(t, x) = F(t, e^x)$, where the function F is defined by

$$F(t, x) = \sup_{\tau \in \mathcal{T}_{0, T-t}} \mathbf{E} e^{-r\tau} f(X_\tau^x)$$

with

$$f(x) = (K - e^x)^+$$

Note that for any real number x, the process $(e^{-rt} F(t, X_t^x))_{0 \leqslant t \leqslant T}$ is the Snell envelope of $(e^{-rt} f(X_t^x))_{0 \leqslant t \leqslant T}$. As such, it is a supermartingale. The supermartingale property has the following analytic interpretation.

Proposition 10.3.7:   *The function F satisfies $(\partial F/\partial t) + AF \leqslant 0$ (in the sense of distributions) in the open set $(0, T) \times \mathbf{R}$, where*

$$AF = \frac{\sigma^2}{2} \frac{\partial^2 F}{\partial x^2} + \mu \frac{\partial F}{\partial x} - rF$$

The inequality $(\partial F/\partial t) + AF \leqslant 0$ in the sense of distributions means that for any C^∞ non-negative function ϕ, with compact support in $(0, T) \times \mathbf{R}$, we have

$$\int_0^T ds \int_{\mathbf{R}} dx F(s, x) \left(-\frac{\partial \phi}{\partial t}(s, x) + \frac{\sigma^2}{2} \frac{\partial^2 \phi}{\partial x^2}(s, x) - \mu \frac{\partial \phi}{\partial x}(s, x) - r\phi(s, x) \right) \leqslant 0$$

Proposition 10.3.7 is proved in Section 10.5. In what follows, all partial derivatives are to be understood in the sense of distributions (see Rudin, 1991, ch. 6).

We are now in a position to state the following theorem.

Theorem 10.3.8:
(1)   *The partial derivatives $\partial F/\partial x$, $\partial F/\partial t$ and $\partial^2 F/\partial x^2$ are locally bounded. More precisely, $\partial F/\partial x$ is uniformly bounded on $[0, T] \times \mathbf{R}$ and there exists a positive constant C_1 such that*

$$\forall t \in [0, T), \quad \left\| \frac{\partial F}{\partial t}(t, \cdot) \right\|_{L^\infty(\mathbf{R})} + \left\| \frac{\partial^2 F}{\partial x^2}(t, \cdot) \right\|_{L^\infty(\mathbf{R})} \leqslant \frac{C_1}{(T-t)^{1/2}}$$

(2)   *The function F solves the following variational inequality*

$$\max\left(\frac{\partial F}{\partial t} + AF, f - F \right) = 0$$

with the terminal condition $F(T, \cdot) = f$.

Proof:
(1)   The estimates for the first-order derivatives of F follow from the Lipschitz properties of P as stated in Proposition 10.3.5 (cf. Remark 10.3.6). For the second-order derivative, we use the convexity of $x \mapsto P(t, x) = F(t, \log(x))$ on the one hand and Proposition 10.3.7 on the other. Indeed, the convexity implies that $(\partial^2 P/\partial x^2)$ is a non-negative measure, hence

$$\frac{\partial^2 F}{\partial x^2} - \frac{\partial F}{\partial x} \geq 0$$

and Proposition 10.3.7 yields

$$\frac{\sigma^2}{2}\frac{\partial^2 F}{\partial x^2} \leq rF - \frac{\partial F}{\partial t} - \mu\frac{\partial F}{\partial x}$$

The estimates for $(\partial^2 F/\partial x^2)$ then follow from the boundedness of F and the estimates for the first-order derivatives.

(2)   Since the partial derivatives $\partial F/\partial x$, $\partial F/\partial t$ and $\partial^2 F/\partial x^2$ are bounded on any set $[0, T'] \times \mathbf{R}$, with $T' < T$, we can apply Itô's formula with generalised derivatives (cf. Krylov, 1980, ch. 2, section 10). We obtain, for every $t \in [0, T)$,

$$e^{-rt}F(t, X_t) = M_t + \int_0^t e^{-rs}\left(\frac{\partial F}{\partial t} + AF\right)(s, X_s)\,ds$$

where

$$M_t = F(0, X_0) + \int_0^t \mu\frac{\partial F}{\partial x}(s, X_s)e^{-rs}\,dB_s$$

Since $\partial F/\partial x$ is bounded, M is a square integrable martingale. Now, we know that the process $U = (e^{-rt}F(t, X_t))_{0 \leq t \leq T}$ is the Snell envelope of $(e^{-rt}f(X_t))_{0 \leq t \leq T}$. Its Doob–Meyer decomposition is given by $U_t = M_t - a_t$, with

$$a_t = \int_0^t e^{-rs}\left(\frac{\partial F}{\partial t} + AF\right)(s, X_s)\,ds$$

It follows from Lemma 10.2.10 that, for every $t \leq T$

$$\int_t^{\tau_t} e^{-rs}\left(\frac{\partial F}{\partial t} + AF\right)(s, X_s)\,ds = 0$$

where $\tau_t = \inf\{s \geq t \mid F(s, X_s) = f(X_s)\}$. Therefore, we must have $(\partial F/\partial t) + AF = 0$ almost everywhere in the open set $C = \{(t, x) \in (0, T) \times \mathbf{R} \mid F(t, x) > f(x)\}$. This proves that F solves the variational inequality.   □

Remark 10.3.9:   The variational inequality

$$\max\left(\frac{\partial F}{\partial t} + AF, f - F\right) = 0$$

with its terminal condition $F(T, \cdot) = f$ can be viewed as the analytic version of the characteristic properties of the Snell envelope given in Theorem 10.2.13. Indeed, the condition $(\partial F/\partial t) + AF \leq 0$ is the analytic counterpart of the supermartingale property, the condition $F \geq f$ is the translation of the condition $U \geq Z$ and the fact that one of the two inequalities has to be an equality (since the maximum is null) is the analytic version of the last condition in Theorem 10.2.13.

A classical approach to optimal stopping for Markovian models consists of starting from the variational inequality, proving existence and uniqueness results and identifying the solution of the variational inequality as the solution of the optimal stopping problem (cf. Bensoussan and Lions, 1981, 1982 for general results). This approach has been applied to American options in Jaillet *et al.* (1990) for diffusion models and in Zhang (1993, 1994, 1997) for the jump-diffusion model. For recent results about American options on multiple assets, see Broadie and Detemple (1997).   □

Corollary 10.3.10:   *The function $(\partial F/\partial x)$ is continuous on the set $[0, T) \times \mathbf{R}$.*

Proof: For the proof see Section 10.5.   □

Remark 10.3.11:   The continuity of $\partial F/\partial x$ is generally referred to as the 'smooth-fit property'. It can be approached via more probabilistic arguments (see El Karoui and Karatzas, 1991; Jacka, 1993).
   The continuity of $(\partial F/\partial t)$ on $[0, T) \times \mathbf{R}$ can be proved with the help of estimates for the other second-order derivatives of F (cf. Friedman, 1975; Kinderlehrer and Stampacchia, 1980, ch. 8).   □

Remark 10.3.12:   The limit of the American put price as T tends to infinity can be computed explicitly. This is the value function of a *perpetual* put. More precisely, if

$$P_\infty(x) = \sup_{\tau \in \mathcal{T}_{0,+\infty}} \mathrm{E}e^{-r\tau}\psi\left(x\exp\left[\left(r - \delta - \frac{\sigma^2}{2}\right)\tau + \sigma B_\tau\right]\right) \tag{10.7}$$

where $\psi(x) = (K - x)^+$ and $\mathcal{T}_{0,+\infty}$ is the set of all stopping times with values in $[0, +\infty)$, one can prove that $P_\infty(x) = K - x$, if $x \leqslant x^*$ and $P_\infty(x) = (K - x^*)(x/x^*)^{-\gamma}$, with $x^* = K\gamma/(1 + \gamma)$ and

$$\gamma = \frac{1}{\sigma^2}\left[\left(r - \delta - \frac{\sigma^2}{2}\right) + \left(\left(r - \delta - \frac{\sigma^2}{2}\right)^2 + 2r\sigma^2\right)^{1/2}\right]$$

These formulae go back to McKean (1965) (see also Lamberton and Lapeyre (1996), ch. 4, section 4, for a more probabilistic presentation). They can be extended to perpetual options on two stocks (see Gerber and Shiu, 1996).   □

10.3.5   The exercise boundary

We now introduce the so-called *critical* stock price for the American put. For $t \in [0, T)$, let

$$s^*(t) = \inf\{x \in [0, +\infty) \mid P(t, x) > \psi(x) = (K - x)^+\}$$

The number $s^*(t)$ is called the critical stock-price at time t. Clearly, $0 \leqslant s^*(t) < K$ for every $t \in [0, T)$. The inequality $s^*(t) < K$ follows from the fact that $P(t, x) > 0$. To see this, recall that

$$P(t, x) \geqslant \mathrm{E}e^{-r(T-t)}\psi\left(x\exp\left[\left(r - \delta - \frac{\sigma^2}{2}\right)(T - t) + \sigma B_{T-t}\right]\right)$$

We also have $s^*(t) \geqslant x^*$, where x^* is defined in Remark 10.3.12. This follows from the inequality $P(t, x) \leqslant P_\infty(x)$.
   It follows from the convexity of $x \mapsto P(t, x)$ that

$$\forall x \leqslant s^*(t), \quad P(t, x) = K - x \quad \text{and} \quad \forall x > s^*(t), \quad P(t, x) > (K - x)^+$$

Therefore, using the variational inequality satisfied by $F(t, x) = P(t, e^x)$ and writing it in terms of P, we see that, for all $(t, x) \in (0, T) \times [0, +\infty)$

$$\frac{\partial P}{\partial t}(t, x) + \frac{\sigma^2}{2} x^2 \frac{\partial^2 P}{\partial x^2}(t, x) + (r - \delta)x \frac{\partial P}{\partial x}(t, x) - rP(t, x) = (\delta x - rK)\mathbf{1}_{\{x \leqslant s^*(t)\}}$$

The function s^* is called the free boundary or exercise boundary.

Now, using Itô's formula in essentially the same way as in the proof of Theorem 10.3.8, we have

$$e^{-rt}P(t, S_t) = P(0, S_0) + \int_0^t e^{-ru}\sigma S_u \frac{\partial P}{\partial x}(u, S_u)\, dB_u + \int_0^t e^{-ru}(\delta S_u - rK)\mathbf{1}_{\{S_u \leqslant s^*(u)\}}\, du$$

Let $t \to T$ and take expectations. We obtain

$$P_e(0, S_0) = P(0, S_0) + \int_0^T e^{-ru}\mathbf{E}\big((\delta S_u - rK)\mathbf{1}_{\{S_u \leqslant s^*(u)\}}\big)\, du$$

where we have set

$$P_e(t, x) = \mathbf{E}e^{-r(T-t)}\psi\left(x\exp\left[\left(r - \delta - \frac{\sigma^2}{2}\right)(T - t) + \sigma B_{T-t}\right]\right)$$

Note that P_e is the value function of the *European* put. Straightforward computations lead to the following formula relating $P(t, x)$ to $P_e(t, x)$

$$P(t, x) = P_e(t, x) + \int_0^{T-t} \big(rKe^{-ru}N(d_1(x, t, u)) - \delta xe^{-\delta u}N(d_2(x, t, u))\big)\, du$$

where N is the standard normal distribution function and

$$d_1(x, t, u) = \frac{\log(s^*(t + u)/x) - (r - \delta - \frac{\sigma^2}{2})u}{\sigma\sqrt{u}}$$

$$d_2(x, t, u) = \frac{\log(s^*(t + u)/x) - (r - \delta + \frac{\sigma^2}{2})u}{\sigma\sqrt{u}}$$

Remark 10.3.13: The formula relating P and P_e seems to have been discovered and used by several people at about the same time (see Kim, 1990; Charretour *et al.*, 1992; Carr *et al.*, 1992; Jacka, 1991). The proof given in Jacka (1991) does not use the partial differential equation satisfied by P. □

Remark 10.3.14: It is clear that s^* is a non-decreasing function on $[0, T)$ (this follows from the fact that P is a non-increasing function of time). It is known that $\lim_{t \to T} s^*(t) = K \wedge (rK/\delta)$ (see Kim, 1990). It can be proved that s^* is differentiable on $[0, T)$ (cf. Friedman, 1975; Kinderlehrer and Stampacchia, 1980). The behaviour of $s^*(t)$ as t approaches T has been studied by Barles *et al.* (1993, 1995) (see also Lamberton, 1995a) in the case $\delta = 0$ (see Aït-Sahlia, 1995, for some results when $\delta \neq 0$ and Pham, 1997, for an extension to the jump-diffusion model). In higher dimensions, the exercise region may exhibit interesting properties (cf. Broadie and Detemple, 1997, and Villeneuve, 1997).

10.4 Numerical methods

Various numerical methods have been developed for the pricing of American options. To clarify our survey of some of the recent literature, we will separate 'probabilistic methods' and 'analytic methods'. Comparisons of some of these methods have been carried out by Geske and Shastri (1985) and Broadie and Detemple (1996a). See also the introduction of Carr and Faguet (1996) for a presentation of various numerical methods and Broadie and Detemple (1996b) for a recent survey.

10.4.1 Probabilistic methods

A natural idea to approximate the American put price in the Black–Scholes model is to restrict the set of admissible stopping times to those with values in a subdivision of the time interval. This method has been used by Parkinson (1977) and Geske and Johnson (1984), but it seems to be applicable only with a small number of discretisation points. An interesting estimate for the error of approximation can be found in Carverhill and Webber (1990). The method has been extended to jump diffusion models in Mulinacci (1996). The set of stopping times can be restricted in other ways, as in Bjerksund and Stensland (1992) or Broadie and Detemple (1996a).

In order to overcome the computational difficulties arising when the discretisation step gets smaller, discrete approximations of the underlying diffusion can be used. The simplest method consists of approximating Brownian motion by a random walk. This method is known as the binomial approximation in the finance literature (see Cox and Rubinstein, 1985). Convergence results in a rather general setting go back to Kushner (1977) and Aldous (1979), and have been applied to financial models by Amin (1993) and Amin and Khanna (1994). An alternative approach to convergence results has been developed by Lamberton and Pages (1990) (see also Mulinacci and Pratelli, 1995, for some results on the weak convergence of Snell envelopes). It seems difficult to derive sharp estimates for the rate of convergence (some estimates can be found in Lamberton, 1995b). Finally we mention the use of Monte-Carlo techniques to compute American options in Broadie and Glasserman (1995).

10.4.2 Analytic methods

We call *analytic* the methods that are related to the partial differential equations satisfied by the price function (variational inequality, free boundary problem). The finite difference method of Brennan and Schwartz (1977) was the first to appear in the financial literature (see Jaillet *et al.*, 1990, for a rigorous justification and Zhang (1994, 1997) for a complete proof of the strong convergence result and an extension to jump-diffusion models). More complex options are treated by Dempster (1994), Dempster and Hutton (1995) and by Barraquand and Pudet (1996) and Barraquand and Martineau (1995).

The quadratic approximation of MacMillan (1986) (see also Barone-Adesi and Whaley, 1987) is a quasi-explicit approximation of the American put price which has some extensions to the jump-diffusion model (see Zhang, 1994). The error term of this approximation has recently been studied by Chevance (1997). The recent paper by Carr and Faguet (1996) can be viewed as a refinement of this method, which leads to more accurate approximations.

Appendix 10.1

Proofs

Proof of Proposition 10.3.4: It suffices to prove the result for $t = 0$. We have

$$C(0, x; K, r, \delta) = \sup_{\tau \in T_{0,T}} \mathbf{E}e^{-r\tau}\left(xe^{(r-\delta)\tau} - K\exp\left[-\sigma B_\tau + \frac{\sigma^2}{2}\tau\right]\right)^+ \exp\left[\sigma B_\tau - \frac{\sigma^2}{2}\tau\right]$$

$$= \sup_{\tau \in T_{0,T}} \mathbf{E}e^{-r\tau}\left(xe^{(r-\delta)\tau} - K\exp\left[-\sigma B_\tau + \frac{\sigma^2}{2}\tau\right]\right)^+ \exp\left[\sigma B_T - \frac{\sigma^2}{2}T\right]$$

where the second equality follows from the martingale property. Let $\tilde{\mathbf{P}}$ be the probability on $(\Omega, \mathcal{F})$ defined by its density with respect to $\mathbf{P}$

$$\frac{d\tilde{\mathbf{P}}}{d\mathbf{P}} = \exp\left[\sigma B_T - \frac{\sigma^2}{2}T\right]$$

We know from Girsanov's theorem (cf. for instance, Karatzas and Shreve, 1988) that, under $\tilde{\mathbf{P}}$, the process $\tilde{B} = (B_t - \sigma t)_{0 \leqslant t \leqslant T}$ is a standard Brownian motion. Now

$$\mathbf{E}e^{-r\tau}\left(x\exp[(r-\delta)\tau] - K\exp\left[-\sigma B_\tau + \frac{\sigma^2}{2}\tau\right]\right)^+ \exp\left[\sigma B_T - \frac{\sigma^2}{2}T\right]$$

$$= \tilde{\mathbf{E}}e^{-r\tau}\left(x\exp[(r-\delta)\tau] - K\exp\left[-\sigma\tilde{B}_\tau - \frac{\sigma^2}{2}\tau\right]\right)^+$$

Taking the sup over all stopping times and using the fact that $-\tilde{B}$ has the same distribution under $\tilde{\mathbf{P}}$ as B under $\mathbf{P}$, we obtain

$$C(0, x; K, r, \delta) = \sup_{\tau \in T_{0,T}} \mathbf{E}e^{-r\tau}\left(x\exp[(r-\delta)\tau] - K\exp\left[\sigma B_\tau - \frac{\sigma^2}{2}\tau\right]\right)^+$$

$$= \sup_{\tau \in T_{0,T}} \mathbf{E}e^{-\delta\tau}\left(x - K\exp\left[\sigma B_\tau + \left(\delta - r - \frac{\sigma^2}{2}\right)\tau\right]\right)^+$$

$$= x\sup_{\tau \in T_{0,T}} \mathbf{E}e^{-\delta\tau}\left(1 - \frac{K}{x}\exp\left[\sigma B_\tau + \left(\delta - r - \frac{\sigma^2}{2}\right)\tau\right]\right)^+$$

and the proposition follows from the last two equalities. □

Proof of Proposition 10.3.5: The first statement is a simple consequence of the scaling property of Brownian motion. Indeed, let $\bar{\mathcal{F}}_s = \mathcal{F}_{(T-t)s}$. It is easy to check that $\tau \in T_{0,T-t}$ if and only if $\tau/(T-t) \in \bar{T}_{0,1}$, where $\bar{T}_{0,1}$ is the set of all stopping times of the filtration $(\bar{\mathcal{F}}_s)_{0 \leqslant s \leqslant 1}$ with values in $[0, 1]$. Therefore

$$P(t, x) = \sup_{\tau \in \bar{T}_{0,1}} \mathbf{E}e^{-r\tau(T-t)}\psi\left(x\exp\left[\left(r - \delta - \frac{\sigma^2}{2}\right)\tau(T-t) + \sigma B_{\tau(T-t)}\right]\right)$$

Now, it suffices to observe that $(\bar{\mathcal{F}}_s)_{0 \leqslant s \leqslant 1}$ is the augmented filtration of $(B_{s(T-t)})_{0 \leqslant s \leqslant 1}$ and that $(B_{s(T-t)})_{0 \leqslant s \leqslant 1}$ has the same distribution as $((T-t)^{1/2} B_s)_{0 \leqslant s \leqslant 1}$.

For the second statement, observe that if $\tau \in \mathcal{T}_{0,T-t}$

$$\left| \psi \left(x \exp \left[\left(r - \delta - \frac{\sigma^2}{2} \right) \tau + \sigma B_\tau \right] \right) - \psi \left(y \exp \left[\left(r - \delta - \frac{\sigma^2}{2} \right) \tau + \sigma B_\tau \right] \right) \right|$$

$$\leqslant |x - y| \exp \left[\left(r - \delta - \frac{\sigma^2}{2} \right) \tau + \sigma B_\tau \right]$$

where we have used the Lipschitz property of ψ. The result follows from the fact that $\delta \geqslant 0$ and $\mathbf{E} \exp[\sigma B_\tau - \frac{\sigma^2}{2} \tau] = 1$.

The proof of the third statement proceeds analogously from Equation (10.5). □

Proof of Proposition 10.3.7: Since $(e^{-rt} F(t, X_t^x))_{0 \leqslant t \leqslant T}$ is a supermartingale, we have, for $0 < s \leqslant t \leqslant T$,

$$e^{-rt}[Q_{t-s} F(t, \cdot)](X_s^x) \leqslant e^{-rs} F(s, X_s^x)$$

almost surely (recall that $(Q_t)_{t \geqslant 0}$ is the semi-group of the diffusion X). Since F is continuous and the support of the distribution of X_s^x is $\mathbf{R}$, this yields

$$e^{-r(t-s)}[Q_{t-s} F(t, \cdot)](y) \leqslant F(s, y)$$

for all $y \in \mathbf{R}$.

Now, let ϕ be a C^∞ non-negative function with compact support in $(0, T) \times \mathbf{R}$. We have, for $h > 0$ small enough,

$$\int_0^T ds \int_{\mathbf{R}} dx \, \phi(s, x) \left(e^{-rh}[Q_h F(s+h, \cdot)](x) - F(s, x) \right) \leqslant 0$$

Using the definition of Q_h, Fubini's theorem and the invariance of the Lebesgue measure under translation, we have

$$\int_{\mathbf{R}} dx \, \phi(s, x)[Q_h F(s+h, \cdot)](x) = \int_{\mathbf{R}} dx \, \phi(s, x) \mathbf{E}(F(s+h, x + \mu h + \sigma B_h))$$

$$= \int_{\mathbf{R}} dx \, \mathbf{E}(\phi(s, x - \mu h - \sigma B_h) F(s+h, x))$$

Hence,

$$\int_0^T ds \int_{\mathbf{R}} dx \, F(s, x) \left(e^{-rh} \mathbf{E}(\phi(s-h, x - \mu h - \sigma B_h)) - \phi(s, x) \right) \leqslant 0$$

Dividing by h and letting h tend to 0, we obtain

$$\int_0^T ds \int_{\mathbf{R}} dx \, F(s, x) \left(-\frac{\partial \phi}{\partial t}(s, x) + \frac{\sigma^2}{2} \frac{\partial^2 \phi}{\partial x^2}(s, x) - \mu \frac{\partial \phi}{\partial x}(s, x) - r\phi(s, x) \right) \leqslant 0$$

for any non-negative test-function ϕ, which proves the proposition. □

Proof of Corollary 10.3.10: Using Theorem 10.3.8 and standard localisation arguments, it suffices to prove that if $U(t, x)$ is a continuous function on $\mathbf{R} \times \mathbf{R}$

such that its partial derivatives $\partial U/\partial x$, $\partial/\partial t$ and $\partial^2 U/\partial x^2$ are bounded functions on $\mathbf{R} \times \mathbf{R}$, then $\partial U/\partial x$ is a continuous function. We will sketch the proof of this classical result (see Ladyzenskaja *et al.*, 1968, ch. 2, lemma 3.1).

Fix $s, t, x_0 \in \mathbf{R}$. For every real number x, we have

$$\int_{x_0}^x \frac{\partial U}{\partial x}(t, y)\,dy - \int_{x_0}^x \frac{\partial U}{\partial x}(s, y)\,dy = U(t, x) - U(t, x_0) - U(s, x) + U(s, x_0)$$

$$= U(t, x) - U(s, x) + U(s, x_0) - U(t, x_0)$$

Therefore

$$\left| \int_{x_0}^x \frac{\partial U}{\partial x}(t, y)\,dy - \int_{x_0}^x \frac{\partial U}{\partial x}(s, y)\,dy \right| \leqslant 2 \left\| \frac{\partial U}{\partial t} \right\|_{L^\infty} |t - s|$$

On the other hand, for $\tau = t, s$, we have

$$\left| \frac{\partial U}{\partial x}(\tau, x_0) - \frac{1}{x - x_0} \int_{x_0}^x \frac{\partial U}{\partial x}(\tau, y)\,dy \right| \leqslant \left\| \frac{\partial^2 U}{\partial x^2} \right\|_{L^\infty} |x - x_0|$$

Hence

$$\left| \frac{\partial U}{\partial x}(t, x_0) - \frac{\partial U}{\partial x}(s, x_0) \right| \leqslant 2 \left\| \frac{\partial U}{\partial t} \right\|_{L^\infty} \frac{|t - s|}{|x - x_0|} + 2 \left\| \frac{\partial^2 U}{\partial x^2} \right\|_{L^\infty} |x - x_0|$$

Choosing x so that $|x - x_0| = (t - s)^{1/2}$, we obtain

$$\left| \frac{\partial U}{\partial x}(t, x_0) - \frac{\partial U}{\partial x}(s, x_0) \right| \leqslant 2(|t - s|)^{1/2} \left(\left\| \frac{\partial U}{\partial t} \right\|_{L^\infty} + \left\| \frac{\partial^2 U}{\partial x^2} \right\|_{L^\infty} \right)$$

Since we also have Lipschitz-continuity in the x-variable (due to the boundedness of $\partial^2 U/\partial x^2$), we can conclude that $\partial U/\partial x$ is continuous.   □

References

Aït-Sahlia, F. (1995) Optimal stopping and weak convergence methods for some problems in financial economics. Ph.D. dissertation, Stanford University.

Aldous, D. (1979) *Extended weak convergence.* Unpublished manuscript.

Amin, K. I. (1993) Jump diffusion option valuation in discrete time, *Journal of Finance*, **48**, 1833–1863.

Amin, K. I. and Khanna, A. (1994) Convergence of American option values from discrete to continuous-time financial models. *Mathematical Finance*, **4**, 289–304.

Barles, G., Burdeau, J., Romano, M. and Sansœn, N. (1993) Estimation de la frontière libre des options américaines au voisinage de l'échéance. *Comptes Rendus de l'Académie des Sciences, Paris, Série I*, **316**, 171–174.

Barles, G., Burdeau, J., Romano, M. and Sansœn, N. (1995) Critical stock price near expiration. *Mathematical Finance*, **5**, 77–95.

Barone-Adesi, G. and Whaley, R. (1987) Efficient analytic approximation of American option values. *Journal of Finance*, **42**, 301–320.

Barraquand, J. and Pudet, T. (1996) Pricing of American path dependent contingent claims. *Mathematical Finance*, **6**, 17–51.

Barraquand, J. and Martineau, D. (1995) Numerical valuation of high dimensional multivariate American securities. *Journal of Financial and Quantitative Analysis* **30**, 383–405.

Bensoussan, A. (1984) On the theory of option pricing. *Acta Applicandae Mathematicae*, **2**, 139–158.

Bensoussan, A. and Lions, J. L. (1981) *Contrôle impulsionnel et inéquations quasi-variationnelles*. Paris: Dunod.

Bensoussan, A. and Lions, J. L. (1982) *Applications of Variational Inequalities in Stochastic Control*. Amsterdam: North-Holland.

Bjerksund, P. and Stensland, G. (1992) Closed form approximation of American options. *Scandinavian Journal of Management*.

Brennan, M. J. and Schwartz, E. S. (1977) The valuation of the American put option. *Journal of Finance*, **32**, 449–462.

Broadie, M. and Detemple, J. B. (1997) The valuation of American options on multiple assets. *Mathematical Finance*, **7**, 241–286.

Broadie, M. and Detemple, J. B. (1996b) American option valuation: new bounds, approximations, and a comparison of existing methods. *Review of Financial Studies*, **9**, 1211–1250.

Broadie, M. and Detemple, J. B. (1996c) Recent advances in numerical methods for pricing derivative securities. In L. C. G. Rogers and D. Talay (Eds). *Numerical Methods in Financial Mathematics*. Cambridge: Cambridge University Press.

Broadie, M. and Glasserman, P. (1995) Pricing American-style securities using simulation. To appear in *Journal of Economic Dynamics and Control*.

Carr, P. and Faguet, D. (1996) Valuing finite lived options as perpetual. Morgan Stanley working paper.

Carr, P., Jarrow, R. and Myneni, R. (1992) Alternative characterization of American put options. *Mathematical Finance*, **2**, 87–106.

Carverhill, A. P. and Webber, N. (1990) American options: theory and numerical analysis. In *Options: Recent Advances in Theory and Practice*. Manchester: Manchester University Press.

Charretour, F., Elliott, R. J., Myneni, R. and Viswanathan, R. (1992) Paper presented at the Oberwolfach meeting on mathematical finance, August 1992.

Chevance, D. (1997) Résolution numérique des EDSR. Ph.D. dissertation, Université de Provence.

Cox, J. and Rubinstein, M. (1985) *Options Markets*. Englewood Cliffs, New Jersey: Prentice-Hall.

Dacunha-Castelle, D. and Duflo, M. (1986) *Probability and Statistics*, 2 volumes. New York: Springer-Verlag.

Dempster, M. A. H. (1994) *Fast numerical valuation of American, exotic and complex options*. Department of Mathematics research report, University of Essex, Colchester, UK.

Dempster, M. A. H. and Hutton, J. P. (1995) *Fast numerical valuation of Amerecan options by linear programming*. Department of Mathematics research report, University of Essex, Colchester, UK.

Duffie, D. (1992) *Dynamic Asset Pricing Theory*. Princeton: Princeton University Press.

El Karoui, N. (1981) Les aspects probabilistes du contrôle stochastique. *Lecture Notes in Mathematics*, **876**, 72–238. New York: Springer-Verlag.

El Karoui, N. (1982) Une propriété de domination de l'enveloppe de Snell des semimartingales fortes. In *Séminaire de Probabilités XVI. Lecture Notes in Mathematics*, **920**, 400–407. New York: Springer-Verlag.

El Karoui, N. and Karatzas, I. (1991) A new approach to the Skorohod problem and its applications. *Stochastics and Stochastics Reports*, **34**, 57–82.

El Karoui, N., Lepeltier, J. P. and Millet, A. (1988) A probabilistic approach of the réduite. Preprint, Université Paris VI.

Friedman, A. (1975) Parabolic variational inequalities in one space dimension and smoothness of the free boundary, *Journal of Functional Analysis*, **18**, 151–176.

Gerber, H. U. and Shiu, E. S. W. (1996) Martingale approach to pricing perpetual American options on two stocks. *Mathematical Finance*, **6**, 303–322.

Geske, R. and Johnson, H. (1984) The American put option valued analytically. *Journal of Finance*, **39**, 1511–1524.

Geske, R. and Shastri, K. (1985) Valuation by approximation: a comparison of alternative option valuation techniques. *Journal of Financial and Quantitative Analysis*, **20**, 45–71.

Grabbe, O. (1993) The pricing of call and put options on foreign exchange. *Journal of International Money and Finance*, **2**, 239–253.

Jacka, S. D. (1991) Optimal stopping and the American put price. *Mathematical Finance*, **1**, 1–14.

Jacka, S. D. (1993) Local times, optimal stopping and semi-martingales. *Annals of Probability*, **21**, 329–339.

Jaillet, P., Lamberton, D. and Lapeyre, B. (1990) Variational inequalities and the pricing of American options. *Acta Applicandae Mathematicae*, **21**, 263–289.

Karatzas, I. (1988) On the pricing of American options. *Applied Mathematics and Optimization*, **17**, 37–60.

Karatzas, I. (1989) Optimization problems in the theory of continuous trading. *SIAM Journal of Control and Optimization*, **27**, 1221–1259.

Karatzas, I. and Shreve, S. E. (1988) *Brownian Motion and Stochastic Calculus*. New York: Springer-Verlag.

Kim, I. J. (1990) The analytic valuation of American options. *Review of Financial Studies*, **3**, 547–572.

Kinderlehrer, D. and Stampacchia, G. (1980) *An Introduction to Variational Inequalities and their Applications*. New York: Academic Press.

Krylov, N. V. (1980) *Controlled Diffusion Processes*. New York: Springer-Verlag.

Kushner, H. J. (1977) *Probability Methods for Approximations in Stochastic Control and for Elliptic Equations*. New York: Academic Press.

Ladyzenskaja, O. A., Solonnikov, V. A. and Ural'Ceva, N. N. (1968) *Linear and Quasi-linear Equations of Parabolic Type*. Translations of mathematical monographs, **23**, A.M.S.

Lamberton, D. (1995a) Critical price for an American option near maturity. *Seminar on Stochastic Analysis*, E. Bolthausen, M. Dozzi and F. Russo (Eds). *Progress in Probability*, **36**, Boston, Barel, Berlin: Birkhauser.

Lamberton, D. (1995b) Error estimates for the binomial approximation of American put options. Prébublications de l'Équipe d'Analyse et de Mathématiques Appliquées, no. 25, Université de Marne-la-Vallée. To appear in *Annals of Applied Probability*.

Lamberton, D. and Lapeyre, B. (1996) *Introduction to Stochastic Calculus Applied to Finance*. London: Chapman and Hall.

Lamberton, D. and Pages, G. (1990) Sur l'approximation des réduites. *Annales de l'Institut H. Poincaré, Probabilités et Statistiques*, **26**, 331–355.

MacMillan, L. (1986) Analytic approximation for the American put price. *Advances in Futures and Options Research*, **1**, 119–139.

McDonald, R. and Schroder, M. (1990) A parity result for American options. Working paper, Northwestern University.

McKean, H. P. (1965) A free boundary problem for the heat equation arising from a problem in mathematical economics. Appendix to a paper by R. Samuelson, *Industrial Management Review*, **6**, 32–39.

Mulinacci, S. (1996) An approximation of American option prices in a jump-diffusion model. *Stochastic Processes and Applications*, **62**, 1–17.

Mulinacci, S. and Pratelli, M. (1995) Functional convergence of Snell envelopes; applications to American options approximations. To appear in *Finance and Stochastics*.

Myneni, R. (1992) The pricing of the American option. *Annals of Applied Probability*, **2**, 1–23.

Neveu, J. (1965) *Mathematical Foundations of the Calculus of Probability*. San Francisco: Holden-Day.

Neveu, J. (1975) *Discrete-Parameter Martingales*. Amsterdam: North Holland.

Parkinson, M. (1977) Option pricing: the American put. *Journal of Business*, **50**, 21–36.

Pham, H. (1997) Optimal stopping, free boundary and American option in a jump-diffusion model. *Applied Mathematics and Optimization*, **35**, 145–164.

Revuz, A. and Yor, M. (1994) *Continuous Martingales and Brownian Motion*. New York: Springer-Verlag.

Rudin, W. (1991) *Functional Analysis*. New York: McGraw-Hill.

Villeneuve, S. (1997) Exercise regions of American options on several assets. *Prépublications de l'Équipe d'Analyse et de Mathématiques Appliquées*, n° 21/97. Université de Marne-la-Vallée.

Zhang, X. L. (1993) Options américaines et modèles de diffusion avec sauts. *Comptes Rendus de l'Académie des Sciences, Paris, Série I*, **317**, 857–862.

Zhang, X. L. (1994) Analyse numérique des options américaines dans un modèle de diffusion avec sauts. Thèse de Doctorat, École Nationale des Ponts et Chaussées, Noisy-le-Grand, France.

Zhang, X. L. (1997) Numerical analysis of American option pricing in a jump-diffusion model. *Mathematics of Operations Research*, **22**, 668–690.

11

Notes on Term Structure Models

Saul D. Jacka

11.1 Introduction

11.1.1 Interest rates

One of the major topics in financial theory is the modelling of interest rates and the pricing of interest rate derivatives. The traditional Black–Scholes model for option pricing assumes a constant spot-rate, r. This is clearly unrealistic and for several years effort has been devoted to stochastic models for interest rates which capture some or all of the characteristics of the yield curve, which is, of course, based on the relatively large number of bonds available in most major financial markets (we note in passing that, nevertheless, there is no well-known model for stock prices which incorporates a stochastic interest rate).

Most recently, Heath *et al.* (1992) and Kennedy (1994) have modelled the whole forward-rate curve: an interesting innovation, reflecting the fact that forward rates are traded and the fact that, because bonds are usually coupon bearing rather than idealised zero-coupon bonds, it can be easier to fit forward-rate models.

We note that, historically, most interest-rate models actually give the dynamics of the spot-rate r_t under an EMM (equivalent martingale measure), from which we may, in principle, deduce the corresponding bond prices.

In this chapter, where necessary, we assume a continuum of bond prices, and make the usual assumptions: that bonds are zero-coupon (i.e. pay no interest) and have no default risk (but see Chapter 12).

11.1.2 An equivalence

We now note a well-known equivalence: a market (of bond prices) is arbitrage-free (in the sense of Chapter 8) iff there exists an EMM under which the discounted bond-prices, $\tilde{P}(t, u) \equiv \exp(-\int_0^t r_s \, ds) P(t, u)$, are all martingales (in the t variable), and in this case, if $\mathbb{Q}$ is an EMM, then

$$\tilde{P}(t, u) = \mathbb{E}_{\mathbb{Q}}\left[\exp\left(-\int_t^u r_s \, ds\right)\middle| \mathcal{F}_t\right] \tag{11.1}$$

The first part of this statement is standard, the second follows immediately from the fact that $P(u, u) = 1$ (so that $\tilde{P}(u, u) = \exp\left(-\int_0^u r_s \, ds\right)$) and $\tilde{P}(\cdot, u)$ is a $\mathbb{Q}$-martingale. The modelling implications of this are fairly obvious: under weak integrability conditions any bond price model can be specified by giving a model for the instantaneous spot-rate – i.e. the dynamics of the spot-rate under a probability measure $\mathbb{Q}$ – then $\mathbb{Q}$ becomes an EMM for the bond price processes.

Consequently, if we wish to price a derivative security which pays 'interest' at a rate ϕ_t, and delivers a terminal payment of ψ_T at time T then, denoting $\exp\left(-\int_0^t r_s \, ds\right)$ by β_t, we have the usual formula for the price of the derivative:

$$D_0 = \mathbb{E}_{\mathbb{Q}}\left[\int_0^T \beta_t \phi_t \, dt + \beta_T \psi_T\right] \tag{11.2}$$

In principle, therefore, to price interest rate derivatives we need to be able to calculate $\mathbb{E}_{\mathbb{Q}}\phi\beta_t$ for a suitable class of random ϕ.

11.2   Spot-rate models

Traditionally, term-structure models have been specified in terms of the probability measure on the underlying instantaneous spot-rate process, as implied by the equivalence noted above. Almost all of the models specify the dynamics of r_t as an Itô process: $dr_t = \sigma_t \, dB_t + \mu_t \, dt$, where B is a Brownian motion under $\mathbb{Q}$. The largest class of such models are the so-called one factor models where σ and μ are functions of r so that r is a (possibly time-inhomogeneous) strong Markov process (see Duffie, 1992).

11.2.1   One factor models

It is not our purpose to give a survey of term-structure models – for an excellent survey see Chapter 7 of Duffie, 1992 – we simply propose to note here an extension of Duffie's one factor framework to the case of discontinuous models. We take as our discontinuous model the jump SDE:

$$dr_t = \sigma_{t-}dB_t + \mu_{t-}dt + \sum_{i=1}^n J_{t-}^i \, dN_t^i \tag{11.3}$$

where B is a Brownian motion, each N^i is a generalised Poisson process with intensity process I_{t-}^i, and each of σ, μ, I^i and J^i are functions of r_t and t. It follows from this that the (space–time) infinitesimal generator of r is $\mathcal{L}$, given by

$$\mathcal{L}f(r, t) = \tfrac{1}{2}\sigma^2 f_{rr} + \mu f_r + f_t + \sum_{i=1}^n I^i(r, t)[f(r + J^i(r, t), t) - f(r, t)]$$

It follows, from the usual generalisation of the Feynman–Kac formula, that $P(t, u) \equiv P^u(r_t, t)$, where

$$\left.\begin{array}{c} \mathcal{L}P^u - rP^u = 0 \\ \text{with} \\ P^u(r, u) \equiv 1 \end{array}\right\} \tag{11.4}$$

More generally, it follows from Equation (11.3) that if $h(r_t, t) \stackrel{\text{def}}{=} \mathbb{E}_\mathbb{Q}\!\left[\exp\left(-\int_t^T r_s \, \mathrm{d}s\right) h(r_T) \mid \mathcal{F}_t\right]$ then h satisfies:

$$\left.\begin{array}{c} \mathcal{L}h - rh = 0 \\ \text{with} \\ h(r, T) \equiv h(r) \end{array}\right\} \tag{11.5}$$

11.2.2 Multi-factor models

We note that a substantial amount of fairly recent work in term-structure has been on multi-factor models. The argument is that extra, explanatory variables have to be added to the model to give a Markov process: so that the assumption is that some vector of processes, $(r_t, \ldots)$, is Markov – see Chapter 14 for a discussion of stochastic volatility and see also the survey on multi-factor models in Duffie and Kas (1994).

11.3 Forward rate models

11.3.1 Forward rates

Forward rate models were introduced by Heath *et al.* (1992) – rather than specify the dynamics of the spot-rate directly they presented a collection of models for the forward (interest) rate processes $f(t, u)$. Just as r_t is the instantaneous rate of interest available at time t, so $f(t, u)$ is the instantaneous rate of interest available at time t for payment at time $u + \mathrm{d}u$ on a loan of \$1 to be made at time u. As they point out, it may, in practice, be easier to model these forward rates, when they exist, rather than the spot-rate. Given the existence of traded forward contracts on interest rates there are fairly compelling practical arguments for the existence of these processes.

Kennedy (1994) introduced a large class of Gaussian sheet models for forward rates, which we discuss in the next sub-section.

In Jacka *et al.* (1996) the authors consider an SPDE model for forward rates and go on to conclude the general structure of all continuous forward rate models which do not admit arbitrage. Jacka (1997) simplifies the presentation and deals with the discontinuous case. We may summarise the results as follows: if we denote $\tilde{P}(t, u)f(t, u)$ by $\tilde{f}(t, u)$ then

$\mathbb{Q}$ is an EMM if and only if $\tilde{f}(\cdot, u)$ is a $\mathbb{Q}$-martingale for Lebesgue a.a. u

$$\tag{11.6}$$

whereupon

$$f(t, u) = \mathbb{E}_\mathbb{Q}\!\left[\exp\left(-\int_t^u r_s \, \mathrm{d}s\right) r_u \;\middle|\; \mathcal{F}_t\right]$$

Condition (11.6) holds iff

$$f_t^u = f_0^u + M_t^u + [M^u, \tilde{M}^u + \xi^u]_t, \quad \text{for each } t, u \qquad (11.7)$$

where $\tilde{M}^u = \int_0^u M^v \, dv$ and $\xi^u = \sum_{s \leqslant t} \frac{(\Delta \tilde{M}^u)^2}{1 - \Delta \tilde{M}^u}$, and M^u is a $\mathbb{Q}$-mg for each u.

11.3.2   Forward rate models

The Heath *et al.* (1992) model (or HJM model) has all the forward rates given by SDEs driven by a d-dimensional BM, and living on the filtration of the BM. The SDEs are:

$$df_t^u = 1_{t \leqslant u} \sigma(t, u) \cdot dB_t + \mu(t, u) \, dt$$

Condition (11.7) says that, under $\mathbb{Q}$, we must have

$$\int_0^t \mu(s, u) \, ds = \left[\int_0^t \sigma(s, u) \cdot dB_s, \int_0^t \int_0^u 1_{s \leqslant v} \sigma(s, v) \cdot dB_s \right]_t = \int_0^t \int_0^u \sigma(s, u)^T \sigma(s, v) \, dv \, dt$$

so that under $\mathbb{Q}$, we must have $\mu(t, u) = \int_0^u \sigma(t, u)^T \sigma(t, v) \, dv$. It follows from the Brownian representation property that any change of measure $\Lambda = d\mathbb{P}/d\mathbb{Q}$ must satisfy $\Lambda_t \equiv \mathbb{E}_{\mathbb{Q}}[\Lambda \mid \mathcal{F}_t] = \exp\left(\int_0^t \phi_s \, dB_s - \frac{1}{2} \int_0^t \phi_s^2 \, ds\right)$, for some predictable process ϕ with $\mathbb{E}_{\mathbb{Q}}\Lambda = 1$, and hence, by the Cameron–Martin–Girsanov transformation, we must have

$$df_t^u = 1_{t \leqslant u} \sigma(t, u) \cdot d\tilde{B}_t + \tilde{\mu}(t, u) \, dt$$

where $\tilde{B}$ is a $\mathbb{P}$-BM and $\tilde{\mu}(t, u) = \int_0^u \sigma(t, u)^T \sigma(t, v) \, dv + \sigma(t, u)^T \phi_t$.

Under the Kennedy (1994, 1997) model: $f(t, u) = W(t, u)$, where W is a continuous Gaussian sheet with filtration $\mathcal{F}_t$ given by $\mathcal{F}_t = \sigma(W_{s,v} : s \leqslant t, s \leqslant v \leqslant T)$, where T is some terminal time. The structure of the Gaussian sheet is specified by the mean and covariance structure: $\mathbb{E}W_{t,u} = \mu_{t,u}$ and $\text{Cov}\,(W(t, u), W(s, v)) = c(s \wedge t, u, v)$, for some function c symmetric in the last two arguments. It is fairly simple to check that condition (11.7) becomes:

$$\mu_{t,u} = \int_0^u (c(t \wedge v, v, u) - c(0, v, u)) \, dv + \mu_{0,u}$$

so that in the case of a Brownian sheet, where $c(t, v, u) = t(v \wedge u)$, $\mu_{t,u} = (t^3/3) + t(u^2 - t^2)/2 + \mu_{0,u}$.

Kennedy (1997) also gives a stationary model which generalises the Ornstein–Uhlenbeck model. In this case $c(s, v, u) = \sigma^2 \exp(\lambda s + (2\mu - \lambda) \times (v \wedge u) - \mu(v + u))$, with the parameter μ satisfying $\mu \geqslant \lambda/2$. As Kennedy observes, this leads to spot rates and forward rates of fixed duration being mean reverting.

11.4   Affine jump models

Duffie, in Chapter 7 of Duffie (1992), discusses in detail the class of so-called affine term-structure models, developing the original idea of Brown and Schaeffer (1991) of classifying all term structure models which give rise to bond prices of the form

$$P(t, u) = \exp(a(t, u) + b(t, u)r_t) \qquad (11.8)$$

He restricts attention to the case of one-factor models where r is an Itô diffusion satisfying

$$dr_t = \sigma(r_t, t) \, dB_t + \mu(r_t, t) \, dt$$

and a and b are functions only of $u - t$, and shows that (under weak conditions on a and b) Equation (11.8) holds if and only if σ^2 and μ are time-dependent affine functions of r, in other words, Equation (11.8) holds iff $\mu = \alpha_1(t) + \alpha_2 r_t$ and $\sigma^2 = \beta_1(t) + \beta_2(t) r_t$.

We note in passing that we have an affine model iff the forward rates are affine (in r_t): this follows from the fact that $f(t, u) = -(\partial/\partial t) \ln P(t, u)$.

In this section we shall extend this to the case where jumps (of various fixed sizes) are allowed.

We assume that, as in Section 11.2.1,

$$dr_t = \sigma_{t-} \, dB_t + \mu_{t-} \, dt + \sum_{i=1}^{n} J_{t-}^i \, dN_t^i \tag{11.9}$$

where B is a Brownian motion, each N^i is a generalised Poisson process with intensity process I_{t-}^i, and each of σ, μ, and I^i are functions of r_t and t, but assume that the possible jump sizes J^i are functions of t alone. With this form for r we obtain the space-time infinitesimal generator $\mathcal{L}$, given by

$$\mathcal{L}f(r, t) = \tfrac{1}{2}\sigma^2 f_{rr} + \mu f_r + f_t + \sum_{i=1}^{n} I^i(r, t)[f(r + J^i(t), t) - f(r, t)]$$

and, substituting the expression (11.8) in (11.4) and dividing by $P(r_{t-}, t, u)$, we obtain:

$$\tfrac{1}{2}\sigma^2 b^2 + \mu b + a_t + b_t r + \sum_{i} I^i(e^{bJ^i} - 1) - r = 0 \tag{11.10}$$

Since Equation (11.10) holds for all t, T and r, it follows that if, for each t, there exist $T_1, \ldots, T_{n+2}$ such that the matrix $M \equiv (m_{ij})$ is of full rank, where $m_{(n+1),j} = b^2(t, T_j)$, $m_{(n+2),j} = b(t, T_j)$ and $m_{ij} = \exp[b(t, T_j)J^i(t)] - 1$ (for $1 \leqslant i \leqslant n$), then we must have $\mu = \alpha_1(t) + \alpha_2 r_t$, $\sigma^2 = \beta_1(t) + \beta_2(t) r_t$ and $I^i = \gamma_1^i(t) + \gamma_2^i(t) r$. Conversely, if this is so, then substituting back into Equation (11.10) we obtain:

$$b_t - 1 + \alpha_2 b + \beta_2 b^2 + \sum_{i=1}^{n} \gamma_2^i(e^{bJ^i} - 1) = 0 \tag{11.11}$$

with boundary condition $b(u, u) = 0$, and

$$a_t + b\alpha_1 + b^2 \beta_1 + \sum_{i=1}^{n} \gamma_1^i(e^{bJ^i} - 1) = 0 \tag{11.12}$$

with boundary condition $a(u, u) = 0$.

We give some examples of time-independent jump affine models.

11.4.1.1   The signed Poisson process

Here $\alpha = \beta = 0$; $J^1 = 1$, $J^2 = -1$, and $\gamma_1^1 = \gamma_1^2 = \lambda$, $\gamma_2^1 = \gamma_2^2 = 0$. This gives $b_t - 1 = 0$

and $a_t = -2\lambda(\cosh(b) - 1)$, so that $b(t, u) = -(u - t)$ and $a(t, u) = a(u - t)$, where $a(\tau) = 2\lambda \int_0^\tau (\cosh(s) - 1)\,ds$.

11.4.1.2 Spot-rate independent

Here, σ, μ and I^i are all independent of r (i.e. $\alpha_2 = \beta_2 = \gamma_2 = 0$). We can deduce that $b(t, u) = -(u - t)$ and

$$a_t = (u - t)\alpha_1(t) - (u - t)^2 \beta_1(t) - \sum_{i=1}^n \gamma_1^i(t)(e^{-(u-t)J^i(t)} - 1)$$

11.4.1.3 Proportional pure jump

Only γ_2 is non-zero, and there is only one jump size J. We get $b_t - 1 = e^{bJ} - 1$, so that $b = -\log c - Jt/J$, with $c = 1 + Ju$ and $a_t = 0$.

11.4.1.4 The jump CIR model

Here $\beta_1 = \gamma_2 = 0$, and we get the same solution for b as in the CIR model. More generally if $\gamma_2 \equiv 0$ then b is unchanged from the corresponding jump-free model.

11.5 Pricing formulae

11.5.1 Pricing formulae for general models

For a discussion of pricing formulae in general one-factor models see Duffie (1992); the basic idea is simply to solve the PDE (Equation (11.5)) numerically; however special models may have more or less explicit pricing schemes.

In the Kennedy model, where the forward rates form a Gaussian sheet, pricing is very straightforward. Since the spot-rates are jointly Gaussian, to price a claim of $h(r_s)$ at time t we need to evaluate $\mathbb{E}_Q[h(r_s)\exp(-\int_0^t r_v\,dv)]$. But we know that r_s and $\int_0^t r_v\,dv$ are jointly Gaussian so we need only specify their means and covariance structure. It is straightforward to deduce that $\mathrm{Var}(\int_0^t r_v\,dv) = 2\int_{w=0}^t \int_{v=0}^w c(v, v, w)\,dv\,dw$ and $\mathrm{Cov}(\int_0^t r_v\,dv), r_s) = \int_0^t c(s \wedge v, v, s)\,dv$ whilst $\mathbb{E}\int_0^t r_v\,dv = \int_0^t \mu_{v,v}\,dv$. The other required terms are specified directly in the Kennedy model.

11.5.2 Pricing formulae in affine models

Using the terminology of Section 11.2.1, three special cases have been analysed fairly fully: in each of these cases there are no jumps, and they are time-independent. They are the Gaussian path-independent model ($\alpha_2 = \beta_2 = 0$), the Vasicek (1977) model, also called the Ornstein–Uhlenbeck model (with $\beta_2 = 0$) and the Cox–Ingersoll–Ross (1985) model (with $\beta_1 = 0$). In the first two cases both r and $\int r_v\,dv$ are Gaussian and a similar analysis to that given above for the Kennedy model can be performed. In the third case, by shifting the origin we may cover all time-independent jump-free affine models (this is the Pearson–Sun model – see Pearson and Sun, 1989) and an explicit solution for the Green's function for the PDE (Equation (11.5)) is given in Chapter 7 of Duffie (1992). From the Green's function we may immediately evaluate the solution to Equation (11.4) (as an integral).

In the remainder of this section we shall discuss the time-dependent affine jump case. The key observation we make is that a certain rich class of functions is stable under the application of the operator $\mathcal{L} - r$.

We suppose that $\mathcal{L}$ has the form given in Equation (11.9) and the function f is of the form $f = p(r, t)e^{a(t) + rb(t)}$ where a and b satisfy Equations (11.11) and (11.12) and p is a polynomial of degree n in r with time-dependent coefficients: then $(\mathcal{L} - r)f$ is of the same form provided the coefficients of p are C^1. This is not hard to see on observing that

$$(\mathcal{L}-r)fg = f(\mathcal{L}-r)g + g(\mathcal{L}-r)f + (\beta_1 + r\beta_2)f_r g_r + \sum_i (\gamma_1^i + r\gamma_2^i)\Delta_{J_i} f \Delta_{J_i} g \quad (11.13)$$

(where $\Delta_{J_i} f \equiv f(r + J_i, t) - f(r, t)$), since if $g = e^{a(t) + rb(t)}$ (so that $(\mathcal{L} - r)g = 0$) and $p = \sum_{k=0}^n c_k(t)r^k$ then $(\mathcal{L} - r)pg = g(\mathcal{L}p + b(\beta_1 + r\beta_2)p_r + \sum_i (e^{bJ_i} - 1)\Delta_{J_i}p)$, and each term in this sum is a polynomial of degree n in r.

We may use this observation to solve Equation (11.5) when $h(r) = r^n$.

Theorem: *In the affine jump model we have*

$$\mathbb{E}_Q r_T^n \exp\left(\int_0^T r_s \, ds\right) = \sum_{k=0}^n c_k^n(t, T)r^k \exp[a(t, T) + r(b(t, T))] \quad (11.14)$$

where (the vector) c^n satisfies

$$\frac{dc^n}{dt} = -A^n(t)c^n \quad (11.15)$$

with boundary conditions $c^n(T) = (0, \ldots, 0, 1)'$, where the upper triangular matrix $A^n = (a_{i,j}^n)$ is given by

$$a_{k,k}^n = k(\alpha_2 + b\beta_2 + \sum_i e^{bJ_i} J_i \gamma_2^i)$$

$$a_{k,k+1}^n = (k+1)(\alpha_1 + \tfrac{1}{2}k\beta_2 + b\beta_1 + \sum_i e^{bJ_i}(\tfrac{1}{2}k\gamma_2^i J_i^2 + \gamma_1^i J_i)); \quad for\ k \leqslant n-1$$

$$a_{k,k+2}^n = \tfrac{1}{2}(k+1)(k+2)\beta_1 + \sum_i e^{bJ_i}\left(\binom{k+2}{3}\gamma_2^i J_i^3 + \binom{k+1}{2}\gamma_1^i J_i^2\right); \quad for\ k \leqslant n-2$$

$$a_{k,l}^n = \sum_i e^{bJ_i}\left(\binom{l}{l-k+1}\gamma_2^i J_2^{l-k+1} + \binom{l}{l-k}\gamma_1^i J_i^{l-k}\right); \quad for\ k \leqslant l-3 \leqslant n-3$$

Proof: The result follows directly from the observation above, on substituting Equation (11.14) into (11.13), and collecting terms in r_k. $\square$

Because A^n is in upper triangular form we may solve Equation (11.15) directly to obtain

$$c_n^n(t, T) = \exp\left(n \int_t^T \left(\alpha_2(s) + b\beta_2(s) + \sum_i e^{bJ_i} J_i \gamma_2^i(s)\right) ds\right)$$

together with the recurrence relation

$$\frac{c_l^n(t, T)}{c_n^n(t, T)} = \sum_{m=l+1}^n \int_t^T a_{l.m}^n(s) \frac{c_m^n(s, T)}{c_n^n(s, T)} ds$$

It should therefore be possible to approximate the solution to Equation (11.5), by first approximating $h(r)$ using a suitable orthogonal sequence of polynomials, $(p_k)_{k \geq 0}$, and then solving Equation (11.5) for each p_k, using our solution to (11.14), although error bounds are hard to obtain.

References

Brown, R. and Schaeffer, S. (1991) Interest rate volatility and the term structure. Unpublished – London Business School.

Cox, J., Ingersoll, J. and Ross, S. (1985) A theory of the term structure of interest rates. *Econometrica*, **53**, 385–408.

Duffie, D. (1992) *Dynamic Asset Pricing Theory*. Princeton, New Jersey: Princeton University Press.

Duffie, D. and Kas, R. (1994) Multi-factor term-structure models. *Philosophical Transactions of the Royal Society London A*, **347**, 577–586.

Heath, D. C., Jarrow, R. A. and Morton, A. (1992) Bond pricing and the term structure of interest rates: a new methodology for contingent claims valuation. *Econometrica*, **60**, 77–105.

Jacka, S. D. (1997) Arbitrage-free term-structure models. University of Warwick, Department of Statistics research report.

Jacka, S. D., Hamza, K. and Klebaner, F. C. (1996) No arbitrage condition in SPDE model for interest rates. University of Melbourne, Department of Statistics research report.

Kennedy, D. P. (1994) The term structure of interest rates as a Gaussian random field. *Mathematical Finance*, **4**, 247–258.

Kennedy, D. P. (1997) Characterizing and filtering Gaussian models of the term structure of interest rates. To appear in *Mathematical Finance*.

Pearson, N. and Sun, T.-S. (1989) A test of the Cox, Ingersoll, Ross model of the term structure of interest rates using the method of moments. Sloane School of Management, MIT.

Vasicek, O. (1977) An equilibrium characterization of the term structure. *Journal of Financial Economics*, **5**, 177–188.

12

Default Risk

Dilip Madan

12.1 Introduction

Default risk, broadly defined, is the risk borne by contract participants delivering value, as per contract, prior to the receipt of the agreed upon quid pro quo. The party delivering value first bears the risk of counterparty default, if the counterparty is unable to deliver the agreed upon value. Typically, these risks are controlled for, by writing contracts for short time periods, and arranging for their automatic renegotiation, on a continuing basis.

For example, contracts for the rendering of services, typically, have the service rendered first, with payment upon completion of service. The company making payment is usually sufficiently well placed financially for the service provider not to be too concerned about receiving this payment, at least for the typical pay period of two weeks. The contract is essentially on automatic renegotiation every two weeks. There are, however, cases when the payment is made first, and the service is rendered later. This occurs, for example, when airline tickets must be fully paid for (well before the commencement of the flight) and then the airlines default on the provision of the flight service, with no refunds to passengers for tickets purchased.

Although default risk is a component of essentially all contracts, we shall be concerned here primarily with financial contracts. In these contracts, both the value delivered and the quid pro quo received are defined by payments of monies by one party to another at various points of time. The actual time of default need not necessarily coincide with the time at which a payment is to be made, but can in general occur at any time. The extent of default depends on the level of any partial payments connected with the default.

Default occurs when the party that has yet to deliver announces its inability to do so, and then arranges to seek a reduction in the payments that it will make. The default time is the time at which the announcement is made. The extent of default depends on the level of partial payments, if any, consequent upon the default announcement. There are thus two components of default risk that may be termed the *timing* and *magnitude* components.

The possibility of default in a financial contract, impacts the *statistical* distribution of payments in a contract, and thereby has an impact on the *market value* of the contract, at all times. Hence, there are two perspectives on each of the two components of default, in a contract. One may evaluate for each contract the statistical probability of default occurrence, and describe the evolution of the statistical density for the magnitude of default, if it were to occur. In addition, one may evaluate the market price of a hypothetical security paying unit face value, only if a default occurs, and also describe the evolution of the market prices of hypothetical securities paying unit face value provided the magnitude of default exceeds pre-specified levels. Both *statistical probabilities* and *market values* provide different but related assessments of default possibilities, in financial contracts.

The literature on default risk can be partitioned on the basis of this focus on statistical probabilities or market valuations. There is an extensive literature concerned with predicting both the occurrence and extent of defaults. Holders of defaultable claims have an interest in assessing these probabilities, especially if they anticipate holding the claims to their maturities. Investors contemplating the sale of such claims, or their management (as part of an investment portfolio) have an interest in assessing the market value of these defaultable claims, and variations thereof. For every concept concerned with valuation there is an analogous concept, with generally different associated numeric values, concerned with statistical probabilities. The focus of this chapter is on the valuation of defaultable claims.

With regard to claim valuation, the question can be addressed, as noted earlier, in the context of each possible contract, and the specifics of contract details will influence the results, for each particular case. Before applications can be considered for contracts involving multiple payment transfers (either bilaterally or multilaterally) at various random or predetermined time points, the issue must be addressed for simple loan contracts, where one party pays another party an initial sum of money for a fixed promised repayment at a future date. Such a contract is the zero coupon bond, in which the face value of the bond is the fixed repayment at the specified future date, termed the maturity of the bond, and the initial payment is the price of the defaultable bond. We shall consider here the problem of valuing default risk in zero coupon bond pricing.

Although it may appear that such a narrow focus is unduly restrictive, in fact it is quite wide, with a considerable agenda for research still open. First, we note that all zero coupon bonds issued by the same borrower are not substitutes for one another, as they may have differing maturities, and they may also be distinguished by their priority in default. Some bonds may be subordinated to others in that they are to receive payment in default only after bonds, with higher priority, have been fully paid off.

The prices of zero coupon bonds are typically represented by yield curves that specify the annualised continuously compounded return that would be earned on the bond, if it were purchased and held to maturity. The relationship between the price, P, the yield, y, and the time to maturity, τ, is $y = -\ln(P)/\tau$, or equivalently $P = e^{-y\tau}$. One would expect, in general, that these yields would depend on: the state of the economy, information specific to the borrower, and the priority level of the bond. If we let x denote a vector of sufficient statistics about the economy, c be a vector of borrower specific information and n be the priority level of the bond, then the problem of pricing zero coupon bonds is the problem of specifying the components of x and c and modelling the multidimensional yield surface $R(x, c, n, \tau)$. As we shall see, the literature is far short of delivering on this objective, with much of the work yet remaining to be done.

Section 12.2 presents a general perspective on the statistical and risk neutral modelling of default risk, and we examine the models of the literature from this vantage point in the sections that follow. Section 12.3 considers the European option theoretic approach of Merton (1974), while Section 12.4 takes up the barrier option framework introduced by Longstaff and Schwartz (1995), Hull and White (1995) and Nielsen *et al.* (1993). In Section 12.5 we examine the use of inaccessible stopping times to describe default, as studied by Artzner and Delbaen (1995), Jarrow and Turnbull (1995), Jarrow *et al.* (1993). Section 12.6 presents the rewriting of defaultable bond pricing models, as if they were default-free, with adjusted discount rates as initiated by Madan and Unal (1994), Duffie *et al.* (1994) and Duffie and Singleton (1996). Section 12.7 presents the defaultable HJM model of Duffie and Singleton (1996). Section 12.8 concludes with a discussion of directions for further research.

12.2 A general perspective on the modelling of default risk

We consider an economy over the time interval $[0, T]$, with uncertainty represented by a probability space $(\Omega, \mathcal{F}, P)$, and the arrival of information modelled by a filtration $\mathbb{F} = \{\mathcal{F}_t \mid 0 \leqslant t \leqslant T\}$ satisfying the usual conditions of being complete, increasing and right continuous. The economy has a money market account paying the continuously compounded annualised interest rate $r(t)$ at time t, for an $\mathbb{F}$ adapted stochastic process $r = [r(t) \mid 0 \leqslant t \leqslant T]$. The return at time t on a dollar invested at time 0 is given by the stochastic process $B = [B(t) \mid 0 \leqslant t \leqslant T]$, where

$$B(t) = \exp\left(\int_0^t r(u)\, du\right) \tag{12.1}$$

In addition to a money market account, there are, trading in the economy, default-free pure discount bonds of all maturities with price paths $\Pi(T) = [\Pi(t, T) \mid 0 \leqslant t \leqslant T]$, and time t prices $\Pi(t, T)$, for the bond maturing at time T, $T \leqslant \mathcal{T}$. Bonds issued by the Government are typically treated in the literature as free of default. There are also trading, in the economy, bonds issued by private companies that are subject to default. We suppose the existence of a set of sufficient statistics on company and bond characteristics, such that two bonds with the same characteristics and maturity have the same bond prices. Let c denote this vector of company specific information, and n be the vector of bond characteristics. There is then in the economy, a defaultable bond price function, of the form $P(c, n, t, T)$, for bonds with company characteristics c and bond characteristics n. The company characteristics could be the level of leverage and the volatility in the evolution of the underlying asset values, while the bond characteristics could specify the priority of the bond or the total sum of monies promised at higher priorities. In general, one would expect that both c and n would be stochastic processes $c = [c(t) \mid 0 \leqslant t \leqslant T]$ and $n = [n(t) \mid 0 \leqslant t \leqslant T]$, though for many applications these could be taken as constant processes for simplification of the problem.

The time of default can, in general, be modelled by a random stopping time $\zeta(c)$, and one may define a default process $\Delta(c) = [\Delta(c, t) \mid 0 \leqslant t \leqslant T]$, where

$$\Delta(c, t) = 1_{\{t \geqslant \zeta(c)\}} \tag{12.2}$$

The process $\Delta(c)$ is zero before the time of default and takes the value unity for all t exceeding or equal to the random stopping time $\zeta(c)$. Note that the time of default does not depend on the bond characteristics or the bond maturity, as it is the company that defaults, and a default on any one of its bonds is technically viewed as a default on all its bonds.

Consequent upon a default occurring, each bond may be entitled to some partial (or complete) payment, and we let $y(c, n, T) = [y(c, n, t, T) \mid 0 \leqslant t \leqslant T]$, be the stochastic process of the partial payment made at time t to creditors holding bonds maturing at T, with characteristics n, for the company with characteristics c, if default occurs at time t, or $\zeta(c) = t$. If t is not a default time then $y(c, n, t, T)$ may be defined by convention to be zero.

We suppose the absence of arbitrage opportunities and assume the existence of a measure Q equivalent to P such that under the measure Q, asset price processes discounted by the money market account are martingales. Such a measure is referred to as a risk neutral measure. The measure Q has a density with respect to P, by virtue of the property of equivalence, that we denote by $\Lambda(T) = \mathrm{d}Q/\mathrm{d}P$. The change of measure density process is the process $\Lambda = [\Lambda(t) \mid 0 \leqslant t \leqslant T]$, where $\Lambda(t) = E^P[\Lambda(T) \mid \mathcal{F}_t]$. We associate with Λ, its instantaneous rate of return process $\lambda = [\lambda(t) \mid 0 \leqslant t \leqslant T]$, such that $\Lambda = \mathcal{E}(\lambda)$, where $\mathcal{E}$ is the stochastic or Doleans-Dadé exponential operator. The two processes, Λ and λ are related by the property that $\mathrm{d}\Lambda = \Lambda(t_-)\,\mathrm{d}\lambda$, where $\Lambda(t_-)$ is the left limit of the process Λ at time t.

A model for default risk under either the statistical measure P, or the risk neutral measure Q, specifies the process for the default time $\Delta(c)$ and the partial payment processes $y(c, n, T)$. One may associate with the process $\Delta(c)$ a unique increasing integrable predictable process H with $H_0 = 0$, such that the process $\Delta(c) - H$ is a uniformly integrable martingale, by the Doob–Meyer decomposition theorem (Jacod and Shiryaev, 1980, p. 32). The process H is called the compensator of the stopping time defining the time of default.

Bond prices are determined by evaluating conditional expectations of money market account discounted pay-offs, under a risk neutral measure. Under a Markovian assumption on the evolution of relevant uncertainties, one has a set of economy wide variates defining the process $x = [x(t) \mid 0 \leqslant t \leqslant T]$ that form sufficient statistics for describing the state transition probabilities of the economy. The bond pricing function then takes the form

$$P(c, n, t, T) = \psi(x, c, n, t, T) \tag{12.3}$$

for some function ψ.

Models of default risk may be compared and contrasted in terms of: the associated default stopping time compensators, H; the specification of the partial payment processes, $y(c, n, T)$; the dynamics of the economy wide information processes, x, employed in the evaluation of the prices of defaultable bonds; and the dynamic structure of the firm and bond character- istics c and n respectively. In principle, one has these entities under both the statistical and risk neutral measures, and the process for change of measure density, Λ, or its return process, λ, provides information on what risks are priced in the economy, and how these prices are determined. The sections that follow conduct an evaluation of well established models in the literature in these terms.

12.3 The European option theoretic approach

This option theoretic approach was initiated by Merton (1974). In this approach, corporate equity is viewed as a European call option written on the underlying asset value of the firm, with the strike of the option given by the face value of the debt, and maturity matching the debt maturity. If the debt is the issue of a single zero coupon bond with maturity T and face value F, then equity holders receive the excess of the asset value over the promised face of F at maturity, and bond holders receive the minimum of F and the firm value. The magnitude of default at maturity is given by the excess of the face value F over the value of the underlying assets, if positive, and is therefore the value of a European put option written on the underlying asset value, with strike equal to the face value of debt, and maturity matching the bond maturity.

The risk neutral asset value process is modelled as a geometric Brownian motion with initial value V, a drift equal to a constant interest rate of r and a volatility of σ. The statistical asset value process has a drift of μ and the same volatility of σ. The value of the default at time t is given by the Black–Scholes value of a European put option. Hence, the value of the put option is

$$p = Fe^{-r\tau}N(-d_2) - VN(-d_1) \qquad (12.4)$$

where $N(x)$ is the standard normal distribution function

$$d_1 = \frac{\ln(V/F)}{\sigma\sqrt{\tau}} + \left(\frac{r}{\sigma} + \frac{\sigma}{2}\right)\sqrt{\tau}$$

$$d_2 = d_1 - \sigma\sqrt{\tau}$$

and

$$\tau = T - t$$

Subtracting Equation (12.4) from the price of the Treasury bond gives the value of defaultable bonds or corporate debt, as

$$P(V, \sigma, F, t, T) = Fe^{-r\tau}N(d_2) + VN(-d_1) \qquad (12.5)$$

Defining $d = Fe^{-r\tau}/V$, one may write the yields $R(r, d, \sigma, t, T) = -(\ln P(V, \sigma, t, T))/(T-t)$ as

$$R(r, d, \sigma, t, T) = r - \frac{1}{\tau}\ln\left[N(h_2(d, \sigma^2\tau)) + \frac{1}{d}N(h_1(d, \sigma^2\tau))\right] \qquad (12.6)$$

where

$$h_1(d, \sigma^2\tau) = \frac{\ln(d)}{\sigma\sqrt{\tau}} - \frac{\sigma\sqrt{\tau}}{2}, \quad \text{and} \quad h_2(d, \sigma^2\tau) = -\frac{\ln(d)}{\sigma\sqrt{\tau}} - \frac{\sigma\sqrt{\tau}}{2}$$

The yields can be seen from Equation (12.6) to depend on firm specific information c, represented by d, the level of leverage, and σ, the volatility of the underlying asset value process. The only bond specific information required is its maturity and the only economy wide variable of relevance is the interest rate r.

The model could easily be extended to account for bonds of differing levels of priority. Let F_n be the face value of debt at priority level n, then debt holders at this priority level would receive at maturity, $\text{Min}((V - \Sigma_{j<n}F_j)^+, F_n)$. This pay-off is that of a bull spread with a lower strike of $K_{n-1} = \Sigma_{j<n}F_j$ and an upper strike of $K_n = K_{n-1} + F_n$. The value of this pay-off is given by

$$P(V, \sigma, F_1, F_2, \ldots, F_N, n, t, T) = VN(d_{1,n-1}) - K_{n-1}e^{-rt}N(d_{2,n-1})$$
$$- VN(d_{1,n}) + K_n e^{-rt}N(d_{2,n}) \quad (12.7)$$

where

$$d_{1,n} = \frac{\ln(V/K_n)}{\sigma\sqrt{\tau}} + \left(\frac{r}{\sigma} + \frac{\sigma}{2}\right)\sqrt{\tau}$$

and

$$d_{2,n} = d_{1,n} - \sigma\sqrt{\tau}$$

Upon taking logarithms of Equation (12.7) and dividing by time to maturity we obtain the yields on defaultable bonds of priority n as $R(r, V, \sigma, F_1, F_2, \ldots, F_N, n, t, T)$.

The market wide information required for the pricing is the interest rate. However, the dynamics of this process are elementary in that interest rates are assumed to be constant over time. The required firm specific information is the debt structure, the value of the underlying asset and its volatility. Bond specific information includes the priority and the maturity.

In terms of the nature of the default times and partial pay-off functions, the time of default is random, but the random variable can take on just one value, and this is T, the maturity of the bond. In fact the process $\Delta(t)$ is identically zero for $t < T$, and jumps to unity at T, if $V(T)$ falls short of F. The compensator of the default time is the predictable process defined by

$$H(t) = \mathbf{1}_{t \geqslant T}\mathbf{1}_{F > V(t \wedge T)} \quad (12.8)$$

The function $\mathbf{1}_{t \geqslant T}$ is predictable as it is easily seen to be the limit of the left continuous functions $\mathbf{1}_{t > T-1/n}$. On the other hand the function $\mathbf{1}_{F > V(t \wedge T)}$ is left continuous itself. Hence H is predictable and $\Delta - H$ is the identically zero martingale. We have in this model a separate default time process for each bond, as bonds of differing maturities have different possible default times. The firm may default on its lower maturity bonds, and yet pay in full the longer maturity bonds.

The partial pay-off process is zero at all times other than T, and at T the partial pay-off is $V(T)$, which in a default state is less than F by definition. More generally, if we consider bonds of varying priorities then the pay-off process is $y(c, n, t, T) = \mathbf{1}_{t=T}\text{Min}((V(T) - \Sigma_{j<n}F_j)^+, F_n)$.

The measure change density is the Black–Scholes measure change density. Explicitly we have that

$$\Lambda t = \exp\left(\frac{r-\mu}{\sigma^2}(\ln V(t) - \mu t + \sigma^2 t/2) - \frac{(r-\mu)^2}{\sigma^4}\frac{t}{2}\right) \qquad (12.9)$$

We see from this representation that the only uncertainty priced is the uncertainty in the asset value process. There is no need to price the uncertainty associated with the default time, or the pay-offs connected with default, as by construction these are derivative securities with prices determined by arbitrage as claims contingent on $V(T)$.

The statistical probability of default occurring in the interval $[0, T]$ may then be determined as the probability that $V(T)$ falls short of F at T, under the statistical measure. This is given by $N(-d_2)$, where d_2 is as defined in Equation (12.4), with μ replacing r in the expression for d_2. For a mean rate of return on asset values of 15%, with a 20% volatility, in a 5% interest rate market, the default probability over a two year period for a firm with asset value 1.5 times face value of debt is 0.009. The corresponding risk neutral default probability is 0.0476. Hence, although the default event is quite remote, its market price can be substantial.

12.4 The barrier option theoretic approach

Two unsatisfactory features of the European option theoretic approach, the presence of separate random default times for different bonds, and the assumption of constant interest rates, are relaxed in the barrier option theoretic approach initiated by Longstaff and Schwartz (1995). Since all the creditors of a firm simultaneously face the risk of default by the firm, holders of bonds of differing maturities or priorities, along with other creditors who may be owed monies for services rendered in the course of business, have their claims adjusted collectively upon default. Defaults on corporate debt liabilities may therefore occur before the time at which payments become due. If the company defaults then this simultaneously affects the claim position of all creditors. As noted in Section 12.3, the European option theoretic approach of Merton (1974), essentially gave each bond its own default time process. With a common default time for all bonds, early default is a possibility and this is modelled in Longstaff and Schwartz (1995). Additionally, the assumption of a constant interest rate is relaxed and, furthermore, correlation between the interest rate and asset value process is allowed for.

Longstaff and Schwartz (1995) model the time of default as occurring at the first time when the value of the underlying assets reaches a default threshold K. The default event can then be likened to a barrier option that pays one dollar if the asset value path reaches the barrier of the threshold level prior to maturity, and pays zero otherwise. One does not have to wait until time T for the default to eventuate. The asset value process is modelled as in Merton (1974), as a risk neutral geometric Brownian motion with a drift equal to the interest rate and a volatility of σ. The corresponding statistical process has a drift of μ and the same volatility.

Unlike Merton (1974), where a specific reading on the magnitude of default was given by the extent of the shortfall of asset value relative to face at maturity, the default magnitude is here modelled as a fixed exogenous percentage write down of the claim. The claim is viewed as having promised F at T and full payment at some time $t < T$ constitutes the receipt of an equal number of default-free bonds with face F and maturity T, which at time t have a value given by $F\Pi(t, T)$. The payment at t is $(1 - w)F\Pi(t, T)$, which is equivalent to the receipt of $(1 - w)F$ at T. Hence the pay-off process is the function $y(c, n, \cdot, T) = (1 - w)F\Pi(t, T)$, where w is the fixed percentage write down.

The interest rate, however, is not a constant, but follows a risk neutral mean reverting Vasicek process, with dynamics given by a long term rate of α/β, a speed of adjustment, β, and a volatility of η. Specifically we have

$$dr = (\alpha - \beta r)\,dt + \eta\,d\tilde{W}_2 \tag{12.10}$$

where $\tilde{W}_2$ is a standard Brownian motion correlated with the standard Brownian motion $\tilde{W}_1$, driving the risk neutral asset value process, with an instantaneous correlation coefficient of ρ.

The defaultable bond is then a path dependent contingent claim receiving the pay-off F if the barrier option defining default pays 0, while on the set of paths where the barrier option defining default pays unity, the defaultable bond receives $(1 - w)F\Pi(t, T)$ at the time of default, or the time at which the asset value process first reaches the default threshold K. The uncertainty in the economy is a two-dimensional Markov process given by the vector process of the firm's underlying asset value and the interest rate. The price of the defaultable bond may then be written in terms of $X = V/K$ and the prices of default-free zero coupon bonds $\Pi(t, T) = \Psi(r, t, T)$, as

$$P(r, X, \sigma, t, T) = F\Psi(r, t, T)(1 - Q(r, X, t, T)) + y(r, F, t, T)Q(r, X, t, T) \tag{12.11}$$

where $y(r, F, t, T)$ is the pay-off $(1 - w)F\Psi(r, t, T)$, and $Q(r, X, t, T)$ is the risk neutral probability of default (or equivalently the forward price of the barrier option that pays unity, if default occurs in the interval $[t, T]$).

The structure of Equation (12.11) matches that of Equation (12.5) in that the first term in both cases accounts for full payment with appropriate probabilities for the occurrence of this event evaluated under the appropriate measures. The second term accounts for partial payment which in the case of Equation (12.11) is simpler as it amounts to $Q(r, X, t, T)$ times the predetermined exogenous pay-off.

The complete answer is obtained on determining the risk neutral default probability, and this is the probability that the first passage time of $\ln X$ to zero occurs before T. Integrating the process for $\ln X$, with the Gaussian interest rate process of the Vasicek substituted into the dynamics of $\ln X$, we observe that $\ln X(t)$ is normally distributed with mean $\ln X + M(t, T)$ and variance $S(t)$. Furthermore, conditional on $\ln X_t = 0$, $\ln X(T)$ is normally distributed with mean $M(T, T) - M(t, T)$ and variance $S(T) - S(t)$. The functions $M(t, T)$ and $S(t)$ are given by

$$M(t, T) = \left(\frac{\alpha - \rho\sigma\eta}{\beta} - \frac{\eta^2}{\beta^2} - \frac{\sigma^2}{2}\right)t + \left(\frac{\rho\sigma\eta}{\beta^2} + \frac{\eta^2}{2\beta^3}\right)\exp(-\beta T)(\exp(\beta t) - 1)$$
$$+ \left(\frac{r}{\beta} - \frac{\alpha}{\beta^2} + \frac{\eta^2}{\beta^3}\right)(1 - \exp(-\beta t)) - \left(\frac{\eta^2}{2\beta^3}\right)\exp(-\beta T)(1 - \exp(-\beta t))$$

$$(12.12)$$

and

$$S(t) = \left(\frac{\rho\sigma\eta}{\beta} + \frac{\eta^2}{\beta^2} + \sigma^2\right)t - \left(\frac{\rho\sigma\eta}{\beta^2} + \frac{2\eta^2}{\beta^3}\right)(1 - \exp(-\beta t))$$
$$+ \left(\frac{\eta^2}{2\beta^3}\right)\exp(-\beta T)(1 - \exp(-2\beta t))$$

It is shown by Buonocore *et al.* (1987) that the probability that $\ln X_t$ is less than or equal to zero is given by

$$N((-\ln X - M(t, T))/S(t)^{1/2})$$

Alternatively, this probability may be written in terms of the first passage to zero at u in $(0, t)$ followed by a return to a value less than or equal to zero. Letting $q(0, u \mid \ln X, 0)$ be the density of the first passage to zero at u from $\ln X$ at 0, one may write the alternative expression for the probability that $\ln X \leqslant 0$ as

$$\int_0^t q(0, u \mid \ln X, 0)N\left(\frac{M(u, T) - M(t, T)}{(S(t) - S(u))^{1/2}}\right)du$$

Equating these expressions yields an integral equation for the first passage density as

$$N\left(\frac{-\ln X - M(t, T)}{S(t)^{1/2}}\right) = \int_0^t q(0, u \mid \ln X, 0)N\left(\frac{M(u, T) - M(t, T)}{(S(t) - S(u))^{1/2}}\right)du$$

$$(12.13)$$

Longstaff and Schwartz (1995) solve for $Q(r, X, t, T)$ by solving for a discretisation of this integral equation. Specifically, partitioning the interval $[0, T]$ into n pieces of length T/n they define the approximation for $Q(r, X, t, T)$ as $Q(r, X, t, T, n) = \Sigma_{i=1}^n q_i$ where q_i is the approximation for $\int_{(i-1)T/n}^{iT/n} q(0, u \mid \ln X, 0)\,du$, for $i = 1, \ldots, n$. The integral Equation (12.13) is approximated by the discrete summation

$$N(a_i) = q_i + \sum_{j=1}^{i-1} q_j N(b_{ij}) \qquad (12.14)$$

where

$$a_i = \frac{-\ln X - M(iT/n, T)}{S(iT/n)^{1/2}}$$

and

$$b_{ij} = \frac{M(jT/n, T) - M(iT/n, T)}{(S(iT/n) - S(jT/n))^{1/2}}$$

The default time is predictable as it is the first time that a continuous process reaches a smooth boundary given by the threshold. We define the compensator for Δ, the default time by

$$H(t) = 1_{\{\exists s \leqslant t (V(s)=K \text{ and } V(u)>K \text{ for } 0 \leqslant u<s)\}} \tag{12.15}$$

$H(t)$ is right continuous but the functions $H(t, n)$ defined by

$$H(t, n) = 1_{\{\exists s<t(V(s)=K+1/n \text{ and } V(u)>K+1/n \text{ for } 0 \leqslant u<s)\}} \tag{12.16}$$

are left continuous with a limit as n tends to infinity given by $H(t)$. The compensated martingale $\Delta - H$ is once again the identically zero martingale.

As already noted, the pay-off process is explicitly given by $FP(r, t, T)(1 - w)$ for an early pay-out at t and is independent of priority. The firm specific information required for valuation is the level of leverage, and the economy wide information is the interest rate that is assumed to follow a Vasicek process.

Both asset value risk and interest rate risk are priced in the model. The explicit change of measure density has the form for λ given by

$$\lambda(t) = (r - \mu)/\sigma W_1(t) + \theta W_2(t) \tag{12.17}$$

where the statistical process for the interest rate is

$$dr = (\zeta - \beta r) dt + \eta dW_2 \tag{12.18}$$

and $\alpha = \zeta + \eta\theta$.

The statistical probability of default is much simpler to evaluate as this is just the first passage distribution of $\ln X$ to zero and $\ln X$ is normally distributed with mean μT and variance $\sigma^2 T$. The market risk premium on asset value risk is $(r - \mu)/\sigma$ and this is time varying for a constant μ, yielding a simple statistical probability of default obtained directly from first passage times of geometric Brownian motion. Alternatively, one could suppose that $(r - \mu)/\sigma$ is constant and μ is accordingly time varying. This would lead to computations for the statistical probability of default along the lines of the Longstaff and Schwartz (1995) risk neutral calculation described above, but with suitably altered parameter values.

Hull and White (1995) have extended this approach to allow for more general default boundaries given by the first time that a functional equation of the form $G(V, x, t)$ is satisfied, where x may denote other state variables relevant to the specification of the default conditions. Hull and White apply their methods to the valuation of default in option and swap contracts. This more general approach is applied by Nielsen *et al.* (1993), to the case of risky zero coupon bonds where the default boundary is allowed to evolve stochastically.

12.5   Inaccessible default times

Jarrow and Turnbull (1995) were the first to model the default time as a compensated jump process with a predictable compensator H, such that the compensated martingale $\Delta - H$ is not identically zero, as it is in all the cases considered in the above sections. This distinction and its properties in modelling default were first noted in Madan and Unal (1994). These methods have been extended and applied in a variety of contexts by Duffie *et al.* (1994), Duffie and Singleton (1996), and Duffie and Huang (1996). An analysis of the relationship between the statistical and risk neutral measures in this context is provided by Artzner and Delbaen (1995), who also study the market structure required for the unique identification of the required measure to change from the statistical to the risk neutral process.

In this approach, attention shifts away from defining the default event, towards just specifying the evolution of event likelihoods. As a result, one may not employ the methods of contingent claims directly to evaluate the pay-offs. If we model the evolution of economic information by continuous processes then the default event, in this approach, is not a contingent claim with a pay-off adapted to the evolution of this information. Jarrow and Turnbull (1995) exploit an interesting analogy of default, with the devaluation of currencies in fixed exchange rate regimes, to introduce the nature of their default time.

Consider a firm that has in circulation fixed maturity debt claims, as well as bearer bonds issued by the firm. The bearer bonds essentially circulate in the economy as private money issued by the firm, which in the absence of default, exchange for one US dollar for each bearer bond promised dollar. The firm at maturity can easily meet its obligations default-free by issuing bearer bonds to the tune of the face value of the fixed maturity debt. The firm therefore has a default-free term structure in terms of bearer bonds, with $v(t, T)$ being the value in time t bearer bonds of a unit-face promise of bearer bonds at time T. The yield curve associated with the default-free term structure $v(t, T)$ can be likened to the term structure in a foreign country. In addition to this term structure at time t, there is a term structure, also at time t, of default-free US dollars promised by the Treasury denoted $\Pi(t, T)$. Further, we have an exchange rate between the bearer bonds of the firm and US dollars of $e(t)$ that specifies the number of US dollars that exchange for a single dollar promised on a bearer bond of the firm. To begin with, we are in a state of no default and $e(t)$ is unity. However, there is a random time ζ such that at time ζ the exchange rate between bearer bonds and US dollars jumps down to a value δ and stays at this level thereafter. Letting Δ be the process of the default time, so that $\Delta(t) = \mathbf{1}_{t \geqslant \zeta}$, then one may write the exchange rate process as

$$e(t) = 1 - (1 - \delta)\Delta(t) \tag{12.19}$$

In Sections 12.3 and 12.4 above we saw that the process for $\Delta(t)$ was itself a predictable process that was the limit of left continuous processes and the compensator for $\Delta(t)$ was $H(t) = \Delta(t)$. Here the process $\Delta(t)$ has a predictable compensator $H(t)$, such that the difference $\Delta(t) - H(t)$ is a non-zero compensated jump martingale and the actual time of default is not predictable. The situation can be likened to the arrival of a Poisson process $N(t)$ with arrival rate ξ, whereby the actual time of arrival is never defined in terms of fundamentals, and remains a surprise time, but the compensated process $N(t) - \xi t$ is a martingale. To know the default time we must know the exchange rate process that is, by construction, not a continuous process. It is in this sense that the default cannot be viewed as a claim contingent on the evolution of continuous processes, as one cannot pinpoint the default time from a knowledge of these continuous processes.

The compensator, $H(t)$, for the default time process itself, however, may be adapted to continuously evolving economic information, and for most applications we suppose that one may write $H(t) = \int_0^t h(u)\, du$, for an $\mathbb{F}$ adapted process $h(t)$. The process $h(t)$ is called the process for the hazard rate of default, with $h(t)\, dt$, being the conditional probability at time t that a default occurs in the interval $[t, t + dt]$.

Regarding the pay-off contingent on default, if default occurs at time $\tau_1 < T$ then at time T the exchange rate is δ and the creditors receive $F\delta$ at time T, or equivalently they receive default-free bonds with face δF at time τ_1, that mature at time T with a time τ_1 value of $F\Pi(\tau_1, T)\delta$. The formulation of this pay-off function is exactly as formulated by Longstaff and Schwartz (1995). The difference between the models of Jarrow and Turnbull (1995) and Longstaff and Schwartz (1995) lies in the modelling of the default time.

The pricing of defaultable claims requires that we specify and identify the risk neutral process for the prices of the assets of the economy. In this case we have to specify: the risk neutral process for default-free bond prices $\Pi(t, T)$; prices of default-free bonds promising to pay in bearer bonds $v(t, T)$ converted to US dollars at the exchange rate $e(t)$, or the spot price for corporate bonds of $P(t, T) = e(t)v(t, T)$; and the price process of the money market account in bearer bonds converted at the exchange rate, $B_1(t)e(t)$. As usual we employ the US dollar money market account $B(t)$, as a discounting asset.

Jarrow and Turnbull (1995) employ a Heath, Jarrow and Morton (1992) (HJM) model for the default-free term structure $\Pi(t, T)$. Discounted bond prices are then positive martingales given by the stochastic exponential of

$$\Pi(t, T)/B(t) = \mathcal{E}\left(-\int_0^t \sigma^*(u, T)'\, d\tilde{W}(u)\right) \tag{12.20}$$

where $\tilde{W}$ is an m-dimensional standard Brownian motion under the risk neutral measure and the bond price volatility is related to forward rate volatilities by

$$\sigma^*(t, T) = \int_t^T \sigma(t, u)\, du \tag{12.21}$$

The risk neutral process for the forward rates is

$$f(t, T) = f(0, T) + \int_0^t \sigma(u, T)'\sigma^*(u, T)\, du + \int_0^t \sigma(u, T)'\, d\tilde{W}(u) \tag{12.22}$$

The price of the defaultable bond at time t, assuming $e(t) = 1$, is given by the valuation principle that discounted values are risk neutral martingales. Since the pay-off at maturity is, by construction, $e(T)$, it follows that the price of the defaultable bond at time t is

$$P(t, T) = E^Q[e(T)/B(T) \mid \mathcal{F}_t]B(t) \tag{12.23}$$

Jarrow and Turnbull (1995) invoke an assumption of risk neutral independence of the process driving interest rates from that driving the default, and hence, write the expectation in Equation (12.23) as the product of two expectations, whereby we have that

$$P(t, T) = \Pi(t, T)E^Q[e(T) \mid \mathcal{F}_t] \qquad (12.24)$$

The expectation in Equation (12.24) can be rewritten in terms of the probability of no default, to obtain an expression comparable to that of Longstaff and Schwartz (1995). Letting $F(t, T)$ denote the probability of no default, one may write

$$P(t, T) = \Pi(t, T)F(t, T) + \delta\Pi(t, T)(1 - F(t, T)) \qquad (12.25)$$

where $F(t, T)$ is comparable to $(1 - Q(r, X, t, T))$ in Longstaff and Schwartz (1995). The only explicit computation for $F(t, T)$ provided in Jarrow and Turnbull (1995) is for the case of a Poisson process driving the default and this yields that

$$F(t, T) = \exp(-\lambda\mu(t, T)) \qquad (12.26)$$

where it is supposed that the statistical hazard rate process $h(t)$ is λt while the risk neutral hazard rate process is $\lambda\mu t$. Substitution of Equations (12.20) into (12.25) yields the defaultable bond pricing function of Jarrow and Turnbull (1995) as

$$P(f(t, \cdot), \lambda\mu, t, T) = \Pi(t, T)e^{-\lambda\mu(T-t)} + \delta\Pi(t, T)(1 - e^{-\lambda\mu(T-t)}) \qquad (12.27)$$

The firm specific information required in pricing corporate bonds is the risk neutral hazard rates of default. However, for the specific model of Equation (12.27) we lose the dependence of corporate bond values on the firm's equity or asset value process and its volatility, as we had in the Merton (1974) and Longstaff and Schwartz (1995) models. This is reintroduced in the context of inaccessible default stopping times in Madan and Unal (1994) and is discussed in the next section. As in Longstaff and Schwartz (1995), all priorities are treated equally in the Jarrow and Turnbull (1995) model. The market wide information required in pricing defaultable bonds incorporates the complete richness of an HJM model for the term structure and, in particular, the complete initial forward rate curve.

For the relationship between the statistical and risk neutral measures, in this context, the reader should consult Artzner and Delbaen (1995). Altzner and Delbaen (1995) show that if the compensator, $H(t)$, for the default time process $\Delta(t)$ is absolutely continuous with respect to Lebesgue measure under the statistical measure P and takes the form $H(t) = \int_0^t h(u) \, du$ for a hazard rate process $h(t)$, then it is also absolutely continuous with respect to Lebesgue measure under any equivalent measure and so it has the form $\tilde{H}(t) = \int_0^t \tilde{h}(u) \, du$ under the risk neutral measure. In particular, if $\Delta - H$ is a P-martingale then $\Delta - \tilde{H}$ is a Q-martingale. Furthermore, one may write that

$$\tilde{h}(t) = h(t)\frac{K(t)}{\Lambda(t_-)} \qquad (12.28)$$

where $\Lambda(t) = E^P[dQ/dP \mid \mathcal{F}_t]$ and $K(t)$ is a positive predictable process such that $E^P[\Lambda(T) \mid \mathcal{F}_\sigma] = K_\sigma$. The process $K(t)/\Lambda(t_-)$ is interpreted in Jarrow and Turnbull (1995) as the process for the market price of the risk of default. This is assumed constant and is μ in Equation (12.27). Artzner and Delbaen (1995) show that if the insurance contracts (paying the default value at the default time against the receipt of a continuous premium payment until the default time) are marketed then the risk neutral compensator is uniquely identified, even if markets, in general, may be incomplete. In particular, they show that the compensated jump martingale $\Delta - \tilde{H}$ lies in the L^2-stable space generated by the discounted money market account or the constant martingale 1, the discounted bond prices, and the cash flows of the discounted insurance contract. The default is therefore spanned by the insurance contract and the market price of default risk is thereby identified.

12.6   Adjusting discount rates for default exposure

The use of inaccessible stopping times for the default time has the flexibility of allowing for very general forms of modelling default probabilities. One may make the hazard rates of default statistically and risk neutrally adapted to the evolution of fairly rich information structures. The initial model in Jarrow and Turnbull (1995) was an example in which bond values did not depend on the firm's asset value or its volatility, a property shared by the prior models of Merton (1974) and Longstaff and Schwartz (1995). Madan and Unal (1994) reintroduce this dependence by making the hazard rate of default dependent on the equity level of the firm and its volatility. Specifically, they employ a function of the relativised equity (the equity level deflated by the level of the money market account). Madan and Unal (1994) also allow for more general pay-off processes and show how to extract, under assumptions of strict priority, information on pay-off distributions from the relative spreads of securities with different priority, that face the same default risk. Furthermore, they show that defaultable claims may be valued as if they were default-free, provided discount rates are adjusted to reflect the exposure to default. The specific adjustment for a bond with no partial pay-out in the event of default is to define an adjusted discount rate as the sum of the interest rate and the hazard rate of default. This result is extended by Duffie *et al.* (1994) to the case where there are partial payments contingent on default.

The default time in Madan and Unal (1994) is an inaccessible stopping time ζ with a predictable compensator for $\Delta(t)$, $H(t)$, which is absolutely continuous with respect to Lebesgue measure and of the form $H(t) = \int_0^t h(u)\,du$. They allow for a general pay-off process $y(t)$ describing pay-outs at time t that are contingent on the occurrence of default at time t. The spot price of the defaultable claim may then be written as the expectation under the risk neutral measure of the money market discounted pay-offs or

$$P(t, T) = B(t)E_t^Q\left[\frac{1}{B(T)}1_{\zeta>T}\right] + E_t^Q\left[\int_t^T \frac{B(t)}{B(u)}y(u)\,d\Delta(u)\right] \quad (12.29)$$

Invoking, as in Jarrow and Turnbull (1995), an independence assumption of the interest rate process from the process driving the default, under the risk-neutral measure, Equation (12.29) may be written as

$$\frac{P(t, T)}{\Pi(t, T)} = E_t^Q[1_{\zeta>T}] + E_t^Q\left[\int_t^T \frac{\Pi(t, u)}{\Pi(t, T)}y(u)\,d\Delta(u)\right] \quad (12.30)$$

The payment of $y(u)$ dollars at time u is then treated as equivalent to the payment of $y(u)B(T)/B(u) = y(T)$ dollars at time T. This equivalent payment at time T is defined as $y(T)$ or the recovery rate. This yields the result that

$$\frac{P(t, T)}{\Pi(t, T)} = E_t^Q[1_{\zeta > T}] + E_t^Q\left[\int_t^T \frac{\Pi(t, u)}{\Pi(t, T)} \frac{B(u)}{B(T)} y(T) \, d\Delta(u)\right] \quad (12.31)$$

Finally, they suppose that $y(T)$ has a stationary distribution with density $q(y)$ with support on the unit interval, and once again, invoking the independence of the default time from the interest rate process, they obtain

$$\frac{P(t, T)}{\Pi(t, T)} = E_t^Q[1_{\zeta > T}] + E_t^Q\left[\int_t^T d\Delta(u)\right] \int_0^1 q(y) \, dy \quad (12.32)$$

Now since the integral $\int_t^T d\Delta(u)$ equals $\Delta(T)$, given that $\Delta(t) = 0$, we may write Equation (12.32) as

$$\frac{P(t, T)}{\Pi(t, T)} = F(t, T) + (1 - F(t, T)) \int_0^1 yq(y) \, dy \quad (12.33)$$

Equation (12.33) may usefully be compared with Equations (12.25), (12.11) and (12.3). All four equations partition the value of a defaultable claim into the sum of full payment multiplied by its probability, plus the value of partial payment, which in Merton (1974) occurs at T, but in the other models may occur earlier; however, it is effectively accounted for, as if it were to occur at T. Essentially, Longstaff and Schwartz (1995), Jarrow and Turnbull (1995) and Madan and Unal (1994) all define recovery in terms of payment relative to the default-free present value of the promise, or equivalently as the payment carried over to maturity relative to the promise at this date. Duffie *et al.* (1994), Duffie and Singleton (1996) and Duffie and Huang (1996) depart from this convention for interesting reasons that we shall shortly come to.

Madan and Unal (1994) then introduce two securities, with one having strict priority over the other in default. The securities are termed senior and junior debt with default pay-outs of $S(y)$ and $J(y)$ respectively. The aggregate pay-out rate is y and $p_S S(y) + (1 - p_S)J(y) = y$ where p_S is the proportion of senior debt calculated on the basis of face values. These conditions, along with the strict priority rule, imply that $J(y)$ is a holding of $1/(1 - p_S)$ call options on y at a strike of p_S, and $S(y)$ is a unit face bond less $1/p_S$ put options on y at a strike of p_S. Defining the forward prices to the senior and junior claims by $V(i, t, T) = P(i, t, T)/\Pi(t, T)$ for $i = S, J$ ($P(i, t, T)$ is the spot price of the claim) one may deduce from Equation (12.33) that

$$\frac{V(S, t, T) - V(J, t, T)}{1 - V(J, t, T)} = \frac{\int_0^1 S(y)q(y) \, dy - \int_0^1 J(y)q(y) \, dy}{1 - \int_0^1 J(y)q(y) \, dy} \quad (12.34)$$

Equation (12.34) forms the basis for estimating the density $q(y)$ for which Madan and Unal (1994) employ a Beta distribution.

Estimates for the risk neutral probability of no default may then be inferred from Equation (12.33) applied to either the senior or junior security on obtaining a proxy for $F(t, T)$ as

$$\hat{F}(t, T) = \frac{V(S, t, T) - \int_0^1 S(y)q(y)\,dy}{1 - \int_0^1 S(y)q(y)\,dy} \qquad (12.35)$$

Specific models for the probability, $F(t, T)$, of no default in the interval $[t, T]$ require a specification of the risk neutral hazard rate of the default process $\tilde{h}(t)$. Jarrow and Turnbull (1995) employed a Poisson process for which $\tilde{h}(t)$ was the constant $\lambda\mu$ of Equation (12.27). Madan and Unal (1994) employ a decreasing function of the level of equity value $S(t)$ relativised by the money market account, or $x(t) = S(t)/B(t)$. They suppose that

$$\tilde{h}(t) = \phi(x(t)) \qquad (12.36)$$

where $x(t)$ is a geometric Brownian motion with zero drift. They generalise the result of Jarrow and Turnbull (1995) for the Poisson process and show that, in general, one obtains $F(t, T)$ as

$$F(t, T) = E^Q\left[\exp\left(-\int_t^T \tilde{h}(u)\,du\right) \mid \mathcal{G}_t\right] \qquad (12.37)$$

where $\mathbb{G} = \{\mathcal{G}_t \mid 0 \leqslant t \leqslant T\}$ is a subfiltration of the $\mathbb{F}$ satisfying the usual conditions and such that the hazard rate process is adapted to $\mathbb{G}$. Equation (12.37) is similar in mathematical structure to equations for pricing risk free bonds with $\tilde{h}(t)$ now playing the role of the spot interest rate process. Madan and Unal (1994) use methods from the term structure literature to obtain explicit solutions for $F(t, T)$. Modelling the risk neutral hazard rate as a function of a Markov process leads, by standard methods, to a partial differential equation for $F(t, T)$ which one solves subject to the boundary condition that $F(t, t)$ is unity.

Equation (12.37), when coupled with Equation (12.33), for the special case of no partial payments, leads to the equation

$$P(t, T) = \Pi(t, T)F(t, T) = E_t^Q\left[\exp\left(-\int_t^T (r(u) + \tilde{h}(u))\,du\right) \mid \mathcal{G}_t\right] \quad (12.38)$$

under the maintained assumption of independence of the interest rate process from the process driving the hazard rate. Equation (12.38) shows that risky debt may be priced as if it were risk free provided we raise the discount rate from $r(t)$ to $r(t) + \tilde{h}(t)$ to account for the exposure to the hazard of default.

This result is generalised in Duffie *et al.* (1994) and Duffie and Singleton (1996) to allow for partial payments and to relax the independence assumption used in Madan and Unal (1994)'s derivation of Equation (12.38). They show that for a defaultable bond one may always obtain a discounting process $R(t)$ such that one may write

$$P(t, T) = E_t^Q\left[\exp\left(-\int_t^T R(u)\,du\right) \mid \mathcal{F}_t\right] \qquad (12.39)$$

The intuition for the construction of the process $R(t)$ is presented in Duffie and Singleton (1996). Consider a defaultable loan over a single period that pays 1 dollar with the risk neutral probability $(1 - \tilde{h})$ and makes the default payment y with probability $\tilde{h}$. The time 0 value of this claim is

$$V = \frac{1}{1+r}[(1 - \tilde{h}) + \tilde{h}y] \qquad (12.40)$$

or equivalently

$$V = \frac{1}{1+R} \qquad (12.41)$$

provided we define

$$R = \frac{r + \tilde{h}(1-y)}{1 - \tilde{h}(1-y)} \qquad (12.42)$$

Now we write Equation (12.42) in annualised terms for a period of length Δt to obtain

$$R\Delta t = \frac{r\Delta t + \tilde{h}\Delta t(1-y)}{1 - \tilde{h}\Delta t(1-y)} \qquad (12.43)$$

Dividing Equation (12.43) by Δt and taking limits as Δt tends to zero leads to the required definition for R of

$$R = r + \tilde{h}(1-y) \qquad (12.44)$$

The special case considered in Madan and Unal (1994) was for $y = 0$ and so the required discount rate was $r + \tilde{h}$. In general, when there is a partial pay-out of y the revised discount rate is reduced by the extent of this pay-out and the discount rate is as in Equation (12.44).

It is important to note the form of the definition of y used in adjusting the discount rate. Suppose the defaultable claim has a value of V_1 at time 1 if there is no default and a pay-out of y contingent on default. The time zero value is then

$$V = \frac{1}{1+r}[V_1(1 - \tilde{h}) + \tilde{h}y] \qquad (12.45)$$

On the other hand the corresponding equation for a non-defaultable claim discounted at R is

$$V = \frac{1}{1+R}V_1 \qquad (12.46)$$

Equating the right-hand sides of Equations (12.45) and (12.46) leads to the definition for R of

$$R = \frac{r + \tilde{h}(1 - y/V_1)}{1 - \tilde{h}(1 - y/V_1)} \qquad (12.47)$$

and annualising and taking limits we obtain the definition

$$R = r + \tilde{h}(1 - \tilde{y}) \qquad (12.48)$$

where the recovery rate $\tilde{y}$ is defined as the ratio of the pay-out y contingent on default to the value of the defaultable claim at time 1, if there were no default.

In all the models considered earlier of Merton (1974), Longstaff and Schwartz (1995), Jarrow and Turnbull (1995) and Madan and Unal (1994), recovery was defined in terms of the ratio of the pay-out to the default-free present value of the promise at maturity. Full payment required the delivery of default-free bonds with a face value and maturity matching that of the promise. Here, full payment amounts to the delivery of defaultable bonds facing the same future default risk structure as that embedded in the current promise. Full payment is then the value that would be obtained if default did not occur. This is an important distinction in the concept of recovery employed in obtaining the default risk adjusted discount rate for the valuation of defaultable claims. Given data on pay-outs, the problem of finding the adjusted discount rates is tantamount to the problem of valuing the defaultable claim. On the other hand, the use of adjusted discount rates makes it possible to study defaultable term structures directly using the method-ologies of modelling default-free term structures.

We close this section with a continuous time perspective leading to the formulation of the discount factor as defined in Equation (12.48). Consider first a unit face zero coupon bond of maturity T that has no partial payment contingent on default. The time T pay-off to this claim may be written as

$$P(t, T)(1 - \Delta(t)) = E^Q\left[\exp\left(-\int_t^T r(u)\,du\right)(1 - \Delta(T)) \mid \mathcal{F}_t\right] \quad (12.49)$$

where we recognise that if $\Delta(t) = 1$ then both sides of the equation are zero, while if $\Delta(t)$ is zero then $P(t, T)$ is the price of the claim by the valuation principle or the martingale condition. Multiplying Equation (12.49) by $\exp(-\int_0^t r(u)\,du)$, and observing that the left-hand side is the conditional expectation of a terminal random variable, we find that

$$U(t) = \exp\left(-\int_0^t r(u)\,du\right)P(t, T)(1 - \Delta(t))$$

is a martingale under the measure Q with respect to the filtration $\mathbb{F}$. Furthermore, as we have accounted for $(1 - \Delta(T))$ explicitly, we may take $P(T, T)$ to be unity. Now suppose that interest rates, asset prices and hazard rates are, in fact, adapted to a subfiltration $\mathbb{G}$ satisfying the usual conditions and generated by a finite-dimensional Markov process. The expectation of $U(t)$ conditional on $\mathcal{G}_t$, i.e. $\overline{U}(t) = E[U(t) \mid \mathcal{G}_t]$, is also a martingale as for $s < t$

$$\overline{U}(s) = E[U(s) \mid \mathcal{G}_s] = E[E[U(t) \mid \mathcal{F}_s] \mid \mathcal{G}_s] = E[U(t) \mid \mathcal{G}_s]$$

$$= E[E[U(t) \mid \mathcal{G}_t] \mid \mathcal{G}_s] = E[\overline{U}(t) \mid \mathcal{G}_s] \quad (12.50)$$

It is shown in Madan and Unal (1994) that $E^Q[(1 - \Delta(t)) \mid \mathcal{G}_t] = \exp(-\int_0^t \tilde{h}(u)\,du)$. This is a generalisation of what one would expect from a knowledge of Poisson processes. We then have that

$$\overline{U}(t) = \exp\left(-\int_0^t (r(u) + \tilde{h}(u))\,du\right)P(t, T)$$

is a martingale. It follows from the Feynmann–Kac theorem (see Duffie, 1988) that P satisfies the partial differential equation

$$DP = (r + \tilde{h})P = 0$$

subject to the boundary condition that $P(T, T)$ equals unity, where D is the infinitesimal generator associated with the Markov process generating the filtration $\mathbb{G}$.

In the presence of partial payments contingent on default, the associated partial differential equation can be shown to be (beginning with the equation for discounted gains as the initial martingale)

$$DP - (r + \tilde{h})P + y\tilde{h} = 0$$

Now we define $\tilde{y} = y/P$ and rewrite this differential equation as

$$DP - (r + \tilde{h}(1 - \tilde{y}))P = 0$$

and deduce by applying Feynmann–Kac once again that

$$P(t, T) = E_t^Q\left[\exp\left(-\int_0^t (r(u) + \tilde{h}(u)(1 - \tilde{y}(u)))\, du\right) \mid \mathcal{G}_t\right] \qquad (12.51)$$

Equation (12.51) shows the explicit adjustment required to the discount rate, if one is to value defaultable claims as if they were default-free.

12.7   The defaultable HJM model

The Heath, Jarrow and Morton (HJM) model for default-free term structures has the attractive feature that the risk neutral dynamics may be specified quite generally, from an empirical and statistical analysis of its martingale component. By using the methods of the previous section to adjust discount rates directly for any default exposure, one may directly model the term structure of defaultable bonds under a risk neutral measure, and thereby formulate a base model for pricing claims contingent on defaultable yield spreads across maturities or assets. The information content of such a model can be quite rich and we shall discuss these matters further in our closing section. Here we present the defaultable HJM model developed by Duffie and Singleton (1996).

Duffie and Singleton (1996) consider two cases for formulating the default pay-out; the first is where a fraction of market value is received at the time of default as discussed in Section 12.6, and the second where recovery is a fraction of an equivalent default-free security. Since, as we have noted earlier, other authors have formulated the problem in the second way, we restrict attention to this case here. Upon default at time t, creditors receive the proportion $y(t)$ of the zero coupon bond with price $\Pi(t, T)$.

The underlying filtration supports an m-dimensional standard Brownian motion and a default time jump process $\Delta(t)$ with risk neutral hazard rate $\tilde{h}(t)$ so that $\Delta(t) - \int_0^t \tilde{h}(u)\, du$ is a risk neutral martingale. The hazard rates and risk free interest rates are however adapted to just the subfiltration generated by the Brownian motions. The defaultable discount function is $P(t, T)$ while the default-free discount function is $\Pi(t, T)$. The two sets of forward rates are $f(t, T)$ for default-free and $g(t, T)$ for the defaultable forward rates. The money market account is $B(t)$ defined by $\exp(\int_0^t r(u)\, du)$ where $r(t) = f(t, t)$.

The risk neutral defaultable forward rate evolution is supposed to satisfy the equation

$$g(t, T) = g(0, T) + \int_0^t \mu(u, T)\,du + \int_0^t \sigma(u, T)\,d\tilde{W}(u) \qquad (12.52)$$

where μ and σ satisfy the usual regularity conditions. The process of discounted gains associated with the defaultable bond is given by

$$\Gamma(t) = \int_0^t \frac{y(u)}{B(u)}\Pi(u, T)\,d\Delta(u) + (1 - \Delta(t))\frac{P(t, T)}{B(t)} \qquad (12.53)$$

Evaluating the *dt* component of Equation (12.53) and setting it to zero we obtain

$$0 = \frac{y(t)}{B(t)}\Pi(t, T)\tilde{h}(t) - \frac{P(t, T)}{B(t)}(\tilde{h}(t) + r(t))$$

$$+ \frac{P(t, T)}{B(t)}\left(-\int_t^T \mu(t, u)\,du + \frac{1}{2}\left(\int_t^T \sigma(t, u)\,du\right)\left(\int_t^T \sigma(t, u)\,du\right)'\right) \qquad (12.54)$$

$$+ \frac{P(t, T)}{B(t)}F(t, t)$$

where $F(t, t)$ is introduced in the usual way to get the drift and diffusion integrals to run from t to T. Equation (12.54) may be rearranged as

$$0 = -\frac{P(t, T)}{B(t)}\tilde{h}(t) + \frac{P(t, T)}{B(t)}\gamma_{t,T} \qquad (12.55)$$

where

$$\gamma_{t,T} = F(t, t) - r(t) - \int_t^T \mu(t, u)\,du$$

$$+ \frac{1}{2}\left(\int_t^T \sigma(t, u)\,du\right)\left(\int_t^T \sigma(t, u)\,du\right)' + y(t)\frac{\Pi(t, T)}{P(t, T)}\tilde{h}(t)$$

Multiplication of Equation (12.55) by $B(t)/P(t, T)$ and the computation of partials with respect to T yields that the partial of $\gamma_{t,T}$ with respect to T must be zero or

$$\mu(t, T) = \sigma(t, T)\left(\int_t^T \sigma(t, u)\,du\right)' + y(t)\tilde{h}(t)\frac{\Pi(t, T)}{P(t, T)}(g(t, T) - f(t, T)) \qquad (12.56)$$

The risk neutral drift in this case is then given by the volatility structure of the defaultable bonds plus the defaultable and default-free discount functions, the recovery function and the risk neutral hazard rate of default. Statistically, one could identify all components, except the product of recovery and hazard rates of default. It may be interesting to investigate if priority may be used, as it was by Madan and Unal (1994), to separate out these components at the instantaneous level. Duffie and Singleton (1996) suggest the use of prices of credit derivatives for this purpose and give an example of a yield spread option.

12.8 Conclusions

The Merton (1974) model incorporated the impact of leverage, asset volatility and interest rates on the values of defaultable equity. The default times matched maturities and interest rates and volatility were constant. Longstaff and Schwartz (1995) extended these methods to allow for stochastic and correlated interest rates as well as accommodating pre-maturity default, thereby tying the default event to the firm as opposed to the security. In these approaches the default event was defined as a contingent claim written on underlying continuous processes. As a result, the stopping time of default was predictable, where the instantaneous likelihoods of default are either zero or unity, depending on whether the conditions defining default are met. Furthermore, variables can influence the likelihood of default only via their impact on the variables used in defining the default conditions.

Jarrow and Turnbull (1995), Artzner and Delbaen (1995), Madan and Unal (1994), Duffie *et al.* (1994), Duffie and Singleton (1996), and Duffie and Huang (1996) all employ the methodology of inaccessible stopping times for modelling the random time of default. In this methodology, default arrival rates (the compensator of the default time process) are modelled as absolutely continuous with respect to Lebesgue measure, and the impact of many variables on default probabilities can be incorporated by allowing them to influence arrival rates directly. Considerable flexibility is therefore gained. This flexibility has, as yet, not been fully exploited, although Madan and Unal (1994) do use it to reintroduce the impact of leverage and asset volatility on default probabilities, in this framework.

The statistical study of default and the methods employed to construct bond ratings contain considerable information on the impact of firm specific and macro-economic variables on default likelihoods. As the risk neutral and statistical measures are equivalent, one would expect many of these variables to be relevant to pricing as well. The strategy of modelling hazard rates directly is well suited to incorporating these impacts, and dynamic evolution need not be considered on all the variables introduced. One could, in the first instance, use cross-sectional information on various financial debt management ratios and model their impact on hazard rates, treating the ratios themselves as firm-specific constants. Admittedly, such an approach is myopic but the aesthetic displeasure of incomplete modelling should not be allowed to induce the exclusion of relevant information.

On the impact of macro-economic variables, certainly the role of interest rates and their correlation with asset values should be incorporated. In this regard, the interaction between the debt markets and other financial markets should be considered. Direct modelling of defaultable yield curves, as infinite dimensional Markov processes, evolving with dynamics completely specified as functions of the current yield curve, is a tempting and attractive strategy, based as it is on the idea that financial markets are so efficient that all variables of relevance to evolution have already made their presence felt in the current yield curve, and all we have to do is back out the required information from this yield curve. Although this may be possible, it may be easier to see the light of day if the variables came to us with their original names as opposed to being transformed into the shape of the defaultable yield curve.

References

Artzner, P. and Delbaen, F. (1995) Default risk insurance and incomplete markets, *Mathematical Finance*, **5**, 187–195.

Buonocore, A., Nobile, A. G. and Ricciardi, L. M. (1987) A new integral equation for the evaluation of first-passage-time probability densities. *Advances in Applied Probability*, **19**, 784–800.

Duffie, D. (1988) *Security Markets: Stochastic Models*. New York: Academic Press.

Duffie, D. and Huang, M. (1996) Swap rates and credit quality. *Journal of Finance*, **51**, 921–950.

Duffie, D., Schroder, M. and Skiadas, C. (1994) Recursive valuation of defaultable securities and the timing of resolution of uncertainty. Working Paper, *Kellogg Graduate School of Management*, Northwestern University.

Duffie, D. and Singleton, K. J. (1996) Modeling term structures of defaultable bonds. Working Paper, *Graduate School of Business, Stanford University*, Stanford CA.

Heath, D., Jarrow, R. and Morton, A. (1992) Bond pricing and the term structure of interest rates: a new methodology for contingent claims valuation. *Econometrica*, **60**, 77–105.

Hull, J. and White, A. (1995) The impact of default risk on the prices of options and other derivative securities. *Journal of Banking and Finance*, **19**, 299–322.

Jacod, J. and Shiryaev, A. N. (1980) *Limit Theorems for Stochastic Processes*. New York: Springer-Verlag.

Jarrow, R. and Turnbull, S. (1995) Pricing derivatives on financial securities subject to credit risk. *Journal of Finance*, **50**, 53–85.

Jarrow, R., Lando, D. and Turnbull, S. (1993) A Markov model for the term structure of credit spreads. Working Paper, *Graduate School of Management, Cornell University*.

Longstaff, F. and Schwartz, E. (1995) A simple approach to valuing risky fixed and floating rate debt, *Journal of Finance*, **50**, 789–819.

Madan, D. and Unal, H. (1994) Pricing the risks of default. To appear in *Review of Derivatives Research*.

Merton, R. C. (1974) On the pricing of corporate debt: the risk structure of interest rates. *Journal of Finance*, **29**, 449–470.

Nielsen, L. T., Saá-Requejo, J. and Santa Clara, P. (1993) Default risk and interest rate risk: the term structure of default spreads. Working Paper, *INSEAD*, 77305 Fontainebleau Cedex, France.

13

Non-parametric Methods and Option Pricing

Eric Ghysels, Éric Renault, Olivier Torrès and Valentin Patilea

13.1 Introduction

Derivative securities are widely traded financial instruments which inherit their statistical properties from those of the underlying assets and the features of the contract. The now famous Black and Scholes (1973) formula is one of the few cases where a call option is priced according to an analytical formula and applies to European-type contracts written on a stock which follows a geometric Brownian motion. The formula can be derived via a dynamic hedging argument involving a portfolio of a riskless bond and the underlying stock (see for example, Duffie, 1996, for more details).

Strictly speaking, the restrictive assumptions underlying the so-called Black–Scholes economy are rarely met. Indeed, among the violations one typically encounters and cites are: (1) volatility is time-varying; (2) trading is not costless or it faces liquidity constraints; (3) interest rates may be stochastic; (4) the stock has dividend payments, etc. It should also be noted that many derivative securities traded on exchanges and over-the-counter are not 'plain vanilla' but feature deviations from the basic European call contract design. In particular, some contracts feature early exercise privileges, i.e. so-called American-type options, some involve caps or floors, some involve multiple securities like swaps or quanto options, while others are written on fixed income securities instead of stocks, etc. Obviously, there are many extensions of the Black–Scholes model which take into account some of the deviations one encounters in practice either in terms of assumptions or contract design. In most cases, however, there is no longer an elegant analytical formula and the contract must by priced via numerical methods. Very often there are limitations to these numerical methods of approximation as well since they remain very specific and inherit many of the aforementioned restrictions which apply to the Black–Scholes model. These are some of the motivating reasons why statistical non-parametric methods are applied to option pricing. Indeed, these methods are appealing for the following reasons: (1) the formula is known but too complex to calculate numerically or (2) the pricing formula is unknown and (3) there is an abundance of data which makes the application of non-parametric methods attractive. The first applies, for instance, to the case of American-type options with stochastic volatility or barriers, to interest rate derivative security pricing, to 'look back' options, etc. The second applies because of incompleteness of markets, trading frictions or else the desire to leave unspecified either the stochastic properties of the underlying asset and/or the attitudes of agents towards risk.

There are a multitude of non-parametric methods, see for example, Silverman (1986) for an introduction to the statistical literature on the subject. In addition there are many ways to tackle the pricing of options via non-parametric methods. Moreover, there are many different types of option contracts, some of which require discussion of special features, like early exercise decisions in the case of American options. Given the large number of possibilities and the multitude of methods, we have to be selective in our survey of methods and applications. The literature is also rapidly growing. Recent papers include Aït-Sahalia (1993, 1996), Aït-Sahalia and Lo (1995), Baum and Barkoulas (1996), Bossaerts *et al.* (1995), Broadie *et al.* (1995, 1996), Ghysels and Ng (1996), Gouriéroux *et al.* (1994, 1995), Gouriéroux and Scaillet (1995), Hutchinson *et al.* (1994), Stutzer (1995), among others. To focus the survey we will restrict our attention only to options on equity.

We may be tempted to exclude any *a priori* economic knowledge from our econometric analysis and rely solely on the brute force of non-parametric techniques. This is the so-called *model-free* approach. These non-parametric methods and their use in the context of option pricing will be presented in Section 13.2. They mainly consist of estimating the relation between the dependent variable (usually the option price) and explanatory variables using non-parametric regression techniques. The function characterising the relationship to be estimated is chosen in a family of loosely defined functions according to an appropriate selection criterion. However, the application of these standard methods in the context of option pricing raises difficulties, some of which are not easy to overcome. This is one of several reasons why we may prefer to introduce some restrictions imposed by economic theory. These restrictions are quite often very mild and appear to be sensible, and they can be of great help to eliminate some of the difficulties met in the model-free approach. From a statistical point of view, these restrictions also call for other types of non-parametric methods, such as the non-parametric specification and estimation of equivalent martingale densities. This will be discussed in Section 13.3.

Despite the fact that the Black–Scholes (henceforth BS) framework fails to describe the behaviour of call prices or the exercise policy of investors when contracts are of American type, it still is the most widely used formula among practitioners. For instance, although it is believed that underlying stock prices are not represented by a log-normal diffusion because of time-varying volatility, the BS formula is used to measure this instantaneous variance, even though a necessary condition for the BS formula to be valid is that volatility is constant. The primary appeal of the BS formula is its simplicity and the belief among practitioners that it captures the variables relevant to price option contracts. Therefore, following the practitioner, an econometrician may find it convenient to use the BS formula as a benchmark for his analysis. In this context, non-parametric statistical techniques provide a way of filling the gap between the Black–Scholes and the real world. As we will show in Section 13.4, these methods can be called upon to 'correct' the BS formula so that it can adequately describe the behaviour of observed series.

13.2 Non-parametric model-free option pricing

In this section we present the simplest of all possible methods, which is probably also the purest in a statistical sense as it involves very little financial theory. Suppose we have a large data set with option prices and the features of the contracts such as strike, time to expiration, etc, and data of the underlying security. Such data sets are now commonly found and distributed to the academic and financial communities. Formally, a call is priced via

$$\Pi_t = f_1(S_t, K, T, t, X_t) \qquad (13.1)$$

where Π_t is the option price and S_t the price of the underlying asset both at time t, K the exercise price of the contract and T its expiration date, and finally X_t is a vector of variables affecting the price of the option contract. The latter may include underlying asset prices prior to t, as for instance in non-Markovian settings and/or latent variables which appear in stochastic volatility models. For the moment we ignore the fact that several contracts are listed on a daily basis which, in principle, should require a panel structure instead of a single time series. In Section 13.3 we will say more about panel structures. The pricing functional f_1 is assumed unknown, only its arguments are suggested by the set-up of the contract.[1] The purpose of applying non-parametric statistical estimation is to recover f_1 from the data. Obviously, this can only be justified if the estimation is applied to a situation where the regularity conditions for such techniques are satisfied. To discuss this let us briefly review the context of non-parametric estimation. In general, it deals with the estimation of relations such as

$$Y_i = g(Z_i) + u_i, \quad i = 1, \dots, n \qquad (13.2)$$

where, in the simplest case, the pair $((Y_i, Z_i), i = 1, \dots, n)$ is a family of i.i.d. random variables, and $E(u \mid Z) = 0$, so that $g(z) = E(Y \mid Z = z)$. The error terms $u_i, i = 1, \dots, n$, are assumed to be independently distributed, while g is a function with certain smoothness properties. Several estimation techniques exist, including kernel-based methods, smoothing splines, orthogonal series estimators such as Fourier series, Hermite polynomials and neural networks, among many others. We will focus here on the kernel-based methods for the purpose of exposition. Kernel smoothers produce an estimate of g at $Z = z$ by giving more weight to observations (Y_i, Z_i) with Z_i 'close' to z. More precisely, the technique relies on a *kernel*

---

[1] Non-parametric techniques for selecting the arguments of a non-parametric regression function have recently been proposed by Aït-Sahalia *et al.* (1995), Gouriéroux *et al.* (1994), Lavergne and Vuong (1996) among others.

function, K, which acts as a weighting scheme (it is usually a probability density function, see Silverman, 1986, p. 38) and a *smoothing parameter* λ which defines the degree of 'closeness' or neighbourhood. The most widely used kernel estimator of g in Equation (13.2) is the Nadaraya–Watson estimator defined by

$$\hat{g}_\lambda(z) = \frac{\sum_{i=1}^n K\left(\frac{Z_i - z}{\lambda}\right) Y_i}{\sum_{i=1}^n K\left(\frac{Z_i - z}{\lambda}\right)} \tag{13.3}$$

so that $(\hat{g}_\lambda(Z_1), \ldots, \hat{g}_\lambda(Z_n))' = W_n^K(\lambda)Y$, where $Y = (Y_1, \ldots, Y_n)'$ and W_n^K is an $n \times n$ matrix with its (i, j)th element equal to $K((Z_j - Z_i)/\lambda)/\sum_{k=1}^n K((Z_k - Z_i)/\lambda)$. W_n^K is called the *influence matrix* associated with the kernel K.

The parameter λ controls the level of neighbouring in the following way. For a given kernel function K and a fixed z, observations (Y_i, Z_i) with Z_i far from z are given more weight as λ increases; this implies that the larger we choose λ, the less $\hat{g}_\lambda(z)$ is changing with z. In other words, the degree of smoothness of $\hat{g}_\lambda$ increases with λ. As in parametric estimation techniques, the issue here is to choose K and λ in order to obtain the best possible fit. A natural measure of the goodness of fit at $Z = z$ is the mean squared error $(\mathrm{MSE}(\lambda, z) = E[(\hat{g}_\lambda(z) - g(z))^2])$, which has a bias/variance decomposition similar to parametric estimation. Of course both K and λ have an effect on $\mathrm{MSE}(\lambda, z)$, but it is generally found in the literature that the most important issue is the choice of the smoothing parameter.[2] Indeed, λ controls the relative contribution of bias and variance to the mean squared error; high λs produce smooth estimates with a low variance but a high bias, and vice versa. It is then crucial to have a good rule for selecting λ. Several criteria have been proposed, and most of them are transformations of $\mathrm{MSE}(\lambda, z)$. We may simply consider $\mathrm{MSE}(\lambda, z)$, but this criterion is local in the sense that it concentrates on the properties of the estimate at point z. We would generally prefer a global measure such as the *mean integrated squared error* defined by $\mathrm{MISE}(\lambda) = E[\int (\hat{g}_\lambda(z) - g(z))^2 \, \mathrm{d}z]$, or the *sup mean squared error* equal to $\sup_z \mathrm{MSE}(\lambda, z)$, etc. The most frequently used measure of deviation is the sample mean squared error $M_n(\lambda) = (1/n) \sum_{i=1}^n [\hat{g}_\lambda(Z_i) - g(Z_i)]^2 \omega(Z_i)$, where $\omega(\cdot)$ is some known weighting function. This criterion only considers the distances between the fit and the actual function g at the sample points Z_i. Obviously, choosing $\lambda = \tilde{\lambda}_n \equiv \operatorname{argmin}_\lambda M_n(\lambda)$ is impossible to implement since g is unknown. The strategy consists of finding some function $m_n(\cdot)$ of λ (and of $((Y_i, Z_i), i = 1, \ldots, n)$) whose argmin is denoted $\hat{\lambda}_n$, such that $|\tilde{\lambda}_n - \hat{\lambda}_n| \to 0$ a.s. as $n \to \infty$. For a review of such functions m_n, see Härdle and Linton (1994, section 4.2).[3] The most widely used m_n function is the *cross-validation* function

---

[2] For a given λ, the most commonly used kernel functions produce more or less the same fit. Some measures of relative efficiency of these kernel functions have been proposed and derived, see Härdle and Linton (1994, p. 2303) and Silverman (1986, section 3.3.2).
[3] See also Silverman (1986, section 3.4), Andrews (1991) and Wand and Jones (1995).

$$m_n(\lambda) = CV_n(\lambda) \equiv \frac{1}{n} \sum_{i=1}^{n} [Y =_i -\hat{f}_\lambda^{(-i)}(Z_i)]^2$$

where $\hat{g}_\lambda^{(-i)}(z)$ is a Nadaraya–Watson estimate of $g(z)$ obtained according to Equation (13.3) but with the ith observation left aside. Craven and Wahba (1979) proposed the *generalised cross-validation* function with

$$m_n(\lambda) = GCV_n(\lambda) \equiv \frac{n^{-1} \sum_{i=1}^{n}}{=} [Y_i - \hat{g}_\lambda(Z_i)]$$

where W_n is the influence matrix.[4]

Another important issue is the convergence of the estimator $\hat{g}_{\lambda_n}(z)$. Concerning the Nadaraya–Watson estimate (Equation (13.3)), Schuster (1972) proved that under some regularity conditions, $\hat{g}_{\lambda_n}(z)$ is a consistent estimator of $g(z)$ and is asymptotically normally distributed.[5] Therefore, when the argmin $\hat{\lambda}_n$ of $m_n(\lambda)$ is found in the set Λ_n (see footnote 5), we obtain a consistent and asymptotically normal kernel estimator $\hat{g}_{\lambda_n}(z)$ of $g(z)$, which is optimal in the class of the consistent and asymptotically Gaussian kernel estimators for the criterion $M_n(\lambda)$.[6]

When the errors are not spherical, the kernel estimator remains consistent and asymptotically normal. The asymptotic variance is affected, however, by the correlation of the error terms. Moreover, the objective functions for selecting λ such as CV_n or GCV_n do not provide optimal choices for the smoothing parameters. It is still not clear what should be done in this case to avoid over- or undersmoothing.[7] One solution that has been suggested consists of modifying the selection criterion (CV_n or GCV_n) in order to derive a constant estimate of M_n. An alternative strategy tries to orthogonalise the error term and apply the usual selection rules for λ. When the autocorrelation function of u is unknown, one has to make the transformation from sample estimates obtained from a first step smooth. In that view, the second alternative seems to be more tractable. Altman (1987, 1990) presents some simulation results which show that, in some situations, the pre-whitening method seems to work relatively well. However, there is no general result on the efficiency of the procedure. See also Härdle and Linton (1994, section 5.2) and Andrews (1991, section 6).

When the observations (Y, Z) are drawn from a stationary dynamic bivariate process, Robinson (1983) provides conditions under which kernel estimators of regression functions are consistent. He also gives some central limit theorems which ensure the asymptotic normality of the estimators. The

---

[4] This criterion generalises CV_n since GCV_n can be written as $n^{-1} \sum_{i=1}^{n} [Y_i - \hat{g}_\lambda^{(-i)}(Z_i)]^2 a_{ii}$, where the a_{ii}s are weights related to the influence matrix. Moreover, GCV_n is invariant to orthogonal transformations of the observations.

[5] The regularity conditions bear on the smoothness and continuity of g, the properties of the kernel function K, the conditional distribution of Y given Z, the marginal distribution of Z, and the limiting behaviour of $\hat{\lambda}_n$. The class of $\hat{\lambda}_n$s which satisfy these regularity conditions is denoted Λ_n.

[6] By definition, the choice $\lambda = \lambda_n^*$ is optimal for the criterion $D(\lambda)$ if $D(\lambda_n^*)/\inf_{\lambda \in \Lambda_n} D(\lambda) \xrightarrow[n \to \infty]{a.s.} 1$.

[7] Altman (1990) shows that when the sum of the autocorrelations of the error term is negative (positive), then the functions CV_n and GCV_n tend to produce values for λ that are too large (small), yielding oversmoothing (undersmoothing).

conditions under which these results are obtained have been weakened by Singh and Ullah (1985). These are mixing conditions on the bivariate process (Y, Z). For a detailed treatment, see Györfy *et al.* (1989). Györfy *et al.* (1989, ch. 6) also discuss the choice of the smoothing parameter in the context of non-parametric estimation from time series observations. In particular, if the error terms are independent, and when $\hat{\lambda}_n = \text{argmin}_{\lambda \in \Lambda_n} CV_n(\lambda)$, then under certain regularity conditions $\hat{\lambda}_n$ is an optimal choice for λ according to the *integrated squared error*, $\text{ISE}(\lambda) = \int [\hat{g}_\lambda(z) - g(z)]^2 \, dz$ (see Györfy *et al.*, 1989, corollary 6.3.1). Although the function $CV_n(\lambda)$ can produce an optimal choice of λ for the criterion $M_n(\lambda)$ in some particular cases (such as the pure autoregression, see Härdle and Vieu, 1992, and Kim and Cox, 1996), there is no general result for criteria such as $\text{MISE}(\lambda)$ or $M_n(\lambda)$. For studies of the performance of various criteria for selecting λ in the context of dependent data, see Cao *et al.* (1993).

In applications involving option price data we have also correlated *non-stationary* data. Indeed S_t, which is one of the arguments of f_1 in Equation (13.1), is usually not a stationary process. Likewise, variables entering X_t may be non-stationary as well. Moreover, characterising the correlation in the data may also be problematic. Indeed, the relevant time scale for the estimation of f_1 is not calendar time, as in a standard time series context, but rather the time to expiration of the contracts which are sampled sequentially through the cycle of emissions. It becomes even more difficult once it is realised that at each time t several contracts are listed simultaneously and trading may take place only in a subset of contracts. Some of these technical issues can be resolved. For instance, while S is non-stationary the variable (S/K) is found to be stationary as exercise prices bracket the underlying asset price process. This suggests an alternative formulation of Equation (13.1) as $\Pi_t = f_2(S_t/K, K, T, t, X_t)$. Moreover, under mild regularity conditions f_1 is homogeneous of degree one in (S, K) (see Broadie *et al.*, 1996, or Garcia and Renault, 1996). Under such conditions we have:

$$\Pi_t/K = f_3(S_t/K, T, t, X_t) \qquad (13.4)$$

A more difficult issue to deal with is the correlation in the data. Indeed, while it is easy to capture the serial correlation in calendar time it is tedious to translate and characterise such dependence in a time to maturity scale (see Broadie *et al.*, 1996, for further discussion). Furthermore, the panel data structure of option contracts even worsens the dependence characteristics. Finally, we also face the so-called 'curse of dimensionality' problem. Non-parametric kernel estimators of regression functions $Y = g(Z)$, where Z is a vector of dimension d, as f_3 in Equation (13.4), are *local* smoothers in the sense that the estimate of g at some point z depends only on the observations (Z_i, Y_i) with Z_i in a neighbourhood $\mathcal{N}(z)$ of z. The so-called curse of dimensionality relates to the fact that, if we measure the degree of localness of a smoother by the proportion of observations (Z_i, Y_i) for which Z_i is in $\mathcal{N}(z)$, then the smoother becomes less local when d increases, in the sense that for a fixed degree of localness $\mathcal{N}(z)$ increases in size as the dimension of Z increases. Consequently, the precision of the estimate deteriorates as we add regressors in g, unless the sample size increases drastically.[8] This problem arises in our context as X_t may contain many variables. A good example is the case of non-parametric estimation of option pricing models with stochastic volatility, a latent variable which requires filtering from past squared returns of the underlying asset (see below for further discussion).

The homogeneity property of option prices helps to reduce the dimension of the pricing function f_1 by eliminating one of its arguments. Moreover, in most pricing models, the expiration date T and the calendar date t affect Π_t through their difference $\tau \equiv T - t$, a variable called time to expiration or alternatively time to maturity. Therefore the non-parametric regression to be estimated becomes:

$$\Pi_t/K = f_3(S_t/K, \tau, X_t) \tag{13.5}$$

Let us elaborate further now on the specification of the vector of variables X_t affecting the option price. Examples of variables that might enter X_t are series such as random dividends or random volatility. Dividend series are observable while volatility is a latent process. This raises a number of issues we need to discuss here as there are fundamental differences between the two cases. In principle, one can filter the latent volatility process from the data, using series on the underlying asset. Obviously we need a parametric model if we were to do this in an explicit and optimal way. This would be incompatible however with a non-parametric approach. Hence, we need to proceed somehow without making specific parametric assumptions. In principle, one could consider a non-parametric fit between $(S/K)_t$ and past squared returns $(\log S_{t-j} - \log S_{t-j-1})^2, j = 1, \ldots, L$, for some finite lag L, resulting in the following $(L+1)$-dimensional non-parametric fit

$$\Pi_t/K = f_3(S_t/K, T, t, (\log S_{t-j} - \log S_{t-j-1})^2, j = 1, 2, \ldots, L) \tag{13.6}$$

It is clear that this approach is rather unappealing as it would typically require a large number of lags, say $L = 20$ with daily observations. Hence, we face the curse of the dimensionality problem discussed before. A more appealing way to proceed is to summarise the information contained in past squared returns (possibly the infinite past). Broadie *et al.* (1996) consider three different strategies using: (*a*) historical volatilities; (*b*) EGARCH volatilities; and (*c*) implied volatilities. Each approach raises technical issues, some of which are relatively straightforward to deal with, while others are more tedious. For example, using GARCH or EGARCH models raises several issues: (1) are equations of EGARCH models compatible with the unspecified asset return processes generating the data? (2) How does parameter estimation of the GARCH process affect non-parametric inference? And (3) how do the weak convergence results also affect the non-parametric estimation? Broadie *et al.* (1996) discuss the details of these issues and illustrate them with an empirical example in which the three aforementioned approaches yield the same results.

The non-parametric regression f_3 discussed so far does not rely, at least directly, on a theoretical financial model. Yet it is possible to use the non-

---

[8] For more details on the curse of dimensionality and how to deal with it, see Hastie and Tibshirani (1990), Scott (1992, ch. 7) and Silverman (1986, pp. 91–94).

parametric estimates to address certain questions regarding the specification of theoretical models, namely questions which can be formulated as inclusion or exclusion of variables in the non-parametric option pricing regression. We may illustrate this with an example drawn from Broadie *et al.* (1996). For European type options there has been considerable interest in formulating models with stochastic volatility (see for example, Hull and White, 1987, among many others) while there has been relatively little attention paid to cases involving stochastic dividends. It is quite the opposite with American type options. Indeed, the widely traded S&P100 Index option or OEX contract has been extensively studied, see in particular Harvey and Whaley (1992), with exclusive emphasis on stochastic dividends (with fixed volatility). This prompted Broadie *et al.* (1996) to test the specification of OEX option pricing using the non-parametric methods described in this section combined with tests described in Aït-Sahalia *et al.* (1995). Hence, they tested the relevant specification of the vector X_t whether it should include dividends and/or volatility (where the latter is measured via one of the three aforementioned proxies). They found, in the case of the OEX contract, that both stochastic volatility and dividends mattered. It implies that either ignoring volatility or dividends results in pricing errors, which can be significant, as Broadie *et al.* show. This is an important illustration on how to use this so-called model-free approach to address specification of option pricing without much financial theory content.

To conclude, we should mention that there are several applications of the techniques discussed here which can be found in Aït-Sahalia and Lo (1995) as well as Broadie *et al.* (1995, 1996). The former study the European option on the S&P500 contract, while the latter study the American contract on the S&P100. By using slightly different techniques, Hutchinson *et al.* (1994) achieve the same objective.

13.3 Non-parametric specification of equivalent martingale measures

An obvious difficulty for the model-free option pricing set-up in the previous section is the so-called panel structure of option prices data. Namely, one typically observes several simultaneously traded contracts (with various exercise prices and maturity dates) so that the option pricing formula of interest must involve two indexes:

$$\Pi_{it} = f(S_t, K_i, T_i, t, X_{it}) \tag{13.7}$$

where $i = 1, 2, \ldots, I_t$ describes the (possibly large) set of simultaneously quoted derivative contracts at time t written on the same asset (with price S_t at time t).[9] In such a case, non-parametric model free option pricing becomes

---

[9] For instance Dumas *et al.* (1995) consider S&P500 Index option prices traded on the Chicago Board Options Exchange (CBOE) during the period June 1988 through December 1993. After applying three exclusionary criteria to avoid undesirable heterogeneity, they find quotes for an average of 44 option series during the last half-hour each Wednesday.

quickly unfeasible since it is not able to capture a large set of crucial restrictions implied by arbitrage. Indeed, as stressed by Merton (1973), any option pricing research must start from deducing a set of restrictions that are necessary conditions for a formula to be consistent with a rational pricing theory. Fitting option pricing formulae using purely (model-free) statistical methodologies therefore foregoes imposing an important feature of derivative asset markets, namely: if a security *A* is dominant over a security *B* (that is the return on *A* will be at least as large as on *B* in all states of the world and exceed the return on *B* for some states), then any investor willing to purchase security *B* would prefer to purchase *A*. A first example of restrictions stressed by Merton (1973) for European call options prices is:

$$K_2 > K_1 \Rightarrow f(S_t, K_2, T, t, X_{it}) \leqslant f(S_t, K_1, T, t, X_{it})$$

and, if no payouts (e.g. dividends) are made to the underlying asset (e.g. a stock) over the life of the option

$$f(S_t, K, T, t, X_t) \geqslant \max[0, S_t - KB(t, T)]$$

where $B(t, T)$ is the price of a riskless pure discount bond which pays one dollar $T - t$ periods from now, date t).

The only way for a pricing scheme to take into account the necessary conditions for a formula to be consistent with a rational pricing theory is to incorporate at a convenient stage the requirement that the derivative asset price $f(S_t, K_i, T_i, t, X_{it})$ has to be related to its terminal pay-off, for instance $\max[0, S_T - K]$ in the case of a European call option. Fortunately, modern derivative asset pricing theory provides us with a versatile tool to do this using equivalent martingale measures. Roughly speaking, the Harrison and Kreps (1979) theory ensures the equivalence between the absence of arbitrage and the existence of a probability measure Q with the property that the discounted price processes are martingales under Q. Hence, for a European option with strike K and maturity T we have

$$f(S_t, K, T, t, X_t) = B(t, T)E^Q[(S_T - K)^+ \mid X_t, S_t] \tag{13.8}$$

where the expectation operator E^Q is defined with respect to the pricing probability measure Q.[10] The vector of variable X_t assumes here again the role it played in the previous section, namely a set of state variables relevant to the pricing of the option. The fundamental difference between the non-parametric methods described in the previous section and those which rely on the Harrison and Kreps theory is that the non-parametric statistical inference focuses on the conditional expectation operator $E^Q[\cdot \mid X_t, S_t]$ instead of the pricing function $f(S_t, K, T, t, X_t)$. It is important to note, of course, that in general the Q-conditional probability distribution of S_T given X_t, S_t, coincides with the Data Generating Process (DGP characterised by a probability

---

[10] For further details, see for example, Duffie (1996). We do not discuss here: (1) the assumptions of frictionless markets which ensure the equivalence between the absence of arbitrage and the existence of such a pricing measure Q, and (2) the interpretation of Q which is unique in the case of complete markets and is often called a 'risk neutral probability' when there is no interest rate risk.

measure denoted P hereafter). Early contributions assumed that:

$$E^Q[\cdot \mid X_t, S_t] = E^P[\cdot \mid X_t, S_t] \tag{13.9}$$

(see for instance Engle and Mustafa, 1992, or Renault and Touzi, 1996) but recently several attempts were made to estimate $E^Q[\cdot \mid X_t, S_t]$ without assuming that Q coincides with P (see for example, Rubinstein, 1994, Abken *et al.*, 1996, and Aït-Sahalia and Lo, 1995). For instance, it is now well-known (see for example, Breeden and Litzenberger, 1978, and Huang and Litzenberger, 1988, p. 140) that there is a one-to-one relationship between a European call option pricing function $f(S_t, \cdot, T, t, X_t)$ as a function of the strike price K and the pricing probability measure $Q \mid X_t, S_t$ via

$$Q\left[\frac{S_T}{S_t} \geqslant \frac{K}{S_t} \mid X_t, S_t\right] = -\frac{1}{B(t, T)} \frac{\partial f}{\partial K}(S_t, K, T, t, X_t) \tag{13.10}$$

Hence, observing European call option prices for any strike K, it is possible to recover the pricing probability measure $Q \mid X_t, S_t$ or the corresponding pricing operator $E^Q[\cdot \mid X_t, S_t]$. This forms the basis for consistent non-parametric estimation of this operator without the restrictive assumption (Equation (13.4)) and using options price data, not only time series ($t = 1, 2, \ldots, T$) but also cross-sections ($i = 1, 2, \ldots, I_t$).

Such inference may be developed within two paradigms: Bayesian (covered in Section 13.3.1) or classical (Section 13.3.2 below). In the former we treat the probability distribution as a random variable, hence the reference to Bayesian analysis. Obviously one needs to find a flexible class that covers a large set of pdfs. In Section 13.3.1 we will present such a class, which enables one to characterise the martingale restrictions of option pricing as well as their panel structure. In Section 13.3.2 we present another approach which builds on the estimation of option pricing formulae presented in the previous section. Both approaches have advantages as well as drawbacks which we will discuss.

13.3.1   Non-parametric Bayesian specification of equivalent martingale measures

The basic ideas presented in this section were introduced by Clément *et al.* (1993) and extended more recently by Renault (1996) and Patilea and Renault (1995).[11] Let us reconsider a European option with price $\Pi_t(K)$ written as:

$$\left.\begin{array}{l} \Pi_t(K) = B(t, T)E^Q[(S_T - K)^+ \mid S_t, X_t] \\[2mm] \qquad = B(t, T)\displaystyle\int_{\mathcal{S}} (s_T - K)^+ Q_t(\mathrm{d}s_T) \end{array}\right\} \tag{13.11}$$

where $\mathcal{S}$ is the set of all possible values of S_t. First, we should note that even when markets are complete, one may not observe the full set of securities which complete the market and therefore one is not able to determine unambiguously the pricing probability measure Q_t. A non-parametric

---

[11] In this section we follow closely the analysis developed in Renault (1996) and Patilea and Renault (1995).

Bayesian methodology views this measure as a random variable defined on an abstract probability space (Ω, a, P), taking values in the set $\mathcal{P}(S)$ of all probability distributions on $(S, \mathcal{B}(S))$. If we denote $Q_t(\cdot, \omega)$ as a realisation of this random variable, then the option pricing formula (Equation (13.11)) becomes

$$\Pi_t(K, \omega) = B(t, T) \int_S (s_T - K)^+ Q_t(ds_T, \omega) \qquad (13.12)$$

A good class of distributions to characterise the random probability Q_t is the Dirichlet process, which is a distribution on $(\mathcal{P}(S), \mathcal{B}(\mathcal{P}(S)))$, introduced by Ferguson (1973). More specifically: a random probability π on $(S, \mathcal{B}(S))$ is called a Dirichlet process with parameter λQ_t^0 where $\lambda > 0$ and $Q_t^0 \in \mathcal{P}(S)$ if, for any measurable partition $B_1, B_2, \dots, B_L$ of S, the random vector $(\pi(B_\ell), 1 \leqslant \ell \leqslant L)$ has a Dirichlet distribution with parameters $(\lambda Q_t^0(B_\ell), 1 \leqslant \ell \leqslant L)$. If π is a Dirichlet process with parameter λQ_t^0, we write hereafter $\pi \rightsquigarrow Di(\lambda Q_t^0)$.[12] We may give an interpretation to Q_t^0 as a mean value of the process and to λ as a precision of the process around this mean value. Large values of λ make the realisations of π more concentrated around Q_t^0. For $\lambda = \infty$ the realisations of the Dirichlet process are, with probability one, equal to Q_t^0. To elaborate on the moment properties and the asymptotic behaviour of Dirichlet processes let us consider a real-valued function f defined on S and integrable w.r.t. Q_0. If $\pi \rightsquigarrow Di(\lambda Q_0)$ we define the random variable:

$$\Pi(f) = \int_S f(s)\Pi(ds, \omega)$$

It can be shown (see Clément *et al.*, 1993) that:

$$
\begin{aligned}
E(\Pi(f)) &= E^{Q_0} f \\
\mathrm{Var}\,(\Pi(f)) &= (1 + \lambda)^{-1} \mathrm{Var}^{Q_0}(f) \\
\mathrm{Cov}\,(\Pi(f_1), \Pi(f_2)) &= (1 + \lambda)^{-1} \mathrm{Cov}^{Q_0}(f_1, f_2)
\end{aligned}
$$

Asymptotic normality has been established by Lo (1987). In particular if $\Pi_n \rightsquigarrow Di(\lambda_n, Q_0)$ and $\lambda_n \to \infty$

$$\sqrt{\lambda_n}(\Pi_n(f) - E^{Q_0}(f)) \overset{L}{\longrightarrow} \mathcal{N}(0, \mathrm{Var}^{Q_0}(f))$$

It proves that for λ sufficiently large, that is for random errors around the basic pricing model (defined by Q_0) relatively small, that one can characterise their distribution by the first two moments, as Clément *et al.* (1993) did.

---

[12] It is worth recalling that the Dirichlet distribution is a multivariate extension of the Beta distribution on $[0,1]$. More precisely, the Dirichlet distribution on the simplex $\{(p_1, p_2, \dots, p_L), p_\ell \geqslant 0, \sum p_\ell = 1\}$ is characterised by the pdf

$$g(p_1, p_2, \dots, p_L) = \frac{\Gamma(\eta_1 + \eta_2 + \dots + \eta_L)}{\Gamma(\eta_1)\Gamma(\eta_2)\dots\Gamma(\eta_L)} p_1^{\eta_1 - 1} p_L^{\eta_L - 1}$$

where $(\eta_1, \eta_2, \dots, \eta_L)$ are given non-negative parameters.

Within this framework, we can introduce random error terms *around* an option pricing model defined by Q_t^0 which provides option prices

$$\tilde{\Pi}_t(K) = B(t, T) \int_S (s_T - K)^+ Q_t^0(ds_T) \tag{13.13}$$

Let us assume for the moment that λ and the parameters defining Q_t^0 are included in the statistician's information set I_t^S at time t while the trader's information set includes (for completeness) both I_t^S and ω. Then, if the pricing probability measure Q_t is described by a Dirichlet model around Q_t^0, i.e. $Q_t \mid I_t^S \leadsto \mathcal{D}i(\lambda Q_t^0)$, then using the properties of Dirichlet processes we know that the expectation of $\pi_t(\omega)$ *with respect to the draw* of ω is

$$\tilde{\Pi}_t(K) = E^\omega \Pi_t(K, \omega) = B(t, T) E^{Q_t^0}(S_t - K)^+$$

Hence, for a set of strike prices $K_j, j = 1, 2, \ldots, J$, the random variables

$$u_t(K_j, \omega) = \Pi_t(K_j, \omega) - \tilde{\Pi}_t(K_j)$$

are zero-mean error terms whose joint probability distribution $(u_t(K_j, \omega))_{1 \leqslant j \leqslant J}$ can be easily deduced from the properties of the Dirichlet process. This can be used to characterise the joint probability distribution of error terms and especially their heteroscedasticity, autocorrelation, skewness, kurtosis, etc, whatever the cross-sectional set of option prices written on the same asset we observe (calls, puts, various strike prices, various maturities . . .). For the sake of simplicity of the presentation let us consider a simple one-period model.[13] First we will randomise the risk-neutral probability around the lognormal distribution of the Black–Scholes model. We suppose that $B(t, T)$ and σ are known. Let us introduce $\tilde{S}_t$ as a latent price of the underlying asset which will appear in the definition of the parameters of the Dirichlet process, more precisely in Q_t^0 (we will justify the use of a latent $\tilde{S}_t$ later). The resulting model then is

$$\left. \begin{aligned} & Q_t^0 = LN(\log \tilde{S}_t - \log B(t, T) - \sigma^2(T - t)/2, \sigma^2(T - t)) \\ & S_T \mid (Q_t(\cdot, \omega), I_t^S, \tilde{S}_t) :\leadsto: Q_t(\cdot, \omega) \\ & Q_t \mid \tilde{S}_t :\leadsto: \mathcal{D}i(\lambda Q_t^0) \end{aligned} \right\} \tag{13.14}$$

In this context S_T is what is usually called a sample of size one of the Dirichlet process π, i.e. the conditional distribution of S_T given the realisation $Q_t(\cdot, \omega)$, is $Q_t(\cdot, \omega)$. It can be shown (see Ferguson, 1973) that the marginal distribution of S_T is Q_t^0. In this model, given $\tilde{S}_t$, the price $\Pi_t(K, \omega)$ of a European call option with maturity date T and strike K is defined as in Equation (13.12). In particular the stock price observed at time t is $S_t = S_t(\omega) = \Pi_t(0, \omega)$. With probability one, the observed price will not coincide with the latent price $\tilde{S}_t$. We introduced $\tilde{S}_t$ not only to make the option pricing formula (Equation (13.12)) coherent but also because it is an interesting variable to be taken into

---

[13] Extensions to more complicated multi-period models appear in Clément *et al.* (1993) and Patilea and Renault (1995).

account.[14] Indeed, if we accept the existence of non-synchronous trading, it is clear that agents willing to buy and sell options have in mind a latent price of the underlying asset. This price $\tilde{S}_t$ can be viewed as a latent factor, which has various interpretations previously encountered in the option pricing literature. Indeed, Manaster and Rendleman (1982) argue for instance that 'just as stock prices may differ, in the short run, from one exchange to another (. . .), the stock prices implicit in option premia may also differ from the prices observed in the various markets for the stock. In the long run, the trading vehicle that provides the greatest liquidity, the lowest trading costs, and the least restrictions is likely to play the predominant role in the market's determination of the equilibrium values of underlying stocks'. Moreover, 'investors may regard options as a superior vehicle' for several reasons, like trading costs, short sales, margin requirements . . . Hence, option prices involve *implicit stock prices* that may be viewed as the option market's assessment of equilibrium stock values and may induce a reverse causality relationship from option market to stock market.[15]

Henceforth, we will write the models in terms of returns because this setting is better suited for dynamic extensions. We can write Equation (13.14) as follows

$$Q_t^0 = LN(-\log B(t, T) - \sigma^2(T - t)/2, \sigma^2(T - t))$$

$$Z_T \stackrel{\text{def}}{=} \frac{S_T}{\tilde{S}_t} \mid (Q_t(\cdot, \omega), I_t^S, \tilde{S}_t) : \leadsto : z_T \mid (Q_t(\cdot, \omega)) : \leadsto : Q_t(\cdot, \omega)$$

$$Q_t \mid \tilde{S}_t : \leadsto : Di(\lambda Q_t^0)$$

The distribution of Z_T given $\tilde{S}_t$ is Q_t^0. We should note that this distribution does not depend on $\tilde{S}_t$ and therefore, the option pricing formula will be homogeneous with respect to $(\tilde{S}_t, K)$, an issue which was deemed important in Section 13.2 to conduct statistical analysis. Indeed, given $\tilde{S}_t$ the price of a call option written in Equation (13.12) becomes

[14] In their paper Clément et al. (1993) neglected the incoherence in their option pricing formula. They considered Q_t^0 based on the observed stock price S_t. Thus, at time t, almost surely, they have two stock prices: S_t and $\Pi_t(0, \omega)$.

[15] See also Longstaff (1995) for a related interpretation. While Manaster and Rendleman (1982) and Longstaff (1995) compute implicit stock prices through the *BS* option pricing formula, Patilea et al. (1995) propose an econometric approach in a Hull and White (1987) (HW) setting which is also based on the concept of stock prices implicit in option prices but without choosing between the above theoretical explanations. Indeed, following the state variables methodology set forth by Renault (1996) they argue that if we observe mainly two liquid option contracts at each date: one near the money and another one more speculative (in or out) we need to introduce two unobserved state variables: the first one is stochastic volatility (to apply HW option pricing) and the second one is an 'implicit' stock price which is taken into account to apply the HW option pricing formula. They show that even a slight discrepancy between S_t and $\tilde{S}_t$ (as small as 0.1%, while Longstaff, 1995, documents evidence of an average discrepancy of 0.5%) may explain a sensible skewness in the volatility smile.

$$\left. \begin{aligned} \Pi_t(K, \omega) = \Pi_t(\tilde{S}_t, K, \omega) &= B(t, T) \int_S (\tilde{S}_t z_T - K)^+ Q_t(\mathrm{d}z_T, \omega) \\ &= B(t, T)\tilde{S}_t \int_S (z_T - K/\tilde{S}_t)^+ Q_t(\mathrm{d}z_T, \omega) \end{aligned} \right\} \quad (13.15)$$

Using the properties of the Dirichlet process functionals we can compute the moments (conditionally on $\tilde{S}_t$) of $\pi_t(K, \omega)$

$$E(\Pi_t(K, \omega)) = BS(\tilde{S}_t, K, \sigma)$$

$$\mathrm{Var}\,(\Pi_t(K, \omega)) = (1 + \lambda)^{-1}\tilde{S}_t^2 B(t, T)^2 [E^{Q_t^0}(f^2) - (E^{Q_t^0}(f))^2]$$

$$\mathrm{Cov}\,(\Pi_t(K_1, \omega), \Pi_t(K_2, \omega)) = (1 + \lambda)^{-1}\tilde{S}_t^2 B(t, T)^2 [E^{Q_t^0}(f_1, f_2) - E^{Q_t^0}(f_1)E^{Q_t^0}(f_2)]$$

where

$$f(z) = \left[z - \frac{K}{\tilde{S}_t}\right]^+ \quad \text{and} \quad f_i(z)\left[z - \frac{K_i}{\tilde{S}_t}\right]^+, \quad i = 1, 2$$

This shows, as noted before, that the heteroscedasticity and the auto-correlation structure of error terms around the BS price depend in a highly complicated non-linear way on the underlying characteristics of the options: strike prices, times to maturity, etc.[16] We also obtain the price of the stock at time t as

$$S_t = S_t(\omega) = B(t, T)\tilde{S}_t \int_S z_T \Pi(\mathrm{d}z_T, \omega) \overset{\text{def}}{=} \tilde{S}_t m_t(\omega)$$

The error term $m_t(\omega)$ is a functional of a Dirichlet process of parameter $\lambda LN(-\sigma^2(T - t)/2, \sigma^2(T - t))$ and

$$E(m_t(\omega)) = 1, \quad \mathrm{Var}\,(m_t(\omega)) = (\lambda + 1)^{-1}(e^{\sigma^2(T-t)} - 1)$$

Moreover, for large values of λ, the observed price S_t, given $\tilde{S}_t$, is approximatively normally distributed with mean equal to $\tilde{S}_t$ and variance $\tilde{S}_t^2 \mathrm{Var}\,(m_t(\omega))$. Hence, one may choose λ in such a way that the variance of S_t does not depend on $T - t$.

A first extension of Equation (13.14) may be obtained by considering a Dirichlet process around the risk-neutral probability of Merton's (1973) model. For this consider:

$$V_t^T = \int_t^T \sigma_u^2 \, \mathrm{d}u$$

Conditionally on V_t^T, one can draw Dirichlet realisations around the lognormal distribution $LN(-\log B(t, T) - V_t^T/2, V_t^T)$ and therefore:

---

[16] Clément *et al.* (1993) provide in detail the explicit formulas and suggest a simulation-based methodology.

$$Q_t^0 = Q_t^0(V_t^T) = LN(-\log B(t, T) - V_t^T/2, V_t^T)$$

$$Z_T = \frac{S_T}{\tilde{S}_t} \mid Q_t(\cdot, \omega), I_t^S, \sigma_t, V_t^T, \tilde{S}_t, \sim V_T \mid Q_t(\cdot, \omega) \rightsquigarrow Q_t(\cdot, \omega)$$

$$Q_t \mid (\sigma_t, V_t^T, \tilde{S}_t) :\rightsquigarrow: \mathcal{D}i(\lambda Q_t^0)$$

One can draw first V_t^T from a conditional distribution, given σ_t (to be specified). As a result we obtain that $Q_t \mid \sigma_t$ is a mixture of Dirichlet processes.[17] The call option formula, given $\tilde{S}_t$, V_t^T and σ_t, does not depend on V_t^T and σ_t and is exactly as in Equation (13.15)

$$\Pi_t(K, \omega) = \Pi_t(\tilde{S}_t, K, \omega) = B(t, T)\tilde{S}_t \int_S (z_t - K/\tilde{S}_t)^+ Q_t(dz_T, \omega) \quad (13.16)$$

The mean of $\Pi_t(K, \omega)$, conditionally on $\tilde{S}_t$ and σ_t, is

$$E(\Pi_t(K, \omega)) = E_{\sigma_t}(BS(\tilde{S}_t, K, V_t^T))$$

Since Equation (13.16) is very similar to several well-known extensions of the BS option pricing model, including the Merton (1973) jump diffusion and Hull and White (1987) stochastic volatility models, we may conclude that the class of mixtures of Dirichlet processes allows us to introduce error terms around any extension of the BS model where unobserved heterogeneity (like stochastic volatility) has been introduced.

To conclude, this analysis suggests that we should distinguish two types of option pricing errors: (1) errors due to a limited set of unobserved state variables like stochastic volatility, stochastic interest rate, $\tilde{S}_t$; and (2) errors to make the model consistent with any data set. In the first approach, state variables are introduced as instruments to define mixtures of Dirichlet processes around the 'structural' option pricing model. Therefore, the suggested approach is not fully model-free since the model is built *around* a structural model defined by Q_t^0, with a parameter λ which controls the level of neighbouring around this model.[18] This is the price to pay to take into account arbitrage restrictions. We have therefore only two solutions. Either we adopt a semiparametric approach by introducing a non-parametric disturbance around a given probability measure Q_t^0 (while this was done above in a Bayesian way, it will be done in Section 13.4 in a classical way through the concept of functional residual plots), or alternatively we consider a genuine non-parametric estimation of the equivalent martingale measure. This is the issue addressed in Section 13.3.2.

So far, the Bayesian approach as discussed in this section is not yet fully explored in empirical work. The only attempt that we know of is the work of

---

[17] See Antoniak (1974) for the definition and the properties of the mixtures of Dirichlet processes. The fact that a mixture of Dirichlet processes is conditionally a Dirichlet process allows one to carry over many properties of Dirichlet processes to mixtures. Moreover, one can show that any random probability on $(S, B(S))$ can be approximated arbitrarily closely in the sense of the weak convergence for distributions, by a mixture of Dirichlet processes. Hence, the richness of the class of mixtures of Dirichlet processes suggests that it enables us to build general models.

[18] This parameter λ must not be confused with the smoothing parameter λ in Section 13.2, which was also devised to control the level of neighbouring, but with a very different concept of 'closeness' or 'neighbourhood'.

Jacquier and Jarrow (1995) who applied Bayesian analysis to BS option pricing models. Their analysis does not, however, take full advantage of the complex error structure which emerged from the Dirichlet process specification.

13.3.2   Non-parametric estimation of state-price densities implicit in financial asset prices

We observed in Equation (13.10) that the risk neutral probability distribution or 'state price density' can be recovered from taking derivatives of European calls with respect to their strike price. Kernel estimation techniques provide an estimate of the pricing function f. Provided that they exist, it is straightforward to recover estimates of the derivatives of f from $\hat{f}$. This is particularly important for option valuation in the context of complete markets with no arbitrage opportunities. In this situation, several of the partial derivatives of the pricing function are of special interest. One of them is the 'delta' of the option defined as $\Delta \equiv \partial\Pi/\partial S$.[19] Another derivative of special interest is $\partial^2\Pi/\partial K^2$ since it is related by Equation (13.10) to the *state price density*.

The use of kernel methods for deriving estimates of Δ and the state price density is due to Aït-Sahalia and Lo (1995). Many of the issues raised in Section 13.2 regarding kernel estimation and the non-stationarity of the data, the dependence of the data etc, apply here as well. In addition, some additional issues should be raised. Indeed, we noted in Section 13.2 that kernel smoothing is based on a certain approximation criterion. This approximation criterion applies to the estimation of the function f but not necessarily its derivatives. The bandwidth selection affects the smoothness of the estimate $\hat{f}$ and therefore indirectly its derivatives. Since the ultimate objective is to estimate the derivatives of the function rather than the function itself it is clear that the choice of objective function and approximation criterion of standard kernel estimation are not appropriate. It is a drawback of this approach that still needs to be investigated in greater detail.

Up to now, we presented two non-parametric approaches to the option valuation problem. The first one, the pure non-parametric pricing, makes very little use of the economic or financial dimensions of the problem and relies almost exclusively on the statistical exploitation of *market* data. The second one incorporates elements of a rational option pricing theory: it exploits the equivalence between the absence of arbitrage assumption and the existence of a risk neutral probability measure to derive the pricing formula. From this relation, it appears that the parameter to be estimated is the risk neutral density. The next section presents a third way which can be seen as a blend of the previous approaches.

---

[19] This quantity is useful as it determines the quantity of the underlying stock an agent must hold in a hedging portfolio that replicates the call option.

13.4   Extended Black and Scholes models and objective driven inference

Practitioners recognise that the assumptions of constant dividends, interest rates and volatility of the Black and Scholes (BS) model are not realistic. The most revealing evidence of this is the systematic use of the BS formula as a pricing and hedging tool by practitioners through the so-called BS implicit volatility; that is, the volatility measure which equates the BS option valuation formula to the observed option price

$$\Pi_t = BS(S_t, \tau, K, r, \sigma_t(S_t, K)) \qquad (13.17)$$

where Π_t denotes the observed call price at time t and $\sigma_t(S_t, K)$ the corresponding BS implicit volatility. This practice can be assimilated in a forecasting rule where the BS formula is used as a *black box* which integrates the time varying and stochastic environment through the volatility parameter. Since the BS formula is inherently misspecified we can think of modifying the underlying model so that it incorporates these new features, such as stochastic volatility. However, such attempts lead to very complex models which usually do not admit a unique and dosed form solution except for some special cases.

Taking into account that (1) analytic extensions of BS often pose many computational difficulties; (2) the BS formula is not a valid *modelling* tool; (3) the BS formula is used as a *prediction* tool, Gouriéroux *et al.* (1994, 1995), henceforth GMT, present the statistical foundations of dealing with a misspecified BS model. We first present the GMT approach in its general formulation and then show how it can be applied to model a modified BS formula.

13.4.1   Kernel *M*-estimators

To discuss the generic set-up of kernel *M*-estimators, let us suppose that we observe a realisation of length T of a stationary stochastic process $\{(Z_t, Y_t) : t \in \mathbf{Z}\}$ and that the parameter vector parametrises the conditional distribution of Y_t given $\mathcal{F}_t = \sigma(Z_s, Y_{s-1}, s \leqslant t)$. The estimation strategy proposed by GMT tries to approximate a *functional* parameter vector $\theta(\mathcal{F}_t)$ implicitly defined as the solution of

$$\min_{\theta} E_0[\psi(U_t, \theta) \mid \mathcal{F}_t] \qquad (13.18)$$

where $U_t = (Z_t, Y_t, Y_{t-1})$, ψ is an objective function, and $E_0(\cdot \mid \mathcal{F}_t)$ denotes the conditional expectation with respect to the unknown true conditional distribution of U_t given $(Z_s, Y_{s-1}, s \leqslant t)$. Let X be a d-dimensional process such that the solution $\theta(X_t)$ of

$$\min_{\theta} E_0[\psi(U_t, \theta) \mid X_t] \qquad (13.19)$$

coincides with that of Equation (13.19). GMT suggest approximating Equation (13.19) by

$$\min_{\theta} \frac{1}{T} \sum_{t=1}^{T} \frac{1}{h_T^d} K\left(\frac{x_t - x}{h_T}\right) \psi(u_t, \theta) \qquad (13.20)$$

where $K(\cdot)$ is a kernel function, and h_T is the bandwidth, depending on T, the sample size. Estimators obtained according to Equation (13.20) are denoted $\hat{\theta}_T(s)$ and called kernel M-estimators, since they are derived by *minimising* an 'empirical' criterion, which is a *kernel* approximation of the unknown theoretical criterion to be minimised appearing in Equation (13.14). Under suitable regularity conditions (see Gouriéroux *et al.*, 1995), $\hat{\theta}_T(x) - \theta(x)$ converges to 0, where $\theta(x)$ denotes the solution of $\min_\theta E_0(\psi(U_t, \theta) \mid X_t = x)$.

Gouriéroux *et al.* (1995) also show how local versions of such estimators may be used to compute *functional residuals* in order to check the hypothesis of a constant function $\theta(\cdot)$. In other words, as is fairly standard in econometrics, error terms (and residual plots) are introduced to examine whether unobserved heterogeneity is hidden in seemingly constant parameters θ. The GMT contribution is to give a non-parametric appraisal of these error terms, which justifies the terminology '*functional* residual plots'. GMT show how their functional residual plots are related to some standard testing procedures for the hypothesis of parameter constancy and how they may be introduced as important tools in a modelling strategy. A Bayesian alternative is considered by Jacquier and Jarrow (1995) who suggest considering draws from the posterior distribution of parameters of an extended model in order to deduce some draws of the residual vector and to perform a Bayesian residual analysis.

13.4.2 Extended Black–Scholes formulations

The methodology proposed by GMT can be applied to the problem of option pricing by extending the Black–Scholes formulation. The search for such a prediction model is made starting from the BS formula, namely we look for models of the form

$$E_0(\Pi_t \mid X_t) = \mathrm{BS}(S_t, K, \tau, \theta_0(X_t)) \qquad (13.21)$$

where θ is the vector of parameters (r, σ) which enters the standard BS formula and X_t is a vector of state variables which is believed to affect the volatility and the interest rate. For instance, we may decide to include in X_t variables such as $S_{t-1}, K, \dots$. This approach is very much in line with the idea of computing BS implicit volatilities except that it involves a statistically more rigorous scheme which also serves as a basis for building new prediction tools. GMT propose choosing the objective ψ function

$$\psi(U_t, \theta) = [\Pi_t - \mathrm{BS}(S_t, K, \tau, \theta)]^2$$

with $U_t = (S_t, \Pi_t)$.[20] Such a choice is motivated by observing that if Equation (13.21) is correct, then under suitable assumptions, the solution $\theta(X_t)$ of

$$\min_\theta E_0[\psi(U_t, \theta) \mid X_t]$$

---

[20] In our case, since we are interested in option pricing or option price prediction, the ψ function is related to a pricing error. However, if option hedging is the main goal, the ψ may be chosen accordingly as a tracking error.

is $\theta_0(X_t)$. Therefore, if the regularity conditions are satisfied, the convergence result of kernel M-estimators ensures that for T large enough, $\hat{\theta}_T(x)$ will be close to $\theta_0(x)$. This is a justification for the use of the modified BS formula $BS(S_t, K, \tau, \hat{\theta}_T(X_t))$ as a predictor of option prices.

One of the conditions to be imposed to obtain the convergence of $\hat{\theta}_T(\cdot)$ is the stationarity of the process (S_t, Π_t). Such an assumption is hardly sustainable in view of the stylised facts concerning the variables entering this process. To remedy this problem, GMT suggest using the homogeneity of degree one in (S, K) of the BS function together with the measurability of S_t with respect to $\sigma(X_t)$. Indeed, this leads to a new pricing relationship:

$$E_0\left[\frac{\Pi_t}{S_t} \mid X_t\right] = \frac{1}{S_t} BS(S_t, \tau, K, \theta(X_t)) = BS^*(\tau, k_t, \theta(X_t)) \qquad (13.22)$$

where $k_t \equiv K/S_t$ is the inverse of the moneyness ratio. In this formulation, all the prices are expressed in terms of the time t underlying asset price. Then, a new objective function is now

$$\psi^*(U_t^*, \theta^*) \equiv \left[\frac{\Pi_t}{S_t} - BS^*(\tau, \theta^*)\right]^2$$

where $U_t^* = \Pi_t/S_t$ and $\theta^* = (r, \sigma, k)$ and the kernel M-estimator, denoted $\hat{\theta}_T^*(x)$, is the solution of:

$$\min_{\theta^*} \frac{1}{T} \sum_{t=1}^{T} \frac{1}{h_T^d} K\left(\frac{x_t - x}{h_T}\right) \psi^*(U_t^*, \theta^*)$$

An issue that arises when implementing the GMT approach is the choice of the variables to be included in X_t. This problem can be seen as a problem of model choice which arises very often in econometrics. Gouriéroux *et al.* (1995) propose a modelling approach based on the use of functional residual plots and confidence bands for these residuals. In a first step, the parameter θ is assumed to be constant and is estimated by θ_{0T}^* obtained by minimising the sample average of $\psi^*(U_t^*, \theta^*)$. Then a state variable X_1 potentially affecting the parameter is introduced. In order to test whether θ^* should be considered as constant or depending on X_1, one computes in a second step *functional residuals*. They are defined as the difference between an approximation of $\hat{\theta}_T^*(x_1)$ near the constancy hypothesis and θ_{0T}^*. Confidence bands on these residuals help to determine whether they depend on X_1. A plot of these residuals against X_1 is also helpful in choosing a parametric form for expressing the relationship between θ^* and X_1. In case of a rejection of the constancy hypothesis, the parameters of this relationship are considered as the new functional parameters to be estimated. The procedure goes on by repeating the previous steps with new state variables.

The approach discussed so far can be extended to models other than the BS. For instance, for American type options one may replace the BS formula with a formula tailored for such options, typically involving numerical approximations. Like the BS formula, they rest on the restrictive assumptions of constant volatility, interest rates, etc. The M-estimators approach suggested by GMT readily extends to such, and other applications.

Acknowledgements

Part of this work was funded by the Fonds FCAR of Québec, the SSHRC of Canada under Strategic Grant 804–96–0027 and the TMR Work Programme of the European Commission under Grant Nr ERB4001GT950641.

References

Abken, P.A., Madan, D. B. and Ramamurtie, S. (1996) Basis pricing of contingent claims: application to Eurodollar futures options. Discussion Paper, Federal Reserve Board of Atlanta.

Aït-Sahalia, Y. (1993) Nonparametric functional estimation with applications to financial models. Ph.D. dissertation, M.I.T.

Aït-Sahalia, Y. (1996) Nonparametric pricing of interest rate derivative securities. *Econometrica*, **64**, 527–560.

Aït-Sahalia, Y. and Lo, A. W. (1995) Nonparametric estimation of state-price densities implicit in financial asset prices. Discussion Paper, Sloan School of Management, M.I.T.

Aït-Sahalia, Y., Bickel, P. and Stoker, T. (1995) Goodness-of-fit tests for regression using kernel methods. Discussion Paper, University of Chicago.

Altman, N. S. (1987) Smoothing data with correlated errors. Technical Report 280, Department of Statistics, Stanford University.

Altman, N. S. (1990) Kernel smoothing of data with correlated errors. *Journal of the American Statistical Association*, **85**, 749–759.

Andrews, D. W. K. (1991) Heteroskedastic and autocorrelation consistent matrix estimation. *Econometrica*, **59**, 817–854.

Antoniak, C. (1974), Mixtures of Dirichlet processes with applications to Bayesian nonparametric problems. *Annals of Statistics*, **2**, 1152–1174.

Baum, C. and Barkoulas, J. (1996) Essential nonparametric prediction of U.S. interest rates. Discussion Paper, Boston College.

Black, F. and Scholes, M. (1973) The pricing of options and corporate liabilities. *Journal of Political Economy*, **81**, 637–659.

Bossaerts, P., Hafner, C. and Härdle, W. (1995) Foreign exchange rates have surprising volatility. Discussion Paper, CentER, Tilburg University.

Breeden, D. and Litzenberger, R. (1978) Prices of state-contingent claims implicit in option prices. *Journal of Business*, **51**, 621–651.

Broadie, M., Detemple, J., Ghysels, E. and Torrès, O. (1995) Nonparametric estimation of American options exercise boundaries and call prices. Discussion Paper, CIRANO, Montréal.

Broadie, M., Detemple, J., Ghysels, E. and Torrès, O. (1996) American options with stochastic volatility and stochastic dividends: a nonparametric investigation. Discussion Paper, CIRANO, Montréal.

Cao, R., Quintela-del-Rió, A. and Vilar-Fernández, J. M. (1993) Bandwidth selection in nonparametric density estimation under dependence: a simulation study, *Computational Statistics*, **8**, 313–332.

Clément E., Gouriéroux, C. and Monfort, A. (1993) Prediction of contingent price measures. Discussion Paper, CREST.

Craven, P. and Wahba, G. (1979) Smoothing noisy data with spline functions, *Numerical Mathematics*, **31**, 377–403.

Duffie D. (1996) *Dynamic Asset Pricing Theory*, 2nd Edition. Princeton: Princeton University Press.

Dumas, B., Fleming, J. and Whaley, R. E. (1995) Implied volatility functions: empirical tests. Discussion Paper, HEC, Paris.

Engle, R. F. and Mustafa, C. (1992) Implied ARCH models from option prices. *Journal of Econometrics*, **52**, 289–311.

Ferguson, T. S. (1973) A Bayesian analysis of some parametric problems. *Annals of Statistics*, **1**, 209–230.

Garcia, R. and Renault, E. (1996) Risk aversion, intertemporal substitution and option pricing. Discussion Paper, CIRANO, Montréal and GREMAQ, Université de Toulouse I.

Ghysels, E., Harvey, A. and Renault, E. (1996) Stochastic volatility. In G. S. Maddala and C. R. Rao (Eds), *Handbook of Statistics – Vol. 14, Statistical Methods in Finance*, Chapter 5, Amsterdam: North Holland.

Ghysels, E. and Ng, S. (1996) A semiparametric factor model of interest rates. Discussion Paper, CIRANO, Montréal.

Gouriéroux, C. and Scaillet, O. (1995) Estimation of the term structure from bond data. Discussion paper 9415, CREST, Paris.

Gouriéroux, C., Monfort, A. and Tenreiro, C. (1994) Kernel *M*-estimators: nonparametric diagnostics for structural models. Discussion Paper 9405, CEPREMAP, Paris.

Gouriéroux, C., Monfort, A. and Tenreiro, C. (1995) Kernel *M*-estimators: nonparametric diagnostics and functional residual plots. Discussion Paper, CREST-ENSAE, Paris.

Györfi, L., Härdle, W., Sarda, P. and Vieu, P. (1989) *Nonparametric Curve Estimation from Time Series*, Lecture Notes in Statistics 60, J. Berger *et al.* (Eds), Heidelberg: Springer-Verlag.

Härdle, W. and Linton, O. (1994) Applied nonparametric methods. In R. F. Engle and D. L. McFadden (Eds), *Handbook of Econometrics*, vol. 4, Amsterdam: North Holland.

Härdle, W. and Vieu, P. (1992) Kernel regression smoothing of time series. *Journal of Time Series Analysis*, **13**, 209–232.

Harrison J.-M. and Kreps, D. (1979) Martingale and arbitrage in multiperiods securities markets, *Journal of Economic Theory*, **20**, 381–408.

Harvey, C. R. and Whaley, R. E. (1992) Dividends and S&P 100 index option valuation. *Journal of Futures Markets*, **12**, 123–137.

Hastie, T. J. and Tibshirani, R. J. (1990) *Generalized Additive Models*. London: Chapman & Hall.

Huang, C.-F. and Litzenberger, R. (1988) *Foundations for Financial Economics*. Amsterdam: North Holland.

Hull J. and White, A. (1987) The pricing of options on assets with stochastic volatilities. *Journal of Finance*, **42**, 281–300.

Hutchinson, J. M., Lo, A. W. and Poggio, T. (1994) A nonparametric approach to pricing and hedging derivative securities via learning networks. *Journal of Finance*, **49**, 851–889.

Jacquier E. and Jarrow, R. (1995) Dynamic evaluation of contingent claim models. Discussion Paper, Cornell University.

Kim, T. Y. and Cox, D. D. (1996) Bandwidth selection in kernel smoothing of time series. *Journal of Time Series Analysis*, **17**, 49–63.

Lavergne, L. and Vuong, Q. (1996) Nonparametric selection of regressors: the nonnested case. *Econometrica*, **64**, 207–219.

Lo, A. Y. (1987) A large sample study of the Bayesian bootstrap. *The Annals of Statistics*, **15**, 360–375.

Longstaff, F. A. (1995) Option pricing and the martingale restriction. *Review of Financial Studies*, **8**, 1091–1124.

Merton, R. C. (1973) Theory of rational option pricing. *Bell Journal of Economics*, **4**, 141–183.

Manaster, S. and Rendleman, R. J. (1982) Option prices and predictors of equilibrium stock prices. *Journal of Finance*, **37**, 1043–1057.

Patilea, V., Ravoteur, M. P. and Renault, E. (1995) Multivariate time series analysis of option prices. Working Paper GREMAQ, Toulouse.

Patilea V. and Renault, E. (1995) Random probabilities for option pricing. Discussion Paper, GREMAQ, Toulouse.

Renault, E. (1996) Econometric models of option pricing errors. Discussion Paper, GREMAQ, Université de Toulouse I.

Renault, E. and Touzi, N. (1996) Option hedging and implied volatilities in a stochastic volatility model. *Mathematical Finance*, **6**, 259–302.

Robinson, P. (1983) Nonparametric estimators for time series. *Journal of Time Series Analysis*, **4**, 185–1207.

Rubinstein, M. (1994) Implied binomial trees. *Journal of Finance*, **49**, 771–818.

Schuster, E. F. (1972) Joint asymptotic distribution of the estimated regression function at a finite number of distinct points. *Annals of Mathematical Statistics*, **43**, 84–88.

Scott, D. W. (1992) *Multivariate Density Estimation: Theory, Practice and Visualization*, New York: John Wiley & Sons.

Singh, R. and Ullah, A. (1985) Nonparametric time series estimation of joint DGP, conditional DGP and vector autoregression. *Econometric Theory*, **1**, 27–51.

Silverman, B. W. (1986) *Density Estimation for Statistics and Data Analysis*, London: Chapman & Hall.

Stutzer, M. (1995) A simple nonparametric approach to derivative security valuation. Discussion Paper, Carlson School of Management, University of Minnesota.

Wand, M. P. and Jones, M. C. (1995) *Kernel Smoothing*. London: Chapman & Hall.

14

Stochastic Volatility

David G. Hobson

14.1 Volatility and the need for stochastic volatility models

14.1.1 Introduction

A common approach in the modelling of financial assets is to assume that the proportional price changes of an asset form a Gaussian process with stationary independent increments. The celebrated (and ubiquitous) Black–Scholes option pricing formula is based on such a premise. The success and longevity of the Gaussian modelling approach depends on two main factors: first the mathematical tractability of the model, and secondly the fact that, in many circumstances, the model provides a reasonable and simple approximation to observed market behaviour.

An immediate corollary of the Gaussian assumption is that the behaviour of the asset price can be summarised by two parameters, namely the mean and the standard deviation of the Gaussian variables. In finance-speak the standard deviation is renamed the volatility. Volatility is a key concept because it is a measure of uncertainty about future price movements, because it is directly related to the risk associated with holding financial securities and hence affects consumption/investment decisions and portfolio choice, and because volatility is the key parameter in the pricing of options and other derivative securities.

This chapter is concerned with the estimation of volatility and the implications of the empirical observation that volatility appears non-constant over time. Some of the evidence for this claim is given in the next subsection. In Section 14.2 we review the continuous-time models that have been introduced to reflect the non-constant volatility phenomenon, including level-dependent and stochastic diffusion models for the volatility. Since the fundamental problem in mathematical finance is to price derivative securities such as options, we focus in Section 14.3 on the implications of these alternative models for derivative pricing. The types of discrete-time models (such as *ARV*, *ARCH* and *GARCH*) favoured by econometricians to model stochastic volatility are the subject of Section 14.4. Some comments and conclusions on the importance of accurate modelling of volatility are given in a conclusion.

14.1.2   Simple models for asset prices

The canonical continuous Gaussian process is Brownian motion. In the same way that the normal law arises as the limit of a normalised sum of independent random variables, so Brownian motion arises as the limit of a random walk as, simultaneously, step sizes are reduced and step frequency increased. The links between Brownian motion and finance are long and illustrious; and indeed (mathematician's) Brownian motion was devised by Bachelier (1900) as a model for French stock prices.

The increments of a Brownian motion are normal random variables. As a consequence, Brownian motion can and does become negative, which makes it an unsatisfactory model for limited liability stocks. Instead, a more reasonable model was proposed by Osborne (1959) and Samuelson (1965) who took the asset price to be an exponential (or geometric) Brownian motion. Thus, the logarithm of the asset price is a Brownian motion.

To be more formal, the standard and classical model for the behaviour of the price of a financial asset, such as a stock, assumes that the price process $(P_t)_{t \geqslant 0}$ is the solution to a stochastic differential equation (SDE)

$$dP_t = P_t(\mu dt + \sigma dB_t) \tag{14.1}$$

where t is measured in units of one year, B_t is a Brownian motion and μ, the mean, and σ, the volatility, are constant parameters of the model. The time convention is chosen to ensure that σ can be interpreted as an annualised volatility. This SDE has solution

$$P_t = P_0 \exp\{\sigma B_t + (\mu - \tfrac{1}{2}\sigma^2)t\} \tag{14.2}$$

For a highly readable introduction to stochastic differential equations see Øksendal (1985). The discrete time analogue of Equation (14.1) based on a daily sequence of observations $(P_n)_{n \geqslant 0}$ is

$$\ln P_n - \ln P_{n-1} \equiv \Delta(\ln P_n) = v + \sigma Z_n \tag{14.3}$$

where (Z_n) is a sequence of independent normal random variables with zero mean and variance $(1/365)$. Again this choice of normalisation ensures that σ can be interpreted as an annualised volatility. The assumption that the innovations have a normal distribution means that the increments have a natural nesting property. For example, the proportional price changes of a weekly time series also have a normal distribution. Thus, the dynamics are not dependent on the choice of timescale.

Two contradictory philosophies are available here. It is possible to view the discrete time series as a δ-skeleton of the underlying continuous Markov process given by Equation (14.1) with the understanding that even tick data provides only an approximation to the inherently unobservable true process. Alternatively, the SDE formulation, whose merit is tractability, can be viewed as a limiting approximation to a discrete stochastic difference equation. Discrete models are suited to qualitative and descriptive analyses, whereas continuous time models provide the natural framework for theoretical option pricing. We shall take the view that the fundamental problem in mathematical finance is the calculation of derivative prices and, in particular, formulae which relate the price of an option to the price of the underlying asset and other key variables. Hence, we shall concentrate on continuous time diffusion models for the price process and volatility.

In principle, if the continuous time model can be observed perfectly (and in continuous time) then it is possible to read off the instantaneous value of the volatility from the asset price. (The square of the volatility is the quadratic variation of the log-price process.) In practice however, the volatility must be estimated from the data. Suppose that the data consist of a series of daily observations of the price of an asset $(P_k)_{k \leqslant N}$. Our first estimate of the volatility, $\hat{\sigma}$, is called the *historic volatility*. At time n, the historic volatility based on the last J days is the maximum likelihood estimator obtained from Equation (14.3) and the data $P_{n-J-1}, \ldots P_n$:

$$\hat{\sigma}_{n,J} \equiv \left(\frac{365}{J-1} \sum_{j=0}^{J-1} \left(\Delta(\ln P_{n-j}) - \frac{1}{J} \sum_{j=0}^{J-1} \Delta(\ln P_{n-j}) \right)^2 \right)^{1/2} \qquad (14.4)$$

The factor of 365 converts daily volatility into an annualised term. Typically, J is taken to be 90 or 180 days. These choices are a compromise between the desire for a large number of observations and a realisation that the dynamics of the price process are unlikely to remain constant over several years.

Figure 14.1 gives a plot of the September 1995 futures price of the FT-SE 100 index (the *Financial Times– Stock Exchange* index of the stock prices of 100 leading UK companies) over the period December 1992 to July 1995. Roughly speaking[1] the futures price is the amount which, it is agreed now, is to be paid in September 1995 for delivery, again in September 1995, of the underlying asset; in this case (the cash value of) the basket of stocks in the *FT-SE* index. The advantage of considering the futures price is that our analysis is not confounded by interest rate and discounting effects.

Figure 14.2 gives an estimate of the 90-day historic volatility based on the above data. This limited evidence supports the contention that stock volatility is not constant and, moreover, that volatility shocks persist through time. This conclusion was reached by Mandelbrot (1963), Fama (1965), Blattberg and Gonedes (1974), and Scott (1987) amongst others. Stochastic volatility models are needed to describe and explain volatility patterns.

---

[1] In fact this is the definition of a forward contract. Futures are closely related instruments which are designed to be traded on an exchange. The main difference between the two is that a forward contract is generally made between two parties, and the terms, and particularly the price to be paid on delivery, are fixed at inception. In contrast, futures contracts are bought and sold on an exchange. Moreover the price to be paid on delivery is adjusted each trading day (to be the new futures price) and compensating payments are made to the holders of contracts in proportion to the number of contracts they own. This has the advantage that all futures contracts with the same maturity have the same terms and thus facilitates trading on an exchange. See, for example, Hull (1993) for a fuller description of these securities.

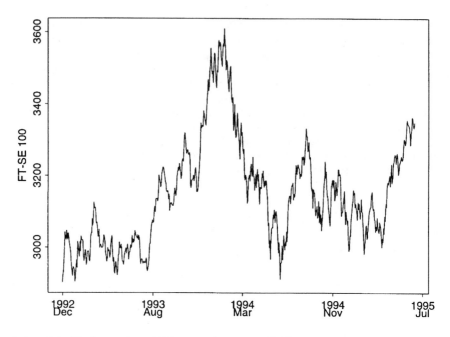

Figure 14.1 The September 1995 futures price for the FT-SE 100 Index

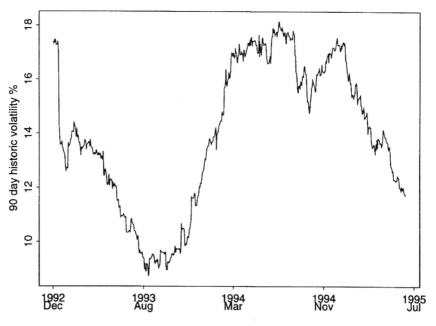

Figure 14.2 90-day historic volatility for the FT-SE 100 Index, based on FT-SE price data in Fig. 14.1

14.1.3   The Black–Scholes paradigm and option pricing

One of the key contributions of mathematics to finance has been the development of formulae for the pricing of options and other derivative securities. Black and Scholes (1973) showed that, subject to certain modelling assumptions, there is a strategy for risklessly hedging options in the sense that it is possible to replicate perfectly the pay-off of the option through dynamic trading. Thus, there is a unique preference-independent rational price for an option. This price corresponds to the fortune needed to purchase the initial portfolio which is required to hedge the option. This observation has revolutionised financial markets and contributed greatly to the explosion in the volume of trading in derivative securities.

The purchaser of a European *call option* on an asset with *strike K* and *expiry T* has the right, but not the obligation, to buy one unit of the asset at time T for a price K. (An American call option conveys the right to buy the asset at any time before T; the option is European style if the right to buy is restricted to the time T alone.) This right will only be exercised if the price P_T of the asset at time T is above K; otherwise at expiry the option is worthless. It is often convenient to think of an option as a derivative security which at time T pays the cash amount $(P_T - K)^+ \equiv \max\{P_T - K, 0\}$. The fundamental problem in mathematical finance is to find the fair price of such an option at a time t prior to expiry.

In order to price this option it is necessary, following Black and Scholes (1973) to make a number of regularity assumptions about the financial market in which the underlying asset is traded. In particular, the market is assumed to be perfect and frictionless, so that there are no transaction costs, there is no taxation, and the underlying asset is available in arbitrary amounts. There is a constant rate of interest r for both borrowing and lending, there are no dividends and there are no restrictions on short selling of stock provided that the net wealth of the trader remains non-negative. In particular, a trader may sell stock or bonds that he does not own provided that, by the end of the trading period, he has repurchased sufficient quantities to cover his obligations. Many of these assumptions can be weakened. For example, it is easy to relax the assumption about constant interest rates to an assumption of deterministic interest rates. Finally, Black and Scholes assume that the asset price process is given by the solution to Equation (14.1) with constant and known parameter values μ and σ.

The Black–Scholes price C of a call option is given by

$$C(P_t, t; K, T; \sigma, r) \equiv C = K\,e^{-r(T-t)}(M_t\Phi(d_1) - \Phi(d_2)) \qquad (14.5)$$

where $M_t \equiv (P_t/K\,e^{-r(T-t)})$ is the *moneyness* of the option and d_1 and d_2 are given by

$$d_1 = \frac{\ln(M_t) + \frac{1}{2}\sigma^2(T-t)}{\sigma(T-t)^{1/2}}$$
$$d_2 = d_1 - \sigma(T-t)^{1/2}$$

respectively. (The term moneyness refers to the fact that if $M_t > 1$ then the option is said to be in-the-money, and if the futures price remains unchanged then the option will make a pay-out on expiry. An option, for which $M_t < 1$, is said to be out-of-the-money since, unless the underlying value of the asset increases, the option pay-out will be zero.) In Equation (14.5) $\Phi(\cdot)$ denotes the cumulative normal distribution function, and K and T are the strike and expiry as before. In the following we will be flexible in deciding which of the quantities P_t, t, K, T, σ are to be considered as variables, and which are fixed parameters.

There are several important remarks that should be made about this formula. First, the justification for calling C the fair price of the option is based on the fact that the quantity C can be used to finance a trading strategy which, at maturity, is guaranteed to match the pay-off of the option. Models with this replication property are said to be complete. Secondly, the drift parameter μ does not appear. Indeed, it is as if the option price was calculated as the discounted expected pay-off of an option on an asset whose dynamics are given by the SDE

$$\frac{dP}{P} = \sigma dW + rdt$$

rather than Equation (14.1). In particular, the call price can be expressed as a conditional expectation given the current price:

$$C \equiv C(P_t, t) = e^{-r(T-t)}\tilde{\mathbb{E}}[(P_T - K)^+ \mid P_t] \qquad (14.6)$$

where $\tilde{\mathbb{E}}$ denotes expectation with respect to the risk-neutral probability measure. This is the measure under which W is a Brownian motion. Thirdly, and as a direct consequence of the second remark, there is a single unknown parameter in the Black–Scholes formula. The strike K and time to expiry $(T - t)$ are part of the specification of the option; the interest rate r is assumed known and the current price P_t is observable. Thus, the price of the option depends solely on the value of the volatility. Moreover the option price depends on the volatility only through the quantity $\sigma^2(T - t)$, which is the integrated squared volatility over the remaining lifetime of the option. This illustrates a more general comment that the Black–Scholes model can easily be adapted to allow for time varying parameter values for the volatility parameter, provided that the behaviour is deterministic, and provided that the term $\sigma^2(T - t)$ is replaced by $\int_t^T \sigma_s^2 \, ds$.

The call option pricing function $C \equiv C(\sigma)$ is an increasing function of the volatility σ. This observation can be verified by differentiation of Equation (14.5), or is immediate from the representation of Equation (14.6); see also Fig. 14.5 later. This means that not only can we calculate the price of an option given a value for the volatility parameter, but also that, given the price of an option, it is possible to deduce the unique value of the volatility that must be substituted into the Black–Scholes formula to obtain the observed option price. We define the *implied volatility* $\tilde{\sigma}$, to be the value of the volatility parameter σ that is consistent with the Black–Scholes formula and the observed call price.

Thus, we have a new measure of volatility. Implied volatility is a market assessment of the expected future volatility over the lifetime of the option. Implied volatility is a useful device because it provides a convenient shorthand for expressing the option price, and because it facilitates price comparisons of options with different characteristics.

Suppose that the assumptions of the Black–Scholes model are satisfied, so that, in particular, the price process of the underlying asset is given by the solution to the SDE, Equation (14.1). Provided that the market prices options using the Black–Scholes formula, then the implied volatility $\tilde{\sigma}$ should be identically equal to the true parameter value σ. For all strikes K and maturities T we can define $\tilde{\sigma}(K, T)$ to be the implied volatility of the call option with maturity $T > t$ and strike K. If the Black–Scholes model is correct then a plot of $\tilde{\sigma}(K, T)$ should yield a constant surface.

Analyses of implied volatility patterns have been attempted by Rubinstein (1985), Skeikh (1991), Fung and Hsieh (1991), Heynen *et al.* (1994) and Xu and Taylor (1994) amongst others. These authors all find systematic biases in the implied volatility surface. In particular there is strong evidence of an implied volatility smile (so that, for the cross-section of the implied volatility surface corresponding to a given maturity, implied volatility is a convex function of the strike) and some evidence of skews (so that added to convexity of the cross-sectional implied volatility, there is an additional linear relationship). Skews are particularly evident in implied volatilities for indices rather than individual stocks; see Wiggins (1987).

Figure 14.3 presents implied volatility data for a set of call options traded on the London Financial Futures Exchange on 18 April 1995. The call options

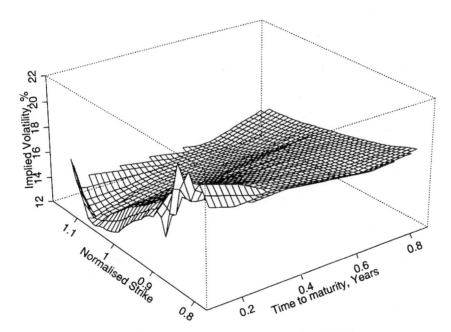

Figure 14.3 Implied volatilities of European call options on the FT-SE 100 Index on 18 April 1995. The normalised strike is the ratio of the true strike and the relevant futures price of the index. The graph has been interpolated by Splus.

are European style options on the FT-SE 100 Index. The plot is presented as a surface parameterised by the date of the expiry of the option and the moneyness of the option. For each of the five expiry dates there are between 13 and 18 options traded with different strikes. There is clear evidence of a skew and some evidence of a volatility smile. These effects become less pronounced as maturity increases. Finally, it appears that the implied volatility of an at-the-money option is an increasing function of time to expiry. A modification of the Black–Scholes model is required to account for these effects and stochastic volatility models provide a potential explanation.

14.2   Non-constant volatility models

In the first section of this chapter we described the standard exponential Brownian motion model for asset prices and noted some of the discrepancies and inconsistencies which arise when this model is compared with market experience. In particular Fig. 14.2 shows time series plots of realised volatility which contradict a stationary normal hypothesis. Moreover, plots of implied volatility in Fig. 14.3 are inconsistent with market belief in Black–Scholes with constant volatility. In this section we aim to outline some of the models which attempt to explain, or at least account for, these inconsistencies. Although there are other potential explanations for the observed biases (for example non-zero transaction costs will require modifications of the Black–Scholes formula and liquidity considerations may inflate the prices of options which are away from the money) we will focus on explanations and models in which the volatility becomes non-constant.

In the 1960s, empirical studies of asset price behaviour by Mandelbrot (1963) and Fama (1965) found leptokurtosis in the distribution of the daily changes in the log-price. These authors were led to suggest an innovations process consisting of random variables with stable Paretian distributions with characteristic exponents between one and two in an attempt to explain the observed fat-tails of the empirical distribution relative to a normal law. Second and higher moments do not exist for such distributions so that the notion of volatility becomes ill-defined, with serious implications for the pricing of options, at least within the Black–Scholes paradigm.

The stable Paretian hypothesis continues to have its proponents (see for example Peters, 1991). However, the tractability that the Black–Scholes model derives from the Gaussian character of its underlying variables allows it to retain its pre-eminence amongst the class of asset price models as the reference model against which others are compared. Instead of rejecting normality, financial economists have searched for alternative explanations for the observed kurtosis and apparent randomness of volatility which rely on modification of the Gaussian framework.

14.2.1   Subordinators and volume effects

One attractive explanation for the apparent randomness of volatility claims that the asset price process *is* an exponential Brownian motion, but only when the time parameter t is interpreted as an *intrinsic clock* rather than real or calendar time. Relative to real time the daily changes are a mixture of normals. This model was proposed by Clark (1973) who argued that the daily proportional price change is a sum of a random number of within-day price changes and that the number of such changes is related positively to the rate of information flow or the volume of trading. Strong supporting evidence for his general thesis was found by Epps and Epps (1976) and Tauchen and Pitts (1983); in particular, a mixture of normals hypothesis was observed to fit the data more accurately than a stable Paretian distribution. Karpoff (1987) documents several studies relating asset price volatility to traded volume.

In general, the model is as follows. Let A_t be a subordinator so that A_t is a non-decreasing Markov process with stationary independent increments. The price process $(P_t)_{t \geqslant 0}$ is a random time-change of an exponential Brownian motion and is given by

$$P_t = P_0 \exp(B_{A(t)} - \mu A_t + vt)$$

for a pair of drift parameters μ and v. For the subordinator $A_t \equiv \sigma^2 t$ we recover the standard Black–Scholes model.

Madan and Seneta (1990) proposed a particular choice of subordinator and termed the resulting stock price model the *Variance–Gamma* model. In this model the subordinator is a Gamma process. The Gamma process is a pure jump process and the price process inherits this property also. Madan and Senata noted that their model has the following desirable properties: first the distribution of proportional price changes is fat-tailed relative to the normal; secondly the distribution has finite moments, at least of lower orders; thirdly the process is consistent with an underlying continuous-time stochastic process; and finally that the model can be extended to a multi-dimensional process. However the model also has one serious disadvantage; the existence of jumps in the asset price makes the pricing and hedging of options very awkward.

14.2.2   Leverage effects and implied volatility skews

A second observed feature of stock price volatility is a correlation between volatility and price level. This relationship is implicit in implied options prices; the market expects volatility to rise as prices fall. One common explanation for this relates volatility to leverage effects.

Imagine a firm with debt whose share value represents the surplus of the firm's assets over this debt. Suppose that the value of the firm's assets fluctuates like an exponential Brownian motion with constant variance whilst the value of the debt remains fixed. Then the magnitude of the proportional changes in the share value is greater than the magnitude of the proportional change in the asset value, although of course the absolute changes are the same. Hence, the volatility of the share price is greater than the volatility of the assets. Moreover, if the value of the assets rises then the ratio of the stock value to asset value approaches unity and fluctuations in the value of assets are directly reflected in the stock price. Conversely, if the value of the assets falls then the debt factor becomes significant and the effect is to magnify changes in the asset value as represented via the stock price. In this way leverage introduces a negative correlation between volatility and price level.

The above arguments provide a direct inspiration for the equity price models of Geske (1979), Rubinstein (1983) and Bensoussan *et al.* (1994). In essence, these models propose that the stock price is the solution of an SDE

$$\frac{dP_t}{P_t} = \sigma(P)\,dB_t + \mu\,dt \tag{14.7}$$

which is a modified form of Equation (14.1) in which the volatility component is allowed to depend on the price level. Geske deduces an explicit form for the function $\sigma(P)$ which is related to the debt structure of the firm, but it is also possible to consider models which begin by specifying arbitrary forms for Equation (14.7). One class of such models is the Constant Elasticity of Variance (CEV) class of models proposed by Cox and Ross (1976) for which $\sigma(P) = \sigma P^{\alpha-1}$ for some $\alpha \in (0, 1)$. In the CEV model there is a negative correlation between the volatility and price level.

The class of models of the form of Equation (14.7) has several desirable features. First, for suitable choices of $\sigma(P)$, the dependence between volatility and price level can be modelled. Secondly, the model is complete and, as in the Black–Scholes model, there is a unique preference independent price for an option.

Given a new model it is illuminating to compare the options prices from this model with those from the Black–Scholes formula. Consider the following exercise; calculate the prices of call options under the alternative model for a range of different strikes and exercise dates. Use these prices to derive the Black–Scholes implied volatility of each option. Finally, plot the resulting implied volatility surface, and compare with market implied volatility data.

Figure 14.4 illustrates the results of such an exercise for the CEV model. Motivated by the calculations of Schroder (1989) we set $\alpha = 2/3$ so that there are simple explicit expressions for the prices of call options. There is a negative skew in the implied volatility surface and this factor dominates any convexity or smile effects. There is also a small increase in implied volatility with maturity.

In general, leverage effects result in a negative skew in the shape of the implied volatility smile, and they may help explain some of the observed biases in market data. However, it is not possible to capture volatility smiles in models that are motivated by leverage considerations alone.

14.2.3   Models of stochastic volatility

Volume and leverage effects can partially account for the observed patterns in volatility. However, these explanations remain incomplete and more sophisticated models for the volatility are required which allow for further random changes in the level of volatility. In response to this need, a series of models for asset price processes was proposed in the late 1980s which took volatility as an exogenous stochastic process.

Scott (1987), Wiggins (1987), Hull and White (1987, 1988), Stein and Stein (1991) and Heston (1993) each proposed models of the form

$$\frac{dP_t}{P_t} = \sigma_t dB_t + \mu dt$$

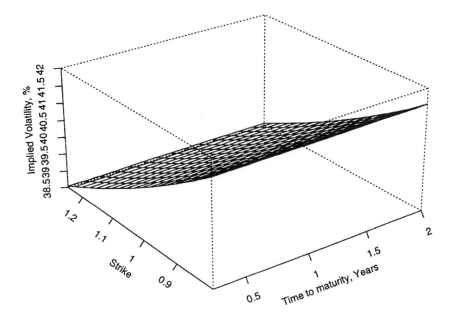

Figure 14.4 Implied volatilities for the CEV model. Note the strong inverse relationship between implied volatility and strike, which decreases only slightly with time.

where σ_t, the stochastic volatility process, is itself the solution of a stochastic differential equation. Several candidate SDEs for the volatility process have been suggested. The candidate models have generally been motivated by intuition, convenience and a desire for tractability, rather than because of an empirical relationship with realised volatility. In particular, the following models have all appeared in the literature:

$$d\sigma_t = \sigma_t(\alpha dt + \gamma dW_t) \tag{14.9}$$

$$d\sigma_t = \sigma_t((\alpha - \beta\sigma_t) dt + \gamma dW_t) \tag{14.10}$$

$$d\sigma_t = \beta(\alpha - \sigma_t) dt + \gamma dW_t \tag{14.11}$$

$$d\sigma_t = \left(\frac{\delta}{\sigma_t} - \beta\sigma_t\right) dt + \gamma dW_t \tag{14.12}$$

In each case W is a Brownian motion, perhaps correlated with the Brownian motion B which forms part of the specification (Equation (14.8)). Denote this correlation by ρ so that $(dB, dW_t) \equiv \rho dt$. We will assume that ρ is a constant with modulus less than one.

Equation (14.9) was introduced by Hull and White (1987) who took $\rho \equiv 0$ and Wiggins (1987) who considered the general case. The volatility is an exponential Brownian motion (or equivalently the logarithm of the volatility is a drifting Brownian motion). Scott (1987) considered the case of Equation (14.10) in which the logarithm of the volatility is an Ornstein-Uhlenbeck (OU) process. The discrete time analogue of an OU process is an $AR(1)$ time series, see Section 14.4.1 below. The models specified by Equations (14.9) and (14.10) have the advantage that the volatility is strictly positive for all time. However, even though the model (14.11) allows the process σ to become negative, this need not be a major handicap since Equation (14.8) remains well defined for negative values of σ, and it is possible to define volatility as the positive square root of the process σ_t^2. This third model was proposed by Scott (1987) and further investigated by Stein and Stein (1991). In both these articles the authors specialised on the case $\rho = 0$. In this model, the volatility process itself is an *OU* process with mean reversion level α. The final model (Equation (14.12)) was proposed by Hull and White (1988) and Heston (1993). In this model the volatility is related to the square-root process of Cox, Ingersoll and Ross (1985) and σ can be interpreted as the radial distance from the origin of a multidimensional *OU* process.

Two other models of note were proposed by Johnson and Shanno (1987) who modelled both the price and volatility as CEV processes, and Melino and Turnbull (1990) who took the price to be a CEV process and the logarithm of the volatility to be an *OU* process.

14.2.4   Transition densities

Consider the model

$$\frac{\mathrm{d}P_t}{P_t} = \sigma_t \mathrm{d}B_t + \mu \mathrm{d}t \tag{14.13}$$

$$\mathrm{d}\sigma_t = \gamma(\sigma_t)\,\mathrm{d}W_t + v(\sigma_t)\,\mathrm{d}t \tag{14.14}$$

where, for the moment, B and W are *independent* Brownian motions. Then σ and B are independent and, conditional on $(\sigma_s)_{0 \leqslant s \leqslant t}$, we have that $\int_0^t \sigma_s\,\mathrm{d}B_s$ is a Gaussian random variable with zero mean and variance $V_t \equiv \int_0^t \sigma_s^2\,\mathrm{d}s$. In particular, from the analogue of the representation of Equation (14.2), we have that $P_t = P_0 \exp\{Z\}$ where Z is a Gaussian random variable with mean $\mu t - \frac{1}{2}V_t$ and variance V_t. Thus, the transition density is a 'mixture of normals', with the mixing distribution depending on the autonomous stochastic process σ. If the value of the volatility is related to the rate of transactions then we recover the volume of transactions model described in Section 14.2.1, with a volume described by a random process.

Thus, for a stochastic volatility model in which the volatility is independent of the Brownian motion which drives the SDE for the price process, it is sufficient to characterise the law of V_t in order to derive the transition law for the price. Stein and Stein (1991) illustrate this result when the volatility process is an *OU* process and give an explicit form for the transition density.

When ρ is non-zero, the interactions between the volatility and the driving Brownian motion complicate the analysis. However, there is strong empirical evidence that ρ is non-zero. A negative value of ρ provides one method of capturing the observed negative correlation between volatility and price. Hence, it is worthwhile to pursue the general case and to resort to numerical methods if necessary (see Johnson and Shanno, 1987; Wiggins, 1987). Heston (1993) has devised an efficient method for calculating options using characteristic functions.

It is possible to recover the level dependent volatility models of, for example, Cox and Ross, and Geske, by taking $|\rho| = 1$ and choosing an appropriate, though potentially unwieldy, specification for the parameters γ and v in Equation (14.14). If $\rho = 1$ then the diffusion (P_t, σ_t) is degenerate and there is a deterministic relationship between the processes P and σ. The model is then similar in spirit to *GARCH* models, see Section 14.4.2. See also Hobson and Rogers (1998) who define a continuous time model of the form of Equation (14.8) in which B and W are perfectly correlated.

14.3   Option pricing for stochastic volatility models

In this section we consider the option pricing implications of diffusion models for the volatility. In particular, it is no longer true that there are unique preference independent options prices. Instead the model is incomplete and economic considerations (such as risk aversion) must be introduced to obtain pricing formulae.

Suppose that P and σ are defined as in Section 14.2.4 above, without the assumption that B and W are independent. Indeed write $W_t \equiv \rho B_t + (1 - \rho^2)Z_t$ for a Brownian motion Z which is independent of B. Suppose that the aim is to price an option, and that the price of that option is given by a (differentiable) function H which depends on the current value of the asset, the current volatility and the time to go. Then we can apply Itô's formula to $H(P_t, \sigma_t, T - t)$ to obtain

$$dH = H_1 dP + H_2 d\sigma + \Lambda dt$$

where suffixes denote partial differentiation with respect to the relevant coordinate of H and

$$\Lambda dt = \tfrac{1}{2} H_{11}(dP)^2 + H_{12}(dP)(d\sigma) + \tfrac{1}{2} H_{22}(d\sigma)^2 - H_3 dt$$
$$= \{\tfrac{1}{2} H_{11} P^2 \sigma^2 + \rho\gamma H_{12} P\sigma + \tfrac{1}{2}\gamma^2 H_{22} - H_3\} dt$$

If volatility were a traded asset then it would be possible to invest in volatility and the stock to form a riskless hedge portfolio for the option. However, this is not the case so there is no riskless hedge and the prices of options will depend on the risk preferences of investors. These preferences may be expressed via a utility function (see Hodges and Neuberger, 1989, or Karatzas *et al.*, 1991), or via a local-risk minimisation criterion (Hofmann *et al.*, 1992, or Platen and Schweizer, 1994).

Substituting for $d\sigma$ we obtain

$$dH = H_1 P \frac{dP}{P} + H_2\left(v dt + \frac{\gamma\rho}{\sigma}\left[\frac{dP}{P} - \mu dt\right] + \gamma(1-\rho^2)^{1/2} dZ\right) + \Lambda dt$$

$$= \left(H_1 P + \frac{\gamma\rho H_2}{\sigma}\right)\frac{dP}{P} + H_2\gamma(1-\rho^2)^{1/2} dZ + \left(H_2\left[v - \frac{\gamma\rho\mu}{\sigma}\right] + \Lambda\right) dt$$

Now define $\Psi = \Psi(P_t, \sigma_t, T - t)$ via

$$\Psi \equiv -\left(H_1 P + \frac{\gamma\rho H_2}{\sigma}\right)$$

Observe that the martingale term of $(dH + \Psi(dP/P))$ only involves dZ so that a portfolio consisting of a call and an amount Φ invested in the stock is uncorrelated with the stock. Asset pricing models imply that the rate of return on this portfolio must be r with an extra return for risk:

$$\mathbb{E}\left[dH + \Psi \frac{dP}{P}\right] = r(H + \Psi)\,dt - \lambda^* H_2 \gamma (1 - \rho^2)^{1/2}\,dt$$

where λ^* is the market price of volatility risk associated with dZ. Typically, the value $H + \Psi$ of the portfolio is negative which explains the sign convention for the market price of risk. Equating finite variation terms we obtain

$$H_2\left(v - \frac{\gamma \rho \mu}{\sigma}\right) + \Lambda = r(H + \Psi) - \lambda^* H_2 \gamma (1 - \rho^2)^{1/2}$$

Finally, some algebraic manipulation of this equation yields the stochastic volatility option pricing partial differential equation for H:

$$\tfrac{1}{2} H_{11} P^2 \sigma^2 + \rho \gamma H_{12} P \sigma + \tfrac{1}{2} \gamma^2 H_{22} - H_3 - rH$$

$$+ rH_1 P + H_2 \left(v - \frac{\gamma \rho (\mu - r)}{\sigma} + \lambda^* \gamma (1 - \rho^2)^{1/2}\right) = 0 \quad (14.15)$$

subject to the boundary condition $H(x, y, 0) = (x - K)^+$. Thus, the price of an option has an interpretation as the expected pay-off of the option under a model in which the price process and the volatility satisfy the SDEs

$$\frac{dP}{P} = \sigma_t dB + r dt$$

$$d\sigma_t = \gamma(\sigma_t) dW + \tilde{v}(\sigma_t)\,dt$$

where

$$\tilde{v}(\sigma) = v(\sigma) - \frac{\gamma(\sigma)\rho(\mu - r)}{\sigma} + \lambda^*(\sigma)\gamma(\sigma)(1 - \rho^2)^{1/2}$$

The option pricing equation (Equation (14.15)) has an analogue in expressions given by Wiggins (1987, Equation (8)), Scott (1987, Equation (4)) and Stein and Stein (1991, Equation (14)). In principle we solve Equation (14.15) to deduce theoretical options prices. Before we comment on the discussion in the literature on the (numerical) solutions of (14.15) some general comments are in order.

First, suppose that $\gamma \equiv 0 \equiv v$ so that the stochastic process (σ_t) is, in fact, a deterministic constant. Then we can view σ as a constant parameter of the model rather than a stochastic variable, and the option pricing Equation (14.15) for the price $C \equiv C(P, u)$ of a call option as a function of the price P of the underlying asset and the time to go u reduces to

$$\tfrac{1}{2} P^2 \sigma^2 C_{PP} - C_u + rPC_P - rC = 0$$

with boundary condition $C(x, 0) = (x - K)^+$. This is the Black–Scholes partial differential equation, and C is the Black–Scholes price of an option.

Secondly, if volatility is stochastic but uncorrelated with the asset (so that $\rho = 0$), then the option price can be expressed as

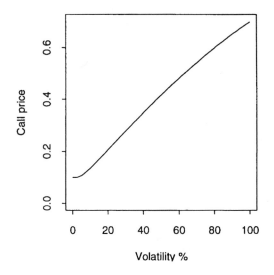

Figure 14.5   The price of an option as a function of volatility. The plot is based on an option with strike $K = 0.9$, and expiry $T = 3$, for an underlying asset whose price is unity. Thus $\sigma_I = 26.5\%$.

$$H(P_t, \sigma_t, T - t) = \tilde{\mathbb{E}}[(P_T - K)^+] = \tilde{\mathbb{E}}[\tilde{\mathbb{E}}[(P_T - K)^+ \mid (\sigma_s)_{t \leqslant s \leqslant T}]]$$
$$= \tilde{\mathbb{E}}\left[C\left(\left(\frac{1}{T-t} \int_t^T \sigma_s^2 \, ds \right)^{1/2} \right) \right]$$

Thus, the option price is an average of Black–Scholes prices. To investigate this relationship further consider the dependence of the Black–Scholes formula on σ. Suppose $t = 0$ and define $\sigma_I = ((2/T) |\ln m|)^{1/2}$ where m is the moneyness of the option. Thus, σ_I is zero for at the money options. Then C is convex in σ for $\sigma \leqslant \sigma_I$, and concave for $\sigma \geqslant \sigma_I$; see Fig. 14.5. Thus, for an at the money option with Black–Scholes implied volatility $\tilde{\sigma}$ it follows from Jensen's inequality that

$$C(\tilde{\sigma}) \equiv \tilde{\mathbb{E}}[C((V_T/T)^{1/2})] \leqslant C(\tilde{\mathbb{E}}((V_T/T)^{1/2})$$

where $V_T = \int_0^T \sigma_s^2 \, ds$. By the monotonicity of the Black–Scholes formula, for an at the money option, the Black–Scholes implied volatility is less than the expected average volatility, under the risk-neutral pricing measure. Conversely, for a far in or out of the money option, then for σ_0 sufficiently small

$$C(\tilde{\sigma}) \equiv \tilde{\mathbb{E}}[C((V_T/T)^{1/2})] \geqslant C(\tilde{\mathbb{E}}((V_T/T)^{1/2}))$$

Thus, we expect that the implied volatility for away from the money options will exceed the expected average volatility, and that there will be an implied volatility smile.

Renault and Touzi (1996) show that, again in the case $\rho = 0$, the volatility smile is symmetric. Consider the Black–Scholes call price C as a function of the moneyness M_t and the time to go u, then Equation (14.5) yields

$$C(M^{-1}, u) = K\,e^{-rT}(M^{-1}\Phi(-d_2) - \Phi(-d_1))$$
$$= \frac{K\,e^{-rT}}{M}((1 - M) + M\Phi(d_1) - \Phi(d_2))$$
$$= M^{-1}C(M, u) + K\,e^{-rT}(M^{-1} - 1)$$

Hence, there is a simple expression relating the prices of in and out of the money calls. Moreover, if we think of the stochastic volatility option pricing function H as a function of moneyness, the current value of the volatility σ_t, and the time to expiry u, then an investigation of the solutions to Equation (14.15) yields that, provided $\rho = 0$.

$$H(M^{-1}, \sigma_t, u) = M^{-1}H(M, \sigma_t, u) + K\,e^{-rT}(M^{-1} - 1)$$

To verify this claim observe that $MH(M^{-1}, \sigma, u) + K\,e^{-rT}(1 - M^{-1})$ also solves Equation (14.15) and the same boundary condition. From this it is a simple exercise to deduce that if the option with moneyness M has an implied volatility $\tilde{\sigma}_M$, so that $H(M, \sigma_t, u) = C(M, u; \tilde{\sigma}_M)$, then also $H(M^{-1}, \sigma_t, u) \equiv C(M^{-1}, u; \tilde{\sigma}_M)$. Now, since $\tilde{\sigma}_{M^{-1}}$ is the value of the implied volatility for which $H(M^{-1}, \sigma_t, u) \equiv C(M^{-1}, u; \tilde{\sigma}_{M^{-1}})$, we must have that $\tilde{\sigma}_M \equiv \tilde{\sigma}_{M^{-1}}$.

In more general situations with non-zero correlation ρ, the picture is more complicated. Several authors have attempted to solve Equation (14.15) in this case. Hull and White (1988) consider solutions which take the form of power series expansions in the volatility of volatility parameter γ:

$$H(M, \sigma_t, u) = C(M, u) + f_0(M, \sigma_t, u) + \gamma f_1(M, \sigma_t, u) + \gamma^2 f_2(M, \sigma_t, u) + \dots$$

Explicit, though complicated, forms can be deduced for the functions $f_0, f_1, f_2 \dots$. See Fig. 14.6 for the predicted implied volatility surface based on an expansion to second order of the Hull and White option pricing series. Alternatively, Johnson and Shanno (1987), Wiggins (1987) and Heston (1993) calculate numerical solutions to Equation (14.15). In each case the authors find that when the correlation is negative, out of the money call options are relatively more expensive under a stochastic volatility model when compared with Black–Scholes prices. This is consistent with the biases found in Rubinstein (1985), Wiggins (1987), Heynen *et al.* (1994), and Fig. 14.3. Wiggins attempts to derive an estimate for ρ. His estimates support the hypothesis that ρ is negative, but the precise estimates vary widely depending on the particular method he uses. However, he does provide evidence that the negative correlation is more pronounced for indices rather than individual stocks.

Continuous-time stochastic volatility provides an attractive and intuitive explanation for observed volatility patterns and for observed biases in implied volatility. In particular, smiles, skews and upward and downward implied volatility term structures arise naturally from a stochastic volatility model. However, the fact that stochastic volatility models fit empirical patterns is not conclusive evidence that those models are correct and the biases in market prices may be the result of other factors such as liquidity problems.

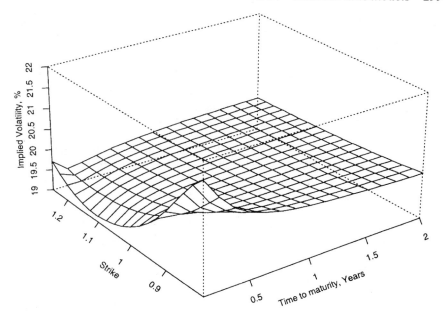

Figure 14.6 Implied volatilities from the Hull–White expansion to second order for an option on an underlying asset whose current price is unity. Note that $\rho = -0.2$ and that for options which are close to maturity there is a pronounced volatility smile, and some evidence of an additional inverse relationship between strike and implied volatility. These effects decrease with maturity.

14.4 Discrete-time models

Whilst continuous time models provide the natural framework for an analysis of option pricing, discrete time models are ideal for the statistical and descriptive analysis of the patterns of daily price changes. There are two main classes of discrete-time models for stock prices with volatility. The first class, the *autoregressive random variance (ARV)* or stochastic variance models are a discrete time approximation to the continuous time diffusion models we outlined in Sections 14.2 and 14.3. The second class of *autoregressive conditional heteroscedastic (ARCH)* models and its descendents are motivated by an attempt to explain volatility clustering and the habit of large price changes to be followed by further large changes.

14.4.1 *ARV* models

Let $Y_n = \ln P_n$ so that Y_n denotes the log price. Then the natural discrete time analogue of Equations (14.13) and (14.14) is to take

$$Y_n = Y_{n-1} + v + \sigma_{n-1} Z_n \qquad (14.16)$$

where $(Z_n)_{n \geq 0}$ is a sequence of independent standard normal variables and σ_n is the solution of a stochastic difference equation. Many authors, including

Chesney and Scott (1989) and Duffie and Singleton (1993), consider a model of the form

$$\ln \sigma_n = \alpha - \phi(\ln \sigma_{n-1} - \alpha) + \theta z_n \qquad (14.17)$$

for parameters α, ϕ, θ and z_n, a sequence of independent identically distributed random variables, such that (Z_n, z_n) forms a bivariate normal sequence with correlation ρ. Equation (14.17) is a direct analogue of Equation (14.10). The model specified by Equations (14.16) and (14.17) is called an *ARV* model (Taylor, 1986).

The *ARV* model is stationary if $|\phi| < 1$ and then $\ln \sigma_t$ has mean α and variance $\beta = \theta^2/(1 - \phi^2)$. Provided that $\rho = 0$, the unconditional distribution of the return $Y_t - Y_{t-1}$ is a mixture of normal distributions, with analogues in the rate of transaction models of Clark (1973) and Tauchen and Pitts (1983).

In the *ARV* model the volatility process is unobservable, which contrasts with the continuous time situation in which the instantaneous value for volatility can be inferred from the quadratic variation of the log-price. As an unfortunate consequence, most *ARV* models lack one-step transition densities for the process Y_n. This means that it is frequently not possible to obtain maximum likelihood estimates for parameter values.

Instead, parameter values are frequently estimated using methods of moments techniques, see Taylor (1986), Melino and Turnbull (1990), and Duffie and Singleton (1993). Of particular interest is the autoregressive coefficient ϕ which governs the persistence of volatility shocks. According to Taylor (1994), most estimates of this parameter, which are based on daily observations, yield values greater than 0.95. Harvey *et al.* (1994) find that a multivariate *ARV* model fits well to exchange rates data and captures movements in volatility, although for certain currencies they are led to suggest a heavy tailed distribution for the innovations process. See Ghysels *et al.* (1996) for a thorough discussion of *ARV* models and their statistical properties.

Since the *ARV* model is an approximation to diffusion models of stochastic volatility there is a correspondence between options prices in an *ARV* model and numerical solutions of the stochastic volatility option pricing equation (Equation (14.15)). Thus, options prices in *ARV* models are preference dependent, and an *ARV* model can account for smiles and skews in implied volatility.

14.4.2   *ARCH* and *GARCH* models

Autoregressive conditional heteroscedastic models were introduced by Engle (1982) in an attempt to model persistence in volatility shocks by assuming an autoregressive structure for the conditional variances. Retaining the convention that $Y_n = \ln P_n$ an *ARCH* model assumes that

$$Y_n = Y_{n-1} + v + \eta_n \epsilon_n \qquad \epsilon_n \text{ i.i.d. } D(0, 1) \qquad (14.18)$$

where D is a general distribution with zero mean and unit variance, and η_n is a function of the past proportional price changes. The simplest *ARCH* model, an *ARCH*(1) combines Equation (14.18) with

$$\eta_n^2 = \alpha + \beta(Y_{n-1} - Y_{n-2} - v)^2 = \alpha + \beta \eta_{n-1}^2 \epsilon_{n-1}^2 \qquad (14.19)$$

ARCH models have the advantage that it is straightforward to write down the log-likelihood and hence to derive maximum likelihood estimators for the parameters.

In empirical applications, higher order $ARCH(q)$ models with a large number of parameters are often needed to characterise the behaviour of financial time series. To circumvent this problem Bollerslev (1986) devised a class of *generalised ARCH* or *GARCH* models which allow the conditional variance to depend directly on previous values. In a $GARCH(1,1)$ model (independently proposed by Taylor, 1986) we have that

$$\eta_n^2 = \alpha + \beta(Y_{n-1} - Y_{n-2} - v)^2 + \gamma\eta_{n-1}^2 = \alpha + \beta\eta_{n-1}^2\epsilon_{n-1}^2 + \gamma\eta_{n-1}^2 \qquad (14.20)$$

Other extensions are also possible, see Bollerslev *et al.* (1994), Harvey *et al.* (1994) or Shephard (1996) for comprehensive surveys.

As defined in Equations (14.19) and (14.20) the updates of the conditional variance depend on the squares of the residual process. Hence these simple models cannot capture leverage effects. However, the exponential *ARCH* model of Nelson (1991) does not treat positive and negative innovations symmetrically and can allow for a correlation between volatility, as expressed by η, and price level.

The natural candidate distribution for $D(0, 1)$ is standard normal. However, some empirical studies of stock prices, including for example Bollerslev (1986) and Bollerslev *et al.* (1994), have found that the standardised innovations process $(Y_n - Y_{n-1} - v)\eta_n^{-1}$ displays excess kurtosis. Taylor (1994) suggests the use of a scaled *t*-distribution or a generalised error distribution. With these choices there are two sources of the kurtosis in the unconditional distribution for the log-price, namely the kurtosis from the price innovations and the changes in the underlying volatility level η.

GARCH models have been extremely successful in the modelling of equity markets. Highly significant test statistics have been reported by Engle and Mustafa (1992) in an analysis of stock returns, and by Schwert (1990) for futures markets. See Bollerslev *et al.* (1992) for an extensive survey of articles reaching similar conclusions. The autoregressive structure imposed by the *GARCH* model for the conditional variances allows volatility to persist over time and captures the observed clustering of large price movements. Note however that Lamoureux and Lastrapes (1990) find that daily trading volume has a significant explanatory power regarding the variance of daily returns and that, furthermore, *ARCH* effects tend to disappear when volume is included in the variance equation.

In an *ARV* model the volatility process is an autonomous process. In contrast, in *GARCH* models the volatility process is a deterministic function of the innovations ϵ_n. Nevertheless, Nelson (1990) has shown that with a judicious choice of parameter values the continuous time limit of a *GARCH* process is a diffusion model with stochastic volatility of the form of Equations (14.8) and (14.11). However, the rate of convergence to the diffusion limit is much slower than that from an *ARV* model. Bollerslev *et al.* (1994) show both how a *GARCH* model can be used to approximate a diffusion and how a diffusion process can be used to approximate a *GARCH* model.

Consideration of the SDE high frequency limit of *GARCH* processes raises the problem of temporal aggregation of *GARCH* processes. The non-linearities of *GARCH* models mean that if a low-frequency sample is taken from a high frequency *GARCH* model, then the resulting time series is not *GARCH*, at least in the sense defined above. Although Drost and Nijman (1993) have introduced the concept of *weak GARCH* models which are stable under temporal aggregation, in general the frequency of observations has an important bearing on the statistical properties of the model. For example an i.i.d. innovations process in the definition of the price process $(Y_n)_{n \geqslant 0}$ will generally result in time dependence of the innovations of $(Y_{kn})_{n \geqslant 0}$ for $k > 1$.

14.4.3 *GARCH* option pricing

Even for discrete time models the pricing of options remains an important issue, and there has been much recent interest in *GARCH* option pricing formulae, which is summarised in the paper by Duan (1995).

Duan assumes a model in which the innovations ϵ_n are normal variables and v takes the form

$$v = r + \lambda^* \eta_{n-1} - \tfrac{1}{2}\eta_{n-1}^2$$

where λ^* is a volatility risk premium, and the $-\tfrac{1}{2}\eta_{n-1}^n$ term ensures that when $\lambda^* = 0$ the discounted price process is a martingale. With this specification, the price process evolves as

$$P_n = P_{n-1} \exp\{r + \lambda^* \eta_{n-1} + \eta_{n-1}\tilde{\epsilon}_n - \tfrac{1}{2}\eta_{n-1}^2\}$$

The concept of a *locally risk neutral valuation relationship* is used to argue that options should be priced as the discounted expected pay-off under a model in which the price and volatility update according to the stochastic difference equations

$$Y_n = Y_{n-1} + r - \tfrac{1}{2}\eta_{n-1} + \eta_{n-1}\tilde{\epsilon}_n$$
$$\eta_n^2 = \alpha + \beta\eta_{n-1}(\tilde{\epsilon}_{n-1} - \lambda^*)^2 + \gamma\eta_{n-1}^2$$

for an i.i.d. sequence $\tilde{\epsilon}_n$ of standard normal variables.

Duan estimates the parameters of the model from market prices and uses Monte Carlo techniques to obtain options prices. He finds implied volatility smiles which become weaker as time to maturity increases. Depending on the initial value of the conditional volatility the term structure of implied volatility of an at the money option can be either downward or upward sloping.

In general, prices from the *GARCH* option pricing model are consistent with the biases found by Rubinstein (1985) and Skeikh (1991). However, although this evidence supports the *GARCH* modelling hypothesis, it cannot guarantee the veracity of the model. Moreover, Monte Carlo techniques are a computationally expensive technique for calculating options prices.

14.5 Conclusions

The Black–Scholes exponential Brownian motion model provides an approximate description of the behaviour of asset prices and a benchmark against which other models can be compared. The volatility parameter is a crucial component of the model and stochastic volatility models aim to reflect the apparent randomness of the level of volatility, as observed in empirical studies.

To this extent, stochastic volatility models are partially successful and, moreover, they can capture, and potentially explain, some of the observed biases in the Black–Scholes formula for options. Both diffusion models and *GARCH* models can account for smiles, skews and term structures which have been observed in market prices for options, and stochastic volatility models are widely used in the financial community as a refinement of the Black–Scholes model. Exotic options are frequently even more sensitive to levels of volatility than standard calls and, as trading in such instruments blossoms, those financial institutions which have models with the ability to price and hedge derivatives, reasonably and consistently, will have a competitive advantage.

References

Bachelier, L. (1900) Théorie de la speculation. *Annales Scientifiques de l'Ecole Normale Supérieure, Troisiéme serie*, **17**, 21–88. Translated in *The Random Character of Stock Market Prices*, Cootner, P. (Ed), MIT Press (1965).

Bensoussan, A., Crouhy, M. and Galai, D. (1994) Stochastic equity volatility and the capital structure of the firm. *Philosophical Transactions of the Royal Society of London, Series A*, **347**, 449–598.

Black, F. and Scholes, M. (1973) The pricing of options and corporate liabilities. *Journal of Political Economy*, **81**, 637–659.

Blattberg, R. C. and Gonedes, N. J. (1974) A comparison of the stable and Student distributions as statistical models for stock prices. *Journal of Business*, **47**, 244–280.

Bollerslev, T. (1986) Generalised autoregressive conditional heteroskedasticity. *Journal of Econometrics*, **31**, 307–327.

Bollerslev, T., Chou, R. Y. and Kroner, K. F. (1992) *ARCH* modelling in finance: a review of the theory and empirical evidence. *Journal of Econometrics*, **52**, 5–59.

Bollerslev, T., Engle, R. F. and Nelson, D. B. (1994) *ARCH* models. In *The Handbook of Econometrics*, Vol. 4, R. F. Engle and D. L. McFadden (Eds), Amsterdam: Elsevier.

Chesney, M. and Scott, L. O. (1989) Pricing European options: a comparison of the modified Black–Scholes model and a random variance model. *Journal of Financial and Quantitative Analysis*, **24**, 267–284.

Clark, P. K. (1973) A subordinated stochastic process model with finite variance for speculative prices. *Econometrica*, **41**, 135–155.

Cox, J. C., Ingersoll, J. E. and Ross, S. A. (1985) An intertemporal general equilibrium model of asset prices. *Econometrica*, **53**, 363–384.

Cox, J. C. and Ross, S. A. (1976) The valuation of options for alternative stochastic processes. *Journal of Financial Economics*, **3**, 145–166.

Drost, F. C. and Nijman, T. E. (1993) Temporal aggregation of *GARCH* processes. *Econometrica*, **61**, 909–927.

Duan, J.-C. (1995) The *GARCH* option pricing model. *Mathematical Finance*, **5**, 13–32.

Duffie, D. and Singleton, K. J. (1993) Simulated moments estimation of Markov models of asset prices. *Econometrica*, **61**, 929–952.

Engle, R. F. (1982) Autoregressive conditional heteroskedasticity with estimates of the variance of United Kingdom inflation. *Econometrica*, **50**, 987–1007.

Engle, R. F. and Mustafa, C. (1992) Implied *ARCH* models from options prices. *Journal of Econometrics*, **52**, 289–311.

Epps, T. W. and Epps, M. L. (1976) The stochastic dependence of security price changes and transaction volumes: implications for the mixture-of-distributions hypothesis. *Econometrica*, **44**, 305–321.

Fama, E. F. (1965) The behaviour of stock market prices. *Journal of Business*, **38**, 34–105.

Fung, W. K. H. and Hsieh, D. A. (1991) Empirical analysis of implied volatility: stocks, bonds and currencies. *Proceedings of the Fourth FORC Conference*, 19–20 July, University of Warwick.

Geske, R. (1979) The valuation of compound options. *Journal of Financial Economics*, **7**, 63–81.

Ghysels, E., Harvey, A. and Renault, E. (1996) Stochastic Volatility. In *Handbook of Statistics, Vol. 14, Statistical Methods in Finance*. Amsterdam: North Holland.

Harvey, A., Ruiz, E. and Shephard, N. (1994) Multivariate stochastic variance models. *Review of Economic Studies*, **61**, 247–264.

Heston, S. L. (1993) A closed-form solution for options with stochastic volatiltiy with applications to bond and currency options. *Review of Financial Studies*, **6**, 327–343.

Heynen, R., Kemma, A. and Vorst, T. (1994) Analysis of the term structure of implied volatility. *Journal of Financial and Quantitative Analysis*, **29**, 31–56.

Hobson, D. G. and Rogers, L. C. G. (1998) Complete models of stochastic volatility. To appear in *Mathematical Finance*.

Hodges, S. D. and Neuberger, A. (1989) Optimal replication of contingent claims under transaction costs. *Review of Futures Markets*, **8**, 222–239.

Hofmann, N., Platen, E. and Schweizer, M. (1992) Option pricing under incompleteness and stochastic volatility. *Mathematical Finance*, **2**, 153–187.

Hull, J. (1993) *Options, Futures and Other Derivative Securities*. Second Edition. Englewood Cliffs, NJ: Prentice Hall.

Hull, J. and White, A. (1987) The pricing of options on assets with stochastic volatilities. *Journal of Finance*, **42**, 281–300.

Hull, J. and White, A. (1988) An analysis of the bias in option pricing caused by a stochastic volatility. *Advances in Futures and Options Research*, **3**, 29–61.

Johnson, H. and Shanno, D. (1987) Option pricing when the variance is changing. *Journal of Financial and Quantitative Analysis*, **22**, 143–151.

Karatzas, I., Lehoczky, J. P., Shreve, S. E. and Xu, G.-L. (1991) Martingale and duality methods for utility maximisation in an incomplete market. *SIAM Journal of Control and Optimisation*, **29**, 702–730.

Karpoff, J. M. (1987) The relation between price changes and trading volume. *Journal of Financial and Quantitative Analysis*, **22**, 109–126.

Lamoureux, C. G. and Lastrapes, W. D. (1990) Heteroskedasticity in stock return data: volume versus *GARCH* effects. *Journal of Finance*, **45**, 221–229.

Madan, D. B. and Seneta, E. (1990) The variance gamma (V.G.) model for share market returns. *Journal of Business*, **63**, 511–524.

Mandlebrot, B. B. (1963) The variation of certain speculative prices. *Journal of Business*, **36**, 394–416.

Melino, A. and Turnbull, S. M. (1990) Pricing foreign currency options with stochastic volatility. *Journal of Econometrics*, **45**, 239–265.

Nelson, D. B. (1990) *ARCH* models as diffusion approximations. *Journal of Econometrics*, **45**, 7–38.

Nelson, D. B. (1991) Conditional heteroskedasticity in asset returns: a new approach. *Econometrica*, **59**, 347–370.

Øksendal, B. (1985) *Stochastic Differential Equations*, Berlin: Springer-Verlag.

Osborne, M. F. M. (1959) Brownian motion in the stock market. *Operations Research*, **7**, 145–173.

Peters, E. E. (1991) *Chaos and Order in the Capital Markets*. New York: Wiley.

Platen, E. and Schweizer, M. (1994) On smile and skewness. Discussion paper No. B–302, University of Bonn.

Renault, E. and Touzi, N. (1996) Option hedging and implied volatilities in a stochastic volatility model. *Mathematical Finance*, **6**, 279–302.

Rubinstein, M. (1983) Displaced diffusion option pricing. *Journal of Finance*, **38**, 213–217.

Rubinstein, M. (1985) Nonparametric tests of alternative option pricing models using all reported trades and quotes on the 30 most active CBOE option classes from Aug. 23 1976 through Aug. 31 1978. *Journal of Finance*, **40**, 455–480.

Samuelson, P. A. (1965) Rational theory of warrant pricing. *Industrial Management Review*, **6**(2), 13–39.

Schroder, M. (1989) Computing the constant elasticity of variance option pricing formula. *Journal of Finance*, **44**, 211–219.

Schwert, G. W. (1990) Stock volatility and the crash of 87. *Review of Financial Studies*, **3**, 77–102.

Scott, L. O. (1987) Option pricing when the variance changes randomly: theory, estimation and an application. *Journal of Financial and Quantitative Analysis*, **22**, 419–438.

Sheikh, A. (1991) Transaction data tests of S&P 100 call option pricing. *Journal of Financial and Quantitative Analysis*, **26**, 459–475.

Shephard, N. (1996) Statistical aspects of *ARCH* and stochastic volatility. In *Time Series Models in Econometrics, Finance and Other Fields*. D. R. Cox, D. V. Hinkley and O. E. Barndorff-Nielsen (Eds), Chapman and Hall Monographs in Statistics and Applied Probability, 65.

Stein, E. M. and Stein, J. C. (1991) Stock price distributions with stochastic volatility: an analytic approach. *Review of Financial Studies*, **4**, 727–752.

Tauchen, G. E. and Pitts, M. (1983) The price variability-volume relationship on speculative markets. *Econometrica*, **51**, 485–505.

Taylor, S. J. (1986) *Modelling Financial Time Series*. Chichester, UK: Wiley.

Taylor, S. J. (1994) Modelling stochastic volatility: a review and comparative study. *Mathematical Finance*, **4**, 183–204.

Wiggins, J. B. (1987) Option values under stochastic volatility. Theory and empirical estimates. *Journal of Financial Economics*, **19**, 351–372.

Xu, X. and Taylor, S. J. (1994) The term structure of volatility implied by foreign exchange options. *Journal of Financial and Quantitative Analysis*, **29**, 57–74.

15

Market Time and Asset Price Movements: Theory and Estimation

Eric Ghysels, Christian Gouriéroux and Joanna Jasiak

15.1 Introduction

The concept of a time deformed or subordinated process to model security prices originated in the work by Mandelbrot and Taylor (1967) and Clark (1973), among others, who argued that since the number of transactions in any time period is random, one may think of asset price movements as the realisation of a process $Y_t = Y_{z_t}^*$, where Z_t is a directing process. This positive non-decreasing stochastic process Z_t can, for instance, be thought of as related to the number of transactions or, more fundamentally, to the arrival of information. This, by now familiar, concept of subordinated stochastic processes, originated by Bochner (1960), was used by Mandelbrot and Taylor (1967) and later refined by Clark (1970, 1973) to explain the behaviour of speculative prices. Originally, it was mostly applied to daily observations since high frequency data were not available. A well-known example in finance is the considerable amount of empirical evidence documenting non-trading day effects. Such phenomena can be viewed as time deformation due to market closure.[1] Obviously, as pointed out by Mandelbrot and Taylor (1967), time deformation is also directly related to the mixture of distributions model of Tauchen and Pitts (1983), Harris (1987), Foster and Viswanathan (1993) among others. More to the point, regarding high frequency data one should mention that, in foreign exchange markets, there is also a tendency to rely on activity scales determined by the number of active markets around the world at any particular moment.

Dacorogna *et al.* (1993a) describe explicitly a model of time deformation along these lines for intra-day movements of foreign exchange rates. Besides

---

[1] Examples of such evidence include Lakonishok and Smidt (1988) and Schwert (1990) who argue that returns on Monday are systematically lower than on any other day of the week, while French and Roll (1986), French *et al.* (1987) and Nelson (1991) demonstrate that daily return volatility on the NYSE is higher following non-trading days.

these relatively simple examples, there are a number of more complex ones. Ghysels and Jasiak (1994) proposed a stochastic volatility model with the volatility equation evolving in an operational time scale. They use trading volume and leverage effects to specify the mapping between calendar and operational time. In Ghysels *et al.* (1995) this framework is extended and applied to intra-day foreign exchange data, providing an alternative to the Dacorogna *et al.* time scale transformation. Madan and Seneta (1990) and Madan and Milne (1991) introduced a Brownian motion evaluated at random (exogenous) time changes governed by independent gamma increments as an alternative martingale process for the uncertainty driving stock market returns. Geman and Yor (1993) also used time-changed Bessel processes to compute path-dependent option prices such as is the case with Asian options. It is also worth noting that there is some research specifically examining the time between trades, see Hausman and Lo (1990) and Han *et al.* (1994) for instance.

Despite the several examples just mentioned, there is no comprehensive treatment of the stochastic process theory and statistical estimation of subordinated processes. The aim of the paper is to describe some of the second-order stochastic properties and the statistical properties of time deformed models. Such models are, in principle, defined in two steps. We first consider the process of interest with respect to intrinsic time Y_z^*, and the time-changing process Z_t, which explains how to pass from calendar time to intrinsic time. Then, the process of interest expressed in calendar time is the subordinated process: $Y_t = Y_{z_t}^*$. Clearly, the observable model (the one corresponding to Y_t) is a dynamic factor model with Z_t as the underlying factor. As is typical in factor models we may distinguish different cases depending on whether Z is assumed to be observable (for instance when it relates a series like transactions volume or number of quotes) or unobservable. In the latter case, it is necessary to specify a latent factor process for Z_t (see Clark, 1970, 1973, for such an approach) and to predict *ex post* the values of the factor.

Sections 15.2 and 15.3 describe the second-order stochastic behaviour of time deformed processes and highlight their use in financial modelling. Sections 15.4 and 15.5 cover the empirical analysis of the processes. Besides estimation we also discuss diagnostic tests which help summarise the potentially vast amounts of data.

15.2   Study of mean and covariance functions

In this section we will compare the second-order properties of the process of interest, when it evolves in calendar time and in intrinsic (or operational) time. We first consider primarily discrete time models and study: (1) second-order stationarity; (2) the conservation of a unit root by time deformation; and (3) the relation between the autocovariance functions of Y and Y^*. A final subsection is devoted exclusively to continuous time models. Distributional and Markov properties will not be considered here, see however Ghysels *et al.* (1996).

15.2.1 Definition of the processes

To set the scene we first introduce some notation.

i) The time changing process, called the directing process by Clark (1973), associates the operational scale with the calendar time. It is a positive strictly increasing process:

$$Z : t \in \mathfrak{I} \longrightarrow Z_t \in \mathcal{Z} \tag{15.1}$$

(ii) The process of interest evolving in the operational time is denoted by:

$$Y^* : z \in \mathcal{Z} \longrightarrow Y_z^* \in \mathcal{Y} \subset \mathbb{R}^M \tag{15.2}$$

(iii) Finally, we may deduce the process in calendar time $t \in \mathfrak{I}$ by considering:

$$Y_t = Y^* \circ Z_t = Y_{z_t}^* \tag{15.3}$$

The introduction of a time scaling process is only interesting if the probabilistic properties of the process of interest become simpler. It explains the introduction of the assumption below which ensures that all the links between the two processes $(Y_t), (Z_t)$ in calendar time come from the time deformation.

Assumption A.15.l: *The two processes Z and Y^* are independent.*

Assumption A.15.1 is not entirely innocent with respect to practical applications. Indeed, if Z is tied to trading volume and Y^* is a return process, for instance, it is clear that the two may not be independent in operational time. However, we would feel more comfortable with letting Y^* be the bivariate process of return and volume and Z the (latent) process of information arrival. Hence, the use of Assumption A.15.1 has to be used judiciously. As noted before, we will proceed with this assumption as it makes the links between Z_t and Y_t result from subordination. It has also to be noted that our formalism allows for the treatment of both discrete and continuous time problems. Indeed one may consider: discrete calendar and operational times with $\mathfrak{I} = \mathcal{Z} = \mathbb{N}$, continuous calendar and operational times with $\mathfrak{I} = \mathcal{Z} = \mathbb{R}^+$ and finally $\mathfrak{I} = \mathbb{N}, \mathcal{Z} = \mathbb{R}^+$ for continuous operational time and discrete calendar time.

15.2.2 Second-order moments

Assuming that both processes are second-order integrable, we consider the first-order moments:

$$\left. \begin{array}{l} m(t) = E(Y_t) \text{ defined on } \mathfrak{I} \\ m^*(z) = E(Y_z^*) \text{ defined on } \mathcal{Z} \end{array} \right\} \tag{15.4}$$

and the autocovariance functions:

$$\left. \begin{array}{l} \gamma(t, h) = E\big[(Y_t - EY_t)(Y_{t+h} - EY_{t+h})'\big], \quad t \in \mathfrak{I}, h \in \mathfrak{I} \\ \gamma^*(z_0, z) = E\big[\big[Y_{z_0}^* - E(Y_{z_0}^*)\big]\big[Y_{z+z_0}^* - E(Y_{z+z_0}^*)\big]'\big], \quad z_0 \in \mathcal{Z}, z \in \mathcal{Z} \end{array} \right\} \tag{15.5}$$

From the definition of the time deformed process, we obtain:

$$m(t) = E(Y_t) = E\big[E\big(Y_{Z_t}^* \mid Z_t\big)\big]$$
$$\gamma(t, h) = E(Y_t Y_{t+h}') - (EY_t)(EY_{t+h})' = E\big[E(Y_t Y_{t+h}' \mid Z_t, Z_{t+h})\big] - (EY_t)(EY_{t+h})'$$
$$\mathrm{Cov}\,(Y_t, Z_{t+h}) = E(Y_t, Z_{t+h}) - EY_t EZ_{t+h} = E[E(Y_t^* \mid Z_t)Z_{t+h}] - EY_t EZ_{t+h}$$

Taking into account the independence assumption between the two processes Z and Y^*, we can establish the following result.

Property 15.2.1.1:　Under Assumption A.15.1

$$m(t) = E[m^*(Z_t)]$$
$$\gamma(t, h) = E[\gamma^*(Z_t, Z_{t+h} - Z_t)] + \mathrm{Cov}\,[m^*(Z_t), m^*(Z_{t+h})]$$
$$\mathrm{Cov}\,(Y_t, Z_{t+h}) = \mathrm{Cov}\,(m^*(Z_t), Z_{t+h})$$

$\square$

15.2.3　Second-order stationarity

It is now possible to discuss some sufficient conditions for the second-order stationarity of the process Y. These conditions are *moment conditions* on the underlying process Y^*, and *distributional conditions* on the directing process Z.

Property 15.2.1.2:　Let us assume the independence Assumption A.15.1. holds. Then the Y process in calendar time is second-order stationary if the following assumptions are satisfied.　$\square$

Assumption A.15.2:　Y^* *is second-order stationary:* $m^*(z) = m^*$, $\forall z$, $\gamma^*(z_0, z) = \gamma^*(z)$, $\forall z_0, z$.

Assumption A.15.3:　*The directing process has strongly stationary increments: the distribution of* $\tilde{\Delta}_h Z_t = Z_{t+h} - Z_t$ *is independent of* t, $\forall h, t$.

15.2.4　Integrated processes

A consequence of Property 15.2.1.2 is that we can have second-order stationarity of the processes Y and Y^* simultaneously. In such a case we get $m(t) = m^*$, $\gamma(t, h) = E[\gamma^*(\tilde{\Delta}_h Z_t)]$, $\mathrm{Cov}\,(Y_t, Z_{t+h}) = 0$, $\forall h$, and in particular we observe no correlation between the series Y and Z, while Y is a (stochastic) function of Z.

Another case of interest is that of a unit root in the calendar time process Y_t. Considering the case $\mathcal{Z} = \mathbb{N}$ we first discuss sufficient conditions for the second-order stationarity of the differentiated process $\tilde{\Delta} Y_t = Y_{t+1} - Y_t$. Let us examine the first and second-order moments of the increments of the underlying process Y^*:

$$E(Y_{z_0+z}^* - Y_{z_0}^*) = \mu^*(z_0, z) \tag{15.6}$$

$$\mathrm{Cov}\,(Y_{z_0+z_1}^* - Y_{z_0}^*, Y_{z_0+z_2}^* - Y_{z_0+z_3}^*) = c^*(z_0, z_1, z_2, z_3) \tag{15.7}$$

Provided the independence Assumption A.15.1 holds, the first and second-order moments of the differentiated process $\tilde{\Delta} Y_t$ are:

$$\mu(t) = E(Y_{t+1} - Y_t) = E\mu^*(Z_t, \tilde{\Delta} Z_t) \qquad (15.8)$$

$$c(t, h) = \text{Cov}(Y_{t+1} - Y_t, Y_{t+h+1} - Y_{t+h})$$

$$= Ec^*(Z_t, \tilde{\Delta} Z_t, \tilde{\Delta}_{h+1} Z_t, \tilde{\Delta}_h Z_t) \qquad (15.9)$$

$$+ \text{Cov}[\mu^*(Z_t, \tilde{\Delta} Z_t), \mu^*(Z_{t+h}, \tilde{\Delta} Z_{t+h})]$$

Both equations yield the following result:

Property 15.2.1.3: Let us assume the independence Assumption A.15.1 holds. Then the process Y_t in calendar time is integrated of order 1, henceforth $I(1)$, and second-order stationary in first differences under the following set of assumptions. ☐

Assumption A.15.4: Y^* is $I(1)$ *and second-order stationary in first differences:*[2]

$$\mu^*(z_0, z) = \mu^*(z), \forall z_0, z, \; c^*(z_0, z_1, z_2, z_3) = c^*(z_1, z_2, z_3), \forall z_0, z_1, z_2, z_3$$

Assumption A.15.5: *The time changes have strongly stationary trivariate increments, i.e. the distribution of* $(\tilde{\Delta} Z_t, \tilde{\Delta}_h Z_t, \tilde{\Delta}_{h+1} Z_t)$ *is independent of t.*

As noted in Property 15.2.1.2, the calendar time process Y is stationary if Y^* is second-order stationary and Assumption A.15.5 is satisfied. From Property 15.2.1.3, however, we can also deduce that for Y to be non-stationary it is necessary that both Y^* and Z are non-stationary.

Finally, we can note that Assumption A.15.5 is satisfied for changing time processes defined by:

$$Z_t = \sum_{i=0}^{t} \eta_i \qquad (15.10)$$

where η_t is a strongly stationary process with positive values.

15.2.5   Continuous time modelling

Continuous time modelling is quite standard in financial theory and therefore it is also important to analyse the consequences of time deformation in such a framework. We will examine two formulations. In the first one the time deformation is assumed to be differentiable and based on a Brownian factor. The second formulation, based on gamma processes (see Madan and Seneta, 1990, Madan and Milne, 1991), takes into account the idea of a large number of discretely spaced trading dates.

---

[2] The conditions on the moments of the differentiated processes might also have been written in terms of the moments of the initial processes. For instance, the condition: $\mu^*(z_0, z) = \mu^*(z)\forall z_0, z$, is equivalent to: $m^*(z_0 + z) - m^*(z_0) = \mu^*(z)\forall z_0, z$. Whenever m^* is continuous, this means that M^* has a linear affine form: $m^*(z) = az + b$.

(i) Differentiable time deformation

The bivariate process (Y^*, Z) is described by the stochastic differential system:

$$\left.\begin{array}{l} dY_z^* = a^*(Y_z^*)\,dz + b^*(Y_z^*)\,dW_z^* \\ dZ_t = \alpha(W_t)\,dt \end{array}\right\} \tag{15.11}$$

where (W_z^*) and (W_t) are two independent Brownian motions, and α is a non-negative function. Since (Z_t) is differentiable, we directly deduce the equivalent system in calendar time:

$$\left.\begin{array}{l} dY_t = a^*(Y_t)\alpha(W_t)\,dt + b^*(Y_t)\alpha(W_t)^{1/2}\,dW_t^1 \\ dZ_t = \alpha(W_t)\,dt \end{array}\right\} \tag{15.12}$$

where (W_t^1) and (W_t) are independent Brownian motions. The evolution of (Y_t) with respect to the information associated with the two processes (Y, Z) depends on the two underlying factors W^1, W.

Since the pair of variables (W_t, dW_t^1) and (W_t, dW_t) have the same distribution, the marginal dynamics of the process (Y_t) alone (i.e. with respect to the information associated with Y) corresponds to the stochastic differential equation:

$$dY_t = a^*(Y_t)\alpha(W_t)\,dt + b^*(Y_t)\alpha(W_t)^{1/2}\,dW_t \tag{15.13}$$

As expected, we obtain a kind of autoregressive-moving average formulation for the dynamics of Y in calendar time, where the drift and volatility parameters both depend on Y_t and W_t. Hence, an effect of time deformation is to increase the lag orders.

(ii) Gamma time deformation

A drawback of the previous formulation is the assumption of differentiable time deformation, while existing markets show a large number of spaced-discretely trading dates. It is then interesting to introduce models for time deformation, such that the paths of (Z_t) may be step functions. Gamma processes have this property, with generally a countable dense set of jumps in their path.

A gamma process with parameter v, denoted by $[C(v)_t, t \in \mathcal{Z} = \mathbb{R}^+]$, is a process with stationary independent increments, such that the distribution of $C(v)_{t_2} - C(v)_{t_1}$ is a gamma distribution $\gamma(v(t_2 - t_1))$, for $t_2 \geqslant t_1 \geqslant 0$, and $C(v)_0 = 0$.

Let us consider the stochastic differential system:

$$\left.\begin{array}{l} dY_z^* = a^*(Y_z^*)\,dz + b^*(Y_z^*)\,dW_z^* \\ dZ_t = \lambda(Z_t)\,dC(v)_t \end{array}\right\} \tag{15.14}$$

where the two processes (W_z^*) and $([C(v)]_t)$ are independent, and $\lambda(Z_t)$ is a non-negative function introduced to create some time dependence in the magnitude and the density of the jumps in the Z process. The first and second-order moments directing processes are:

$$E(dZ_t \mid Z_t) = v\lambda(Z_t)\,dt \\ V(dZ_t \mid Z_t) = v\lambda^2(Z_t)\,dt \Bigg\} \tag{15.15}$$

Before characterising the corresponding system in calendar time, we have to introduce the composed process $(W[C(v)]_t)$. The process $(C(v), W[C(v)])$ has strong stationary independent increments, and the marginal distribution of $W[C(v)]_t$ is examined in the Appendix. The stochastic differential system in calendar time is then directly derived as:

$$dY_t = a^*(Y_t)\lambda(Z_t)\,dC(v)_t + b(Y_t)\lambda(Z_t)^{1/2}\,dW[C(v)]_t$$

To conclude, it is worth noting that, in the Appendix, we also provide a formulation of Itô's lemma for diffusions with gamma distributed subordination.

15.3 Examples

We noted in the introduction to Section 15.2 that time deformed processes are only interesting if we can tackle complex structures via simpler ones thanks to the rescaling of time. It is therefore important to have 'workable' examples which can be used in mathematical finance or in empirical estimation of discrete and/or diffusion processes. The examples described in this section will also serve as illustrations of the results established in the previous section. In this section, we will elaborate on several examples, beginning with time changed Bessel processes in Section 15.3.1, Ornstein–Uhlenbeck processes in Section 15.3.2 and, last but not least, the time deformed random walk with drift.

15.3.1 Time deformed Bessel processes

This first class of processes has been studied extensively by Yor (1992a,b). While we omit all the specific details here, as they are treated by Yor, we would like to use the example of Bessel processes to clarify further the relation between Equations (15.13) and (15.15). The initial model is:

$$dY_z^* = \left(\gamma + \frac{\sigma^2}{2}\right)(Y_z^*)^{-1}\,dz + \sigma dW_z^* \\ dZ_t = \exp 2(\sigma \tilde{W}_t + \gamma t)\,dt \Bigg\} \tag{15.16}$$

where Y_z^* follows a Bessel process. System (15.16) is similar to that defined in Equation (15.12) except that there are parametric restrictions which will be exploited shortly. Then, using Equation (15.13) we obtain:

$$dY_t = \left(\gamma + \frac{\sigma^2}{2}\right)\exp 2(\sigma W_t + \gamma t)Y_t^{-1}\,dt + \sigma \exp(\sigma W_t + \gamma t)\,dW_t$$

A solution of this stochastic differential equation can be written as:

$$Y_t = \exp\left[\sigma W_t + \gamma t\right] \Leftrightarrow dY_t = \left(\gamma + \frac{\sigma^2}{2}\right) Y_t \, dt + \sigma Y_t \, dW_t$$

which corresponds to a geometric Brownian motion with drift.

This example illustrates how simplifications arise because of the strong links introduced between the parameters defining the evolution of Y^* and the evolution of Z in Equation (15.16). As noted before, this process has some useful applications in finance in the pricing of options. See in particular Geman and Yor (1993) and Leblanc (1994). The former study the pricing of Asian options while the latter examines option pricing in a stochastic volatility context.

15.3.2   Time deformed Ornstein–Uhlenbeck process

The Ornstein–Uhlenbeck process is of course the simplest example of a stationary continuous time process satisfying a diffusion equation. It will therefore be ideal for illustrating the properties discussed in Section 15.2. Moreover, it is worth noting that this type of process appears in continuous time finance applications particularly in stochastic volatility models. Ghysels and Jasiak (1994) for instance used a subordinated Ornstein–Uhlenbeck process to analyse the Hull-White stochastic volatility model with a time deformed evolution of the volatility process. We will first examine the autocovariance structure of a subordinated Ornstein–Uhlenbeck process and show how time deformation affects the temporal dependence of the process. Typically, in discrete calendar time such processes have an ARMA representation with uncorrelated, and yet non-linearly dependent, innovations. We therefore also examine the non-linear dependencies.

15.3.2.1.   *Definition of the process*

We consider the one-dimensional case: $n = 1$. The process Y^* is defined as the stationary solution of the stochastic differential equation:

$$dY_z^* = k(m - Y_z^*) \, dz + \sigma \, dW_z^*, \quad k > 0, \quad \sigma > 0 \qquad (15.17)$$

where W^* is a Brownian motion indexed by $\mathcal{Z} = \mathbb{R}^+$, independent of the directing process. It is well known that Y^* is a Markov process of order one, and that the conditional distribution of $Y_{z+z_0}^*$ given $Y_{z_0}^*$ has a Gaussian distribution, with conditional mean:

$$E(Y_{z+z_0}^* \mid Y_{z_0}^*) = m + \rho^z(Y_{z_0}^* - m) \qquad (15.18)$$

and conditional variance:

$$V(Y_{z+z_0}^* \mid Y_{z_0}^*) = \sigma^2 \frac{1 - \rho^{2z}}{1 - \rho^2} \qquad (15.19)$$

with $\rho = \exp{-k}$. Let us now assume again that the independence Assumption A.15.1 holds and that $\mathfrak{J} = \mathbb{N}$. Then the previous properties may be rewritten in calendar time as:

$$Y_t = m + \rho^{\Delta Z_t}(Y_{t-1} - m) + \left\{ \sigma^2 \frac{1 - \rho^{2\Delta Z_t}}{1 - \rho^2} \right\}^{1/2} \epsilon_t \qquad (15.20)$$

where $\epsilon_t \sim I.I.N\,(0, 1)$ and independent of Z, with $\Delta Z_t = Z_t - Z_{t-1}$. Moreover, we also have a similar relation for lag h:

$$Y_t = m + \rho^{\Delta_h Z_t}(Y_{t-h} - m) + \left\{ \sigma^2 \frac{1 - \rho^{2\Delta_h Z_t}}{1 - \rho^2} \right\}^{1/2} \epsilon_{h,t} \qquad (15.21)$$

where $\epsilon_{h,t} \sim N\,(0, 1)$ and is independent of Z.

15.3.2.2. *The autocovariance function*

Now that we have formally defined the process, let us study its second-order properties. This entails, of course, a study of the temporal dependence of the process as measured by the autocovariance function. We will study several cases where we can compare the temporal dependence of Y^* and that of Y. From Property 15.2.1.1, and the fact that $m^*(z) = m$, $\gamma^*(z) = (\sigma^2 \rho^z)/(1 - \rho^2)$, we directly obtain that:

$$m(t) = m$$

$$\gamma(t, h) = \frac{\sigma^2}{1 - \rho^2} E(\rho^{\tilde{\Delta}_h Z_t})$$

It should be noted, however, that the second-order properties of the Y and Y^* processes may be rather different. To clarify this let us first examine a particular case in which they are similar. This is accomplished via the following result.

Property 15.3.2.1: If Z is a strong random walk, independent of the Ornstein–Uhlenbeck process Y^*, then Y has a linear autoregressive representation of order 1. ☐

Proof: From the definition of the autocorrelation we have:

$$\rho(h) = \frac{\gamma(h)}{\gamma(0)} = E(\rho^{\tilde{\Delta}_h Z_t}) = E[\rho^{\Delta Z_{t+h} + \dots + \Delta Z_{t+1}}] = E(\rho^{\Delta Z_{t+h}}) \dots E(\rho^{\Delta Z_{t+1}})$$

$$= [E(\rho^{\Delta Z_{t+1}})]^h = r^h, \quad \text{where } r = E(\rho^{\Delta Z_{t+1}}) \qquad ☐$$

Moreover, from the convexity inequality and the restriction $0 < \rho < 1$, we have:

$$0 \leqslant r = E(\rho^{\Delta Z_{t+1}}) \leqslant \rho^{E(\Delta Z_{t+1})} < 1$$

Hence, the process in calendar time is weakly stationary with an autoregressive coefficient which is smaller than ρ, if $E\rho^{\Delta Z_{t+1}} < 1$. In fact the value of r depends on the distribution of ΔZ_{t+1}. To illustrate this, let us consider increments with a Pascal distribution with parameter $\pi, 0 < \pi < 1$:

$$P[\Delta Z_{t+1} = n] = (1 - \pi)\pi^{n-1}, \quad n \geqslant 1$$

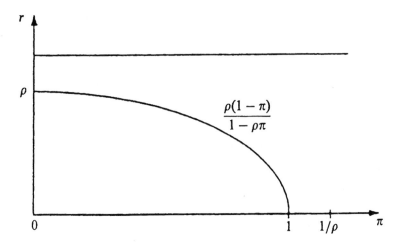

Figure 15.1 First-order calendar autocorrelation of time deformed Ornstein–Uhlenbeck process with the directing process a strong random walk with Pascal distributed increments (π is the parameter of Pascal distribution)

So, that:

$$r = \sum_{n \geqslant 1} \rho^n (1 - \pi)\pi^{n-1} = \frac{\rho(1 - \pi)}{1 - \pi\rho}$$

The effect of changing time is summarised in Fig. 15.1, where the autoregressive coefficient in calendar time is given as a function of π.

So far we have focused on a situation where Z_t is a (strong) random walk, as assumed in Property 15.3.2.1. Let us now consider a situation where Z_t is no longer a strong random walk, but ΔZ_t is still strongly stationary. We can then still characterise the asymptotic behaviour of the autocorrelation coefficient $\rho(\cdot)$. To do so we denote:

$$\Gamma^2 = \gamma(\Delta Z_t) + 2\sum_{h=1}^{\infty} \mathrm{Cov}\,(\Delta Z_t, \Delta Z_{t+h}) \qquad (15.22)$$

For h large, using a central limit argument we have:

$$\tilde{\Delta}_h Z_t = \Delta Z_{t+1} + \ldots + \Delta Z_{t+h} \simeq N[hE(\Delta Z_t), h\Gamma^2]$$

Exploiting this property yields:

$$\rho(h) = E[\rho^{\tilde{\Delta}_h Z_t}] \simeq E(\rho^{hE(\Delta Z_t)+\sqrt{h}\Gamma u}), \quad \text{where } u \sim N(0, 1)$$
$$= r_\infty^h$$

where:

$$r_\infty = \exp\left[E(\Delta Z_t)\log\rho + \frac{\Gamma^2(\log\rho)^2}{2}\right] \qquad (15.23)$$

Hence, for large h, the process Y has approximately the same properties as an autoregressive process of order 1, with autoregressive coefficient r_∞. In particular, we have a larger long range dependence in calendar time than in operational time if:

$$r_\infty > \rho \Leftrightarrow E(\Delta Z_t) - 1 + \frac{\Gamma^2}{2} \log \rho < 0 \tag{15.24}$$

This condition is automatically satisfied for $E(\Delta Z_t) < 1$, but it may also hold for $E(\Delta Z_t) > 1$, in particular if the variance and covariances $\mathrm{Cov}(\Delta Z_t, \Delta Z_{t+h})$ are sufficiently large.

We conclude this section by noting that the behaviour of the entire autocorrelation function can only be obtained under some simplifying assumptions regarding the temporal dependence of the ΔZ_t process. Let us, for instance, consider that ΔZ_t is a Markov chain, with a transition matrix M whose elements are:

$$m_{ij} = P[\Delta Z_t = j \mid \Delta Z_{t-1} = i], \quad i, j = 1, \dots, J \tag{15.25}$$

We then obtain a model with a qualitative factor whose alternatives define J regimes. This model is quite similar to the stochastic switching regime in Hamilton (1989), except that here the effect of the factor is non-linear. Suppose we denote by μ the invariant probability distribution associated with M, then the autocorrelation function is as follows:

$$\begin{aligned}
\rho(h) &= E(\rho^{\tilde{\Delta}_h Z_t}) = E[\rho^{\Delta Z_{t+h} + \dots + \Delta Z_{t+1}}] \\
&= \sum_{v_1,\dots,v_h} (\rho^{v_1 + \dots + v_h}) m_{v_{h-1}.v_h} m_{v_{h-2}.v_{h-1}} \cdots m_{v_1.v_2} \mu(v_1) \\
&= \sum_{v_1,\dots,v_h} \rho^{v_1} \mu(v_1) \rho^{v_2} m_{v_1.v_2} \cdots \rho^{v_h} m_{v_{h-1}.v_h}
\end{aligned}$$

Consider now the matrix $M(\rho)$ whose general element is of the form: $[M(\rho)]_{i,j} = \rho^j m_{i,j}$, while $\mu(\rho)$ is the vector whose general component is: $\rho^i \mu(i)$, then:

$$\rho(h) = M(\rho)^h \mu(\rho) \tag{15.26}$$

Whenever the matrix $M(\rho)$ has distinct eigenvalues $\lambda_1, \dots, \lambda_J$ we can write the autocorrelation as:

$$\rho(h) = \sum_{j=1}^{J} \alpha_j \lambda_j^h, \quad \alpha_j, \lambda_j \in \mathbb{C}$$

This implies that the process Y has a linear ARMA$[J, J-1]$ representation, with autoregressive polynomial:

$$\Phi(L) = \prod_{j-1}^{J} (1 - \lambda_j L) = \det [Id - M(\rho)L]$$

which implies that the time deformed process has longer lags than the process Y^* expressed in intrinsic time.

15.3.2.3.   *The conditional moments*

In the previous subsection we described how, under some circumstances, the process Y may have a linear ARMA representation. Yet, the innovations corresponding to such representation are generally uncorrelated but not white noise. In such a case it is of interest to have some information on the conditional distribution of Y_t given $Y_{t-1}, Y_{t-2,...}$ to capture the non-linear dependencies.

To do this we shall focus on a situation where Z is a strong random walk. Let us first consider the conditional expectation:

$$E(Y_t \mid Y_{t-1}) = E[E(Y_t \mid Y_{t-1}, \Delta Z_t) \mid Y_{t-1}] = E[m + \rho^{\Delta Z_t}(Y_{t-1} - m) \mid Y_{t-1}]$$

$$= m + E(\rho^{\Delta Z_t})(Y_{t-1} - m) = m + r(Y_{t-1} - m)$$

The latter implies that the optimal prediction coincides with the linear regression. However, let us study the conditional variance:

$$V(Y_t \mid Y_{t-1}) = V[E(Y_t \mid Y_{t-1}, \Delta Z_t) \mid Y_{t-1}] + E[V(Y_t \mid Y_{t-1}, \Delta Z_t) \mid Y_{t-1}]$$

$$= V[m + \rho^{\Delta Z_t}(Y_{t-1} - m) \mid Y_{t-1}] + E\left[\sigma^2 \frac{1 - \rho^{2\Delta Z_t}}{1 - \rho^2}\right]$$

$$= (Y_{t-1} - m)^2 V(\rho^{\Delta Z_t}) + \sigma^2 \frac{1 - E(\rho^{2\Delta Z_t})}{1 - \rho^2}$$

Hence, we note that, contrary to the underlying process Y^*, the process in calendar time features conditional heteroscedasticity. This was first noted by Stock (1988), who compared the behaviour of the time deformed Ornstein–Uhlenbeck process in discrete time with ARCH processes. This feature of the Ornstein–Uhlenbeck process makes it, of course, particularly attractive since financial time series are known to exhibit volatility clustering. In the next section we will, in fact, examine a related feature, namely that of leptokurtic asset return distributions as a result of time deformation.

15.3.3   The subordinated random walk with drift

The last class of processes we would like to study as an explicit example is random walks. Again, to facilitate our discussion we divide the section into several subsections. Section 15.3.3.1 covers the continuous time case which is exploited in Section 15.3.3.2 to illustrate applications in finance.

15.3.3.1.   *Definition of the process*

We assume that the initial process is a (multivariate) random walk with drift:

$$dY_z^* = a^* dz + B^* dW_z \tag{15.27}$$

where W_z is a standard Brownian motion. We immediately deduce from this that:

$$Y_t - Y_{t-1} = Y_{Z_t}^* - Y_{Z_{t-1}}^* = a^*(Z_t - Z_{t-1}) + B^*(W_{Z_t} - W_{Z_{t-1}})$$

so that the first differenced process can be written as:

$$\Delta Y_t = a^* \Delta Z_t + (\Delta Z_t)^{1/2} B^* \epsilon_t \qquad (15.28)$$

where $\epsilon_t \sim I.I.N\,[0, Id]$.

Moreover, the first and second-order moments of Y can directly be obtained from Equation (15.28), namely:

$$E(\Delta Y_t) = a^* E(\Delta Z_t), \quad \mathrm{Cov}\,(\Delta Y_t, \Delta Y_{t+h}) = a^* a^{*'} \mathrm{Cov}\,(\Delta Z_t, \Delta Z_{t+h} + E(\Delta Z_t) B^* B^{*'} \delta_0(h))$$

$$(15.29)$$

where $\delta_0(h)$ is the Kronecker symbol, and finally:

$$\mathrm{Cov}\,(\Delta Y_t, \Delta Z_{t+h}) = a^* \mathrm{Cov}\,(\Delta Z_t, \Delta Z_{t+h})$$

15.3.3.2. *Portfolio allocation*

We consider an optimal portfolio allocation problem and show how it depends on the information regarding the directing process Z_t. In particular, if Z_t is latent it will be shown that the optimal allocation rule will resemble one where there is no time deformation but where attitudes toward risk have been changed. To simplify the presentation, we will restrict ourselves to the case where Z_t is a random walk with drift.

Let us assume that the components of Y_t are the log-prices of a set of financial assets, and that the short term interest rate is equal to zero. We may determine two mean–variance optimal portfolios depending on whether or not we have information on time deformation. These optimal allocations are respectively:

$$\alpha_t(Z) = [V(\Delta Y_t \mid Z)]^{-1} E(\Delta Y_t \mid Z)$$
$$\alpha = V(\Delta Y_t)^{-1} E(\Delta Y_t)$$

Since the former is a function of Z_t it corresponds to the case where the portfolio allocation is an explicit function of an (observable) directing process. Replacing the moments by their explicit expressions, we have for the allocation rules using Z_t:

$$\alpha_t(Z) = (B^* B^{*'})^{-1} a^*$$

yielding a fixed composition of the optimal portfolio which is also equal to the composition in intrinsic time.

Now, without the information on the directing time process we have the following allocation rule:

$$\alpha = [a^* a^{*'} V(\Delta Z_t) + E(\Delta Z_t) B^* B^{*'}]^{-1} a^* E(\Delta Z_t)$$

$$= \left[a^* a^{*'} \frac{V(\Delta Z_t)}{E(\Delta Z_t)} + B^* B^{*'} \right]^{-1} a^*$$

$$= \left[(B^* B^{*'})^{-1} - \frac{V(\Delta Z_t)}{E(\Delta Z_t)} \frac{(B^* B^{*'})^{-1} a^* a^{*'} (B^* B^{*'})^{-1}}{1 + \frac{V(\Delta Z_t)}{E(\Delta Z_t)} a^{*'} (B^* B^{*'})^{-1} a^*} \right] a^*$$

$$= \left[1 + \frac{V(\Delta Z_t)}{E(\Delta Z_t)} a^{*'} (B^* B^{*'})^{-1} a^* \right]^{-1} (B^* B^{*'})^{-1} a^*$$

While this portfolio is proportional to $\alpha_t(Z)$ it can be seen that, to correct for the lack of information, the agent has to modify his risk aversion. Suppose the risk aversion coefficient is η, when the information on Z is available. Then to obtain the same optimal portfolio allocation without information requires a risk aversion coefficient equal to:

$$\eta^* = \eta \left[1 + \frac{V(\Delta Z_t)}{E(\Delta Z_t)} a^{*'} (B^* B^{*'})^{-1} a^* \right]$$

15.4   Statistical inference for subordinated stochastic processes

We turn our attention now to statistical issues involving the estimation of subordinated stochastic processes making inference from discrete data in calendar time. In the next subsection we describe the role played by the different parameters in a generic model with time deformation. The discussion of estimation is divided into two cases, one where the directing process Z_t is observable, which is treated in Section 15.4.2, and one where Z_t is latent. The latter is treated in Section 15.4.3.

15.4.1   Parameters of interest

The analysis presented in the previous sections reveals that a generic model contains two types of parameters: (1) those characterising the evolution of the directing process in intrinsic time; and (2) those corresponding to the time deformation. It is important to note that the knowledge of these two types of parameters is important in practice. Indeed, let us for instance consider a problem of option pricing. Consider a European call in intrinsic time, with maturity H, strike price K and hence cash-flow $(Y^*_{z+H} - K)^+$. Furthermore, assume that the complete model is given by the stochastic differential system (Equation (15.12)). This system is driven by two independent Brownian motions, which will result in an incompleteness of the market, if only the price of the underlying asset Y^* is observed. To resolve this problem we may assume that the price of the option depends only on the current and past values of W^*, and not on the randomness specific to the time deformation. In such a case, we have a unique price at z for this option: $P(z, Y^*_z, H, K)$, which will only depend on the parameters appearing in $a^*(\cdot)$ and $b^*(\cdot)$. Yet, we are interested, of course, in the pricing option in calendar time and not in intrinsic time. It is clear that the price of a European style call option $(Y^*_{t+H} - K)^+$ is necessarily $P[Z_t, Y_t, Z_{t+H} - Z_t, K]$. This price cannot be computed, however, when the directing process Z is unobserved. It will only be possible to approximate this price if we know the distribution of Z_t, Z_H, i.e. the parameters of the second equation in Equation (15.12). In summary, this example stresses the importance of estimating all the parameters of the latent model and not just some subset.

It will be rather obvious that the estimation methods will depend on the information available regarding the process Z. We will distinguish two cases; in the first case the set of observable variables contains some variables in deterministic relationship with Z, while in the second case no such variables will be available, resulting in Z being a completely unobservable factor.

15.4.2   Time deformation as a parametric function of observable processes

We will first look at processes where the time deformation is governed by a parametric function which is known up to some unknown parameters involving an observable process X_t. Namely, let us assume that:

$$Z_t = g_t(X_t; b) \tag{15.30}$$

where b is a parameter and X_t is a set of series like trading volume, bid-ask spreads, number of quotes, etc. Once the directing process is specified as in Equation (15.30) we can proceed with estimating the vector b as well as the parameters governing the process Y_z^*. One can think of at least two estimation methods for estimating the parameters. The first one only exploits the second-order properties of subordinated processes while the second one is based on a full characterisation of the distributional properties via the maximum likelihood principle. A subsection is devoted to each method.

15.4.2.1.   Estimation from empirical second-order moments

In analogy with Section 15.2 we first consider estimation only involving the second-order moments of subordinated processes. Sufficient conditions for weak stationarity of subordinated processes were given in *Property 15.2.1.2*, allowing us to estimate parameters through matching empirical and theoretical moments. To illustrate this, let us consider a time deformed Ornstein–Uhlenbeck process discussed in Section 15.3.2. In particular, from Section 15.3.2.2 we know that for the process defined by Equation (15.17) with parameters m, γ and $\rho = \exp -k$, we have the following theoretical first and second moments for the marginal process Y_t in calendar time:

$$m(t) = m \tag{15.31}$$
$$\gamma(t, h) = \sigma^2(1 - \rho^2)^{-1} E(\rho^{\tilde{\Delta}_h g_t(X_t; b)}) \tag{15.32}$$

Hence, with a sufficient number of lags h we can identify the parameters m, σ, ρ as well as b. Consequently, using the empirical mean of Y_t and the empirical autocovariances, we can estimate the aforementioned parameters.

15.4.2.2.   Maximum likelihood estimation

Let us suppose now that we provide a complete specification of the distributional properties to produce parameter estimates. In particular, let us assume that the two processes Y^* and X are independent and Markovian of order one. In such a case we have for discrete variables:

$$P[Y_t = y_t, Z_t = z_t \mid Y_{t-1} = y_{t-1}, Z_{t-1} = z_{t-1}]$$
$$= P[Y_{z_t}^* = y_t \mid Y_{z_{t-1}}^* = y_{t-1}] P[Z_t = z_t \mid Z_{t-1} = z_{t-1}]$$

and a similar decomposition of the conditional p.d.f. holds for continuous variables:

$$\ell_t(y_t, z_t \mid y_{t-1}, z_{t-1}) = \ell_t^*(y_t \mid y_{t-1}; z_t, z_{t-1}) \tilde{\lambda}_t(z_t \mid z_{t-1})$$

where ℓ^* corresponds to the conditional distribution of Y^* and $\tilde{\lambda}$ to the conditional distribution of Z. Furthermore, we assume again that the available data are described by Y_t and X_t where the latter defines Z through Equation (15.30). The process (Y_t, X_t) is Markovian of order one with its transition function given by:

$$\ell_t^*(y_t \mid y_{t-1}; g_t(x_t), g_{t-1}(x_{t-1}))\lambda_t(x_t \mid x_{t-1})$$

where λ_t is the conditional distribution of X. The model is completed by introducing a parametric specification for ℓ_t^*, λ_t and g. To characterise the likelihood function, let α, β and b denote the parameter vectors describing respectively ℓ_t^*, λ_t and g. Then, we have:

$$L_T(\theta) = \prod_{t=1}^{T} \ell_t^*(y_t \mid y_{t-1}; g(x_t; b), g(x_{t-1}; b); \alpha) \prod_{t=1}^{T} \lambda_t(x_t \mid x_{t-1}; \beta) \quad (15.33)$$

From Equation (15.33) we note that the log likelihood function is the product of functions of (α, b) and of β. Therefore, the β parameter can be estimated using observation on X alone, and the ML estimators of the two subsets of parameters will be asymptotically independent.

We can proceed further with an illustrative example which, for the purpose of comparison, is the same as in Subsection 15.4.2.1. Namely, consider again the Ornstein–Uhlenbeck process and suppose that:

$$g_t(X_t; b) = b_0 t + b_1 X_t \quad (15.34)$$

Therefore, $\Delta Z_t = b_0 + b_1 \Delta X_t$, and the evolution of Y_t conditional on X_t is summarised by:

$$Y_t = m + \rho^{b_0 + b_1 \Delta X_t}(Y_{t-1} - m) + \left[\sigma^2 \frac{1 - \rho^{2(b_0 + b_1 \Delta X_t)}}{1 - \rho^2}\right]^{1/2} \epsilon_t \quad (15.35)$$

where ϵ_t is standard Gaussian white noise. We observe immediately that the parameters are not identifiable, and that we must impose some identifying constraint, such as $b_0 = 1$. Then the conditional likelihood becomes:

$$\ell_t^*(y_t \mid y_{t-1}; g_t(x_t; 1, b_1), g_{t-1}(x_{t-1}; 1, b_1), \alpha)$$

$$= (2\pi)^{-1/2}\left[\sigma^2 \frac{1 - \rho^{2(1 + b_1 \Delta X_t)}}{1 - \rho^2}\right]^{-1/2} \exp -\frac{1}{2}\frac{[y_t - m - \rho^{1 + b_1 \Delta X_t}(y_{t-1} - m)]^2}{\sigma^2 \dfrac{1 - \rho^{2(1 + b_1 \Delta X_t)}}{1 - \rho^2}}$$

Finally, it is also worth noting that the corresponding log likelihood function can easily be concentrated with respect to m, σ^2.

15.4.3 Estimation with latent directing processes

It should come as no surprise that the task of estimating subordinated stochastic processes with latent directing processes is considerably more difficult. We no longer assume that Z_t is observable through X_t via the parametric mapping $g_t(\cdot, b)$. Instead we must uncover Z_t through the sample behaviour of Y_t. Once again we can draw a distinction between a method of moments approach, although not necessarily limited to second-order properties, and a maximum likelihood approach. Since we are dealing with latent processes it might be more useful to organise our discussion on the basis of a different attribute. Indeed, we will first study a class of estimators which do not involve simulations of the latent Z process. Such is, for example, the case for the continuous time generalised method of moments (henceforth GMM) approach proposed by Hansen and Scheinkman (1995) and recently adapted by Conley *et al.* (1994) to subordinated diffusions. We shall review this method and, in particular, show the limitations it imposes on the class of time deformed processes we can possibly estimate with such a method. In fact, the continuous time GMM procedure seems only to apply to a restrictive set of circumstances where Z is only governed by a deterministic drift. To estimate a wider class, containing many processes of interest in finance, we must entertain the possibility of simulating the process Z and using simulation-based methods discussed in Duffie and Singleton (1993), Gouriéroux *et al.* (1993), Gallant and Tauchen (1996) and Gouriéroux and Monfort (1994). The next subsection is devoted to the continuous time GMM estimator of Conley *et al.* (1994) while Section 15.4.3.2 covers the simulation-based estimators for subordinated processes.

15.4.3.1. Method of moments using infinitesimal operators

Hansen and Scheinkman (1995) proposed estimating continuous time diffusions through the GMM principle. We will first discuss the principle of the estimation procedure and then elaborate on the identification of parameters. Finally, we will concentrate on a very special case where the directing process is predetermined, i.e. its path is not affected by the randomness of a Brownian motion. The discussion of identification issues will show that it is the latter rather restrictive case only which can be treated by the GMM.

(a) Moment conditions for diffusions

To describe the generic set-up of the continuous time GMM estimator, let us consider the following multivariate system of diffusion equations:

$$\mathrm{d}y_t = \mu_\theta(y_t)\,\mathrm{d}t + \sigma_\theta(y_t)\,\mathrm{d}W_t \tag{15.36}$$

where W_t is a standard n-dimensional Brownian motion and $y_t \in \mathbb{R}^n$. The parameters in Equation (15.36) are described by the vector $\theta \in \mathbb{R}^p$. Hansen and Scheinkman (1995) consider the infinitesimal operator A defined for a class of square integrable functions $\varphi : \mathbb{R}^n \to \mathbb{R}^d$ as follows:

$$A_\theta\varphi(y) = \frac{\mathrm{d}\varphi(y)}{\mathrm{d}y'}\mu_\theta(y) + \frac{1}{2}\mathrm{Tr}\left(\sigma_\theta(y)\sigma_\theta'(y)\frac{\mathrm{d}^2\varphi(y)}{\mathrm{d}y\,\mathrm{d}y'}\right) \tag{15.37}$$

Because the operator is defined as a limit, namely:

$$A_\theta\varphi(y) = \lim_{t\to 0} t^{-1}[\mathbb{E}(\varphi(y_t) \mid y_0 = y) - y]$$

it does not necessarily exist for all square integrable functions φ but only for a restricted domain D. A set of moment conditions can now be obtained for this class of functions $\varphi \in D$. Indeed, as shown for instance by Revuz and Yor (1991), the following equalities hold:

$$EA_\theta\varphi(y_t) = 0 \qquad (15.38)$$

$$E[A_\theta\varphi(y_{t+1})\tilde{\varphi}(y_t) - \varphi(y_{t+1})A_\theta^*\tilde{\varphi}(y_t)] = 0 \qquad (15.39)$$

where A_θ^* is the adjoint infinitesimal operator of A_θ for the scalar product associated with the invariant measure of the process y.[3] By choosing an appropriate set of functions, Hansen and Scheinkman exploit moment conditions (15.38) and (15.39) to construct a GMM estimator of θ.

Conley *et al.* (1994) extended the previous approach to deal with subordinated processes. In particular, let us consider the system of diffusions described in Section 15.2.3. To simplify the presentation let us only concentrate on the set of marginal moment conditions defined in Equation (15.38), leaving aside those in Equation (15.39). The infinitesimal operator argument applied to the joint process $y_t = (Y_t, Z_t)'$ yields:

$$A_\theta\varphi\binom{y}{z} = \begin{bmatrix} a_\theta^*(y)\alpha_\theta(z) \\ \alpha_\theta(z) \end{bmatrix}\begin{bmatrix} \dfrac{\partial\varphi}{\partial y}(y, z) \\ \dfrac{\partial\varphi}{\partial z}(y, z) \end{bmatrix}$$
$$+ \tfrac{1}{2}\mathrm{Tr}\left\{\begin{bmatrix} b_\theta^{*2}(y)\alpha_\theta(z) & a_\theta^*(y)\beta_\theta^2(z) \\ a_\theta^*(y)\beta_\theta^2(z) & \beta_\theta^2(z) \end{bmatrix}\begin{bmatrix} \dfrac{\partial^2\varphi}{\partial y^2}(y, z) & \dfrac{\partial^2\varphi}{\partial y\partial z}(y, z) \\ \dfrac{\partial^2\varphi}{\partial y\partial z}(y, z) & \dfrac{\partial^2\varphi}{\partial z^2}(y, z) \end{bmatrix}\right\} \qquad (15.40)$$

For subordinated diffusions this is not the only infinitesimal operator we can (and should) introduce. Indeed, we can define an infinitesimal operator for the marginal process Y_t in calendar time as soon as it marginally satisfies a univariate diffusion equation (see for instance with $\beta(Z_t) = 0$ as in Equation (15.15)) or even an operator associated with Y_z^*, i.e. with the operational time diffusion. From each of the infinitesimal operators associated with the joint process, as in Equation (15.40), or the marginal process in calendar time, or the operational time diffusion Y_z^*, we can define a set of moment conditions similar to Equation (15.38) (and of course also Equation (15.39) not considered here) and all these conditions may be combined.

(b)   Moment conditions and parameter identification
While parameter estimation via GMM is relatively straightforward, there is the common and well-known point that moment conditions may pose identification problems. In a continuous time GMM framework we construct moment conditions via an appropriate choice of functions φ belonging to the domain of the operator. However, further restrictions on φ must be imposed when the diffusion y_t is only partly observable. As emphasised by Gouriéroux and Monfort (1994), for a large class of diffusions encountered in finance, particularly stochastic volatility models, one often cannot identify the latent parameters governing the dynamics of y. Indeed, to construct moment conditions with an empirical counterpart we must restrict the choice of φ to functions only involving observable transformations of y. Since we are dealing

---

[3] Please note that associated with A_θ^* is a domain D^* so that $\varphi \in D$ and $\tilde{\varphi} \in D^*$ in Equation (15.39).

with a situation where Z_t is latent, this problem is, of course, also encountered here. Consider the moment conditions:

$$E\left[A_\theta\varphi\left(\frac{Y_t}{Z_t}\right)\right] = 0 \tag{15.41}$$

where it is assumed that the functions φ are independent of the parameter θ. We consider the ones where $A_\theta\varphi\left(\frac{Y_t}{Z_t}\right)$ depend only on Y_t for any θ. As soon as the parameterisation does not introduce links between the functions a_θ^*, b_θ^*, α_θ and β_θ defining the diffusions, we deduce from Equation (15.40) that we must restrict the class of functions to the ones where:

$$a_\theta^*(y)\alpha_\theta(z)\frac{\partial\varphi}{\partial y}(y, z), \quad \alpha_\theta(z)\frac{\partial\varphi}{\partial z}(y, z), \quad b_\theta^{*2}(y)\alpha_\theta(z)\frac{\partial^2\varphi}{\partial y^2}(y, z)$$

$$a_\theta^*(y)\beta_\theta^2(z)\frac{\partial^2\varphi}{\partial y\partial z}(y, z), \quad \beta_\theta^2(z)\frac{\partial^2\varphi}{\partial z^2}(y, z) \text{ are all } \textit{independent of } z$$

This yields the following restrictions on the class of admissible functions.

(1) Since

$$\left[b_\theta^{*2}(y)\alpha_\theta(z)\frac{\partial^2\varphi}{\partial y^2}(y, z)\right] \Big/ \left[a_\theta^*(y)\alpha_\theta(z)\frac{\partial\varphi}{\partial y}(y, z)\right]$$

has to be independent of z, we deduce that $\frac{\partial}{\partial y}\log\frac{\partial\varphi}{\partial y}(y, z)$ has also to satisfy this condition. Therefore:

$$\varphi(y, z) = G(y)f(z) + C(z) \tag{15.42}$$

(2) Furthermore since $\alpha_\theta(z)\partial\varphi(y, z)/\partial y$ has to depend only on y one obtains from Equation (15.42) that $f(z) = k(\alpha_\theta(z))^{-1}$ and therefore:

$$\varphi(y, z) = kG(y)(\alpha_\theta(z))^{-1} + C(z) \tag{15.43}$$

(3) Similarly, $\alpha_\theta(z)\partial\varphi(y, z)/\partial z$ must be a function of y only and hence:

$$\varphi(y, z) = -kd\alpha_\theta(z)/dzG(y)(\alpha_\theta(z))^{-1} + \alpha_\theta(z)\,dC(z)/dz \tag{15.44}$$

Using the arguments in (1) through (3) one constrains the choice of φ. Two cases may be distinguished.

(i) If $G(y)$ is not constant, it is necessary to choose $C(z)$ constant, and this choice is only valid if $[\alpha_\theta(z)]^{-1}\,d\alpha_\theta(z)/dz$ is constant.

(ii) If $G(y)$ is a constant, it is necessary that $[\alpha_\theta(z)]^{-2}\,d\alpha_\theta(z)/dz$ is independent of θ.

These constraints are extremely restrictive since they impose conditions on the dynamics of the underlying processes. Therefore, it seems difficult to construct moment conditions that will identify all elements of the parameter vector θ, except in some very special circumstances.

(c) Predetermined latent directing processes
One special case, the one implicitly treated by Conley *et al.* (1994), is where the directing process Z_t satisfies:

$$dZ_t = \alpha_\theta(Z_t)\,dt \tag{15.45}$$

and hence $\beta_\theta(Z_t)$. Recall from the discussion in Section 15.2.3 that, in such a case, one can also derive a diffusion for the marginal process (Y_t) as described by Equation (15.15). Now the moment conditions (Equation (15.40)) greatly simplify and amount to:

$$EA_\theta\varphi\begin{pmatrix}Y_t\\Z_t\end{pmatrix}=E\left[\alpha_\theta(Z_t)\left[a_\theta^*(Y_t)\frac{\partial\varphi}{\partial y}(Y_t,Z_t)+\frac{\partial\varphi}{\partial z}(Y_t,Z_t)+\tfrac{1}{2}(b_\theta^*(Y_t))^2\frac{\partial^2\varphi}{\partial y^2}(Y_t,Z_t)\right]\right]=0$$
(15.46)

Following Conley *et al.* (1994), let us now consider functions separable in y and z, i.e. $\varphi(y,z)=\varphi_0(y)\varphi_1(z)$. Then Equation (15.45) further simplifies to:

$$\left.\begin{aligned}EA_\theta\varphi\begin{pmatrix}Y_t\\Z_t\end{pmatrix}&=E\left[\alpha_\theta(Z_t)\varphi_1(Z_t)\left\{a_\theta^*(Y_t)\frac{d\varphi_0(Y_t)}{dy}+\tfrac{1}{2}(b_\theta^*(Y_t))^2\frac{d^2\varphi_0(Y_t)}{dy^2}\right]\right.\\&\quad+E\left[\alpha_\theta(Z_t)\frac{d\varphi_1(Z_t)}{dz}\varphi_0(Y_t)\right]\end{aligned}\right\}$$
(15.47)

From the infinitesimal operator associated with the changing time process in Equation (15.45) we obtain that

$$E\left[\alpha_\theta(Z_t)\frac{d\varphi_1(Z_t)}{dz}\right]=0$$
(15.48)

for all φ_1 belonging to the appropriate domain.

Therefore we deduce from Equation (15.48), that

$$E\left[\alpha_\theta(Z_t)\frac{d\varphi_1(Z_t)}{dz}\varphi_0(Y_t)\right]=E\left\{\alpha_\theta(Z_t)\frac{d\varphi_1}{dz}(Z_t)E(\varphi_0(Y_t)\mid Z_t)\right\}=0$$

Then condition (15.47) implies:

$$E\left[\alpha_\theta(Z_t)\varphi_1(Z_t)\left(a_\theta^*(Y_t)\frac{d\varphi_0(Y_t)}{dy}+\frac{1}{2}[b_\theta^*(Y_t)]^2\frac{d^2\varphi_0(Y_t)}{dy^2}\right)\right]=0,\quad\forall\varphi_0,\varphi_1$$

which is equivalent to:

$$E\left[a_\theta^*(Y_t)\frac{d\varphi_0(Y_t)}{dy}+\frac{1}{2}[b_\theta^*(Y_t)]^2\frac{d^2\varphi_0(Y_t)}{dy^2}\,\middle|\,Z_t\right]=0,\quad\forall\varphi_0$$

and by integrating over Z_t

$$E\left[a_\theta^*(Y_t)\frac{d\varphi_0(Y_t)}{dy}+\frac{1}{2}[b_\theta^*(Y_t)]^2\frac{d^2\varphi_0(Y_t)}{dy^2}\right]=0,\quad\forall\varphi_0$$
(15.49)

15.4.3.2.   *Simulation-based estimators*

In general, the estimation problem is much more complicated with a latent directing process Z_t because the observable log likelihood, corresponding to $Y_1\ldots Y_T$ is now derived by integrating out the unobservable path of Z:

$$L_T^*(y_1,\ldots,y_T\mid y_0,z_0)=\int\ldots\int\prod_{t=1}^T(\ell_t^*(y_t\mid y_{t-1};z_t,z_{t-1})\tilde\lambda_t(z_t\mid z_{t-1})dz_t)\quad(15.50)$$

The presence of such multiple integrals inside the likelihood function is common in many empirical models for financial data. The best examples are stochastic volatility models. Statistical inference for such processes can be based on simulated estimation methods (Duffie and Singleton, 1993; Gouriéroux *et al.*, 1993; Gallant and Tauchen, 1996; Gouriéroux and Monfort, 1994).

In recent years, considerable advances have been made in this area. Since simulation of a subordinated process with latent Z_t is a special case of the estimation problems treated by this class of simulation-based estimators it is a relatively straightforward application of the available theory. It may be noted here that Ghysels and Jasiak (1994) provide a specific example of such an estimator applied to a subordinated stochastic volatility model.

15.5   Testing the hypothesis of time deformation

In this final section we treat the problem of hypothesis testing, specifically focusing on testing for time deformation. In Section 15.4 we noted that there is an important distinction to be made between a situation where Z_t is latent and one where it is not. We will therefore distinguish these two cases when discussing hypothesis testing.

15.5.1   Parametric models with observable directing processes

Let us consider the maximum likelihood estimator discussed in Section 15.4.2.2. The likelihood function as formulated in Equation (15.33) has a parameter vector $\theta = (\alpha, b, \beta)$ where b determines the mapping between the observable series X_t and the directing process Z_t. For the purpose of hypothesis testing, let us specify the time deformation (Equation (15.30)) such that:

$$g_t(X_t; b)\,|_{b=0} = t \qquad (15.51)$$

This is, for instance, the case in the illustrative example given in Equation (15.34), $g_t(X_t; b) = b_0 t + b_1 X_t$, with the identifying restriction $b_0 = 1$. The test of the hypothesis $Z_t = t$ may be performed by a Lagrange multiplier procedure based on the score: $[\partial \log L_T^*(\theta)/\partial b]_{\theta = \hat{\theta}_0}$, where $\hat{\theta}_0$ is the constrained *ML* estimator and $\theta \equiv (\alpha, b, \beta)$. As an illustration let us consider again the time deformed Ornstein–Uhlenbeck process and $g_t(X_t, b) = t + bX_t$. It can be shown that:

$$T^{-1} \frac{\partial \log L_T^*(\theta)}{\partial b}\bigg|_{\theta = \hat{\theta}_0} \approx \frac{1}{\hat{\gamma}_0^2} \operatorname{Cov}_e(\Delta x_t, \hat{\epsilon}_{ot}^2) + \frac{1 - \hat{\rho}_0^2}{\hat{\rho}_0 \hat{\gamma}_0^2} \operatorname{Cov}_e(\Delta x_t (y_{t-1} - \hat{m}_0), \hat{\epsilon}_{ot})$$

$$(15.52)$$

where

$$\hat{m}_0, \hat{\rho}_0, \quad \hat{\epsilon}_{ot} = y_t - \hat{m}_0 - \hat{\rho}_0 (y_{t-1} - \hat{m}_0), \quad \hat{\sigma}_0^2 = \frac{1}{T} \sum_{t=1}^{T} \hat{\epsilon}_{ot}^2$$

are the constrained *ML* estimators and the constrained residuals. Consequently, the score statistic contains two different terms: the first one $\text{Cov}_e(\Delta x_t, \hat{\epsilon}_{ot}^2)$ is useful for testing the presence of conditional heteroscedasticity in the direction Δx_t, the second one $\text{Cov}_e(\Delta x_t(y_{t-1} - \hat{m}_0), \hat{\epsilon}_{ot})$ for testing the omission of $\Delta x_t(y_{t-1} - m)$ in the conditional mean. This double local effect of time deformation is easy to understand intuitively when we study the expansion of the regression model of Y_t given $Y_{t-1}, \Delta X_t$ in a neighbourhood of the null hypothesis, i.e. when b is small. Indeed we have, from Equation (15.35):

$$Y_t \approx m + (\rho + \rho b \Delta X_t \log \rho)(Y_{t-1} - m) + \left\{ \frac{\sigma^2}{1-\rho^2}(1 - \rho^2[1 + 2b\Delta X_t \log \rho]) \right\}^{1/2} \epsilon_t$$

$$\approx m + \rho(Y_{t-1} - m) + b\rho \log \rho \Delta X_t(Y_{t-1} - m) + \left[\sigma^2 - 2\frac{b\sigma^2 \rho^2 \log \rho}{1-\rho^2} \Delta X_t \right]^{1/2} \epsilon_t$$

Hence, the test combines both effects due to time deformation in the case of an Ornstein–Uhlenbeck model.

15.5.2 Parametric models with latent directing processes

We will concentrate most of our attention on testing the hypothesis of time deformation when one uses the simulation-based estimators described in Section 15.4.3.2. Some observations will also be made about testing when the continuous time GMM estimator is used. Since we discuss primarily simulation-based estimators, let us introduce an analogue to Equation (15.30) to describe the dynamic of the changing time process in discrete time, namely:

$$Z_t = h_t(Z_{t-1}, \varepsilon_{zt}; b) \tag{15.53}$$

where ε_{zt} is I.I.N $(0,1)$. Again, for the purpose of discussion we assume that:

$$h_t(Z_{t-1}, \varepsilon_{zt}; b)|_{b=0} = t \tag{15.54}$$

The score principle was advanced for testing $b = 0$ when the directing process was tied to an observable process X_t through g_t in Equation (15.30). The same score principle can be applied to cases where Z_t is latent. Let us assume again that the parameter vector is $\theta = (\alpha, b)$ and that we estimate the model (via simulation) under the null restriction $b = 0$, yielding $\hat{\theta}_T^0 = (\hat{\alpha}_T^0, 0)$ for a sample of size T. Consider paths for the directing process simulated under the alternative $b \neq 0$. For any choice of b we obtain $[Z_t^s(b)]_{t-1}^T]_{s=1}^S$. Taking the parameter estimates $\hat{\alpha}_T^0$ under the null we can simulate for any alternative b the process $[(y_t^s(\hat{\alpha}_T^0, b))_{t=1}^T]_{s=1}^S$. For the Ornstein–Uhlenbeck example this would amount to:

$$y_t^s(\hat{\alpha}_T^0, b) = \hat{m}_{0T} + \hat{\rho}_{0T}^{(1+b_1\Delta Z_t^s(b))}(y_{t-1}^s(\hat{\alpha}_T^0, b) - \hat{m}_{0T}) + \hat{\sigma}_{0T}\left(\frac{1 - \hat{\rho}_{0T}^{2(1+b_1\Delta Z_t^s(b))}}{1 - \hat{\rho}_{0T}} \right)^{1/2} \varepsilon_t^s$$

where $\hat{\alpha}_T^0 = (\hat{m}_{0T}, \hat{\rho}_{0T}, \hat{\sigma}_{0T})$ and b_1 is an element of the parameter vector b.

Acknowledgement

Eric Ghysels would like to acknowledge the financial support of CREST as well as the SSHRC of Canada and the Fonds F.C.A.R. of Québec.

Appendix 15.1

Gamma time deformation

In the Appendix we provide some of the details of the gamma time deformation presented in Section 15.2.5. We examine the distribution of $W[C(v)]_t$, some of its moments and finally derive an Itô lemma for functional transformations in calendar time of subordinated processes with gamma time deformation.

(i)   *Distribution of* $W[C(v)]_t$
    This distribution may be defined through its characteristic function. We get:

$$E(\exp iu W[C(v)]_t) = EE[\exp iu W[C(v)]_t \mid C(v)_t]$$

$$= E\exp\left[-\frac{u^2}{2}C(v)_t\right]$$

$$= \left(1 + \frac{u^2}{2}\right)^{-vt}$$

(ii)   *Moments of* $W[C(v)]_t$

$$E(W[C(v)]_t^{2p}) = EE[W[C(v)]_t^{2p} \mid C(v)_t]$$

$$= \frac{(2p)!}{2^p p!} E[C(v)_t^p]$$

$$= \frac{\Gamma(vt + p)}{\Gamma(vt)} \frac{(2p)!}{2^p p!}$$

In particular:

$$EW[C(v)_t]^2 = vt$$

$$EW[C(v)_t]^4 = 3vt(vt + 1)$$

We deduce that the kurtosis of this distribution is equal to:

$$k[W[C(v)]_t] = \frac{3vt(vt + 1)}{(vt)^2} = 3 + \frac{3}{vt}$$

and we directly note the leptokurticity due to time deformation.

(iii)   *Itô's formula for the stochastic system in calendar time.*
    Such a formula may be easily derived in the following way. Let us consider a function $g(z, y^*)$. Because of the independence between the two processes

$(Z_t), (Y_z^*)$, we can apply directly the usual Itô formula to the first equation of Equation (15.14). We get:

$$dg(z, Y_z^*) = [a^*(Y_z^*)\frac{\partial g}{\partial y^*}(z, Y_z^*]) + \frac{\partial g}{\partial z}(z, Y_z^*) + \frac{1}{2}b^2(Y_z^*)\frac{\partial^2 g}{\partial y^{*2}}(z, Y_z^*)]\,dz$$

$$+ b(Y_z^*)\frac{\partial g}{\partial y^*}(z, Y_z^*)\,dW_z^*$$

$$dZ_t \qquad = \lambda(Z_t)\,dC(v)_t$$

Consequently, we get the formulation in calendar time for the transformed variables: $g(Z_t, Y_t), Z_t$:

$$dg(Z_t, Y_t) = [a^*(Y_t)\frac{\partial g}{\partial y^*}(Z_t, Y_t]) + \frac{\partial g}{\partial z}(Z_t, Y_t) + \frac{1}{2}b^2(Y_t)\frac{\partial^2 g}{\partial y^{*2}}(Z_t, Y_t)]\lambda(Z_t)\,dC(v)_t$$

$$+ b(Y_t)\frac{\partial g}{\partial y^*}(Z_t, Y_t)\lambda(Z_t)^{1/2}\,dW[C(v)]_t$$

$$dZ_t \qquad = \lambda(Z_t)\,dC(v)_t$$

References

Bochner, S. (1960) *Harmonic Analysis and the Theory of Probability*. Berkeley: University of California Press.

Clark, P. (1970) A subordinated stochastic process model of cotton futures prices. Ph.D. Dissertation, Harvard University.

Clark, P. (1973) A subordinated stochastic process model with finite variance for speculative price. *Econometrica*, **41**, 135–156.

Conley, T., Hansen, L., Luttmer, E. and Scheinkman, J. (1994) Estimating subordinated diffusions from discrete time data. Presented at the CIRANO-C.R.D.E. conference on Stochastic Volatility, Montréal.

Dacorogna, M., Müller, U., Nagler, R., Olsen, R. and Pictet, O. (1993). A geographical model for the daily and weekly seasonal volatility in the foreign exchange market. *Journal of International Money and Finance*, **12**, 413–438.

Duffie, D. and Singleton, K. (1993) Simulated moments estimation of Markov models of asset prices. *Econometrica*, **61**, 929–952.

Foster, D. and Viswanathan, S. (1993) The effect of public information and competition on trading volume and price volatility. *Review of Financial Studies*, **6**, 23–56.

French, K. and Roll, R. (1986) Stock return variances: the arrival of information and the reaction of traders. *Journal of Financial Economics*, **17**, 5–26.

French, K. R., Schwert, G. W. and Stambaugh, R. F. (1987) Expected stock returns and volatility. *Journal of Financial Economics*, **19**, 3–29.

Gallant, A. R. and Tauchen, G. (1996) Which moments to match? *Econometric Theory*, **12**, 657–681.

Geman, H. and Yor, M. (1993) Bessel processes, Asian options and perpetuities. *Mathematical Finance*, **3**, 349–375.

Ghysels, E. and Jasiak, J. (1994) Stochastic volatility and time deformation: an application to trading volume and leverage effects. Paper presented at the 1994 Western Finance Association Meetings, Santa Fe.

Ghysels, E., Gouriéroux, C. and Jasiak, J. (1995) High frequency financial time series data: some stylized facts and models of stochastic volatility. In C. Dunis and B. Zhou (Eds) *Nonlinear Modelling of High Frequency Financial Time Series*, New York: John Wiley.

Ghysels, E., Gouriéroux, C. and Jasiak, J. (1996) Kernel autocorrelogram for time deformed processes. Discussion paper CIRANO.

Gouriéroux, C. and Monfort, A. (1994) Simulation based econometric methods. CORE Lectures Series, Louvain-la-Neuve.

Gouriéroux, C., Monfort, A. and Renault, E. (1993) Indirect inference. *Journal of Applied Econometrics*, **8**, S85–S118.

Hamilton, J. (1989) A new approach to the economic analysis of non-stationary time series and the business cycle. *Econometrica*, **57**, 357–384.

Han, S., Kalay, A. and Rosenfeld, A. (1994) The information content of non trading. Discussion paper, University of Utah.

Hansen, L. and Scheinkman, J. (1995) Back to the future: generating moment implications for continuous-time Markov processes. *Econometrica*, **63**, 767–804.

Harris, L. (1987) Transaction data tests of the mixture of distributions hypothesis. *Journal of Financial and Quantitative Analysis*, **22**, 127–141.

Hausman, J. and Lo, A. (1990) A continuous time discrete state stochastic process for transactions stock prices: theory and empirical evidence. Discussion paper, MIT.

Lakonishok, J. and Smidt, S. (1988) Are seasonal anomalies real? A ninety year perspective. *Review of Financial Studies*, **1**, 403–425.

Leblanc, B. (1994) Une approche unifiée pour une forme exacte du prix d'une option dans les différents modèles à volatilité stochastique. Discussion paper, CREST.

Madan, D. B. and Milne, F. (1991) Option pricing with V.G. martingale components. *Mathematical Finance*, **1**, 39–55.

Madan, D. B. and Seneta, E. (1990) The V.G. model for share market returns. *Journal of Business*, **63**, 511–524.

Mandelbrot, B. and Taylor, H. (1967) On the distribution of stock prices differences. *Operations Research*, **15**, 1057–1062.

Nelson, D. B. (1991) Conditional heteroskedasticity in asset returns: a new approach. *Econometrica*, **59**, 347–370.

Revuz, A. and Yor, M. (1991) *Continuous Martingales and Brownian Motion*, Berlin: Springer-Verlag.

Schwert, G. W. (1990) Indexes of U.S. stock prices from 1802 to 1987. *Journal of Business*, **63**, 399–438.

Stock, J. H. (1988) Estimating continuous time processes subject to time deformation. *Journal of the American Statistical Association*, **83**, 77–84.

Tauchen, G. E. and Pitts, M. (1983) The price variability volume relationship on speculative markets. *Econometrica*, **51**, 485–505.

Yor, M. (1992a) Sur certaines fonctionnelles exponentielles des mouvements Browniens réels. *Journal of Applied Probability*, **29**, 202–208.

Yor, M. (1992b) On some exponential functionals of Brownian motion. *Advances in Applied Probability*, **24**, 509–531.

Index

Page numbers in **bold** refer to figures; those in *italics* refer to tables

Life
cycle/permanent income hypothesis (LCH/
PIH) 157, **157**, 165
insurance, with profit, liabilities 25
tables, actuarial 49
Linear models
distributions
binomial 43
Gamma 43
normal 43
Poisson 43
marine insurance 42
motor insurance 42
Linear programming 94–6
Link functions, canonical 43
Liquidity
capital markets 189
constraints 161
Literature
default risk 240
non-parametric methods 262
LM curve 139–40, **140**
Loans
corporate 79
demands, banks 153
Local-risk minimisation 295
Location model 86
Logistic regression, cf. discriminant analysis 90

Macroeconomics 8
introducing credit 153–6, **154, 155, 156**
Magnitude, default 239–40, 246
Mail-order customers, credit scoring 94
Management
information 116–19
acceptance rate 116
expected score 116
overrides 116
and ownership 188
of uncertainty 41
Marginal utilities, ratio of 190
Marine insurance
claim frequency 59
linear models 42
Marketing initiatives, sampling 111–12
Markets
correlation 18
disequilibrium 162–3
frictionless 177–84
portfolios 19
values, default risk 240
Martingale
calculus 175–6
pricing 193–4
Masking functions 98
Matching, principle of 28, 29, *29*
Maximum likelihood estimation, time
deformation 321–2
Mean–variance models 9–18

developing 26–7
generalising 27–30
limitations 22–4
Measuring
money supply 127–8
risk 106
Microeconomics, consumer credit 156–65, **157,
159, 160, 162**, *164*
Missing data, credit scoring 74, **74**, 114
Model-free approach 262
option pricing 263–8
Models
credit scoring 107
profitability 107, 109
Modern portfolio theory 9
Mondex card 70
Monetary business cycles 144
Money
equilibrium **154**
supply
Bank of England 136–7
changes in 128
and credit 136–7
definition 136
expected 149
increasing 146
Keynesian theory 128
measuring 127–8
nominal 145
and output 147–50
and outstanding credit 150
unexpected 149–50
Moneyness 287–8
Mortality rates, policy duration 51
Mortgages 69
Motor insurance 42
claim frequency 59
claim severity 61
premium rating 59
Multi-factor models 233
Multiple state models 53–6, **54**, *55*

Nearest neighbour 94
Neural networks 96–9, **97**
consumer credit 98
corporate lending 98
definition 96
Neutral default probability, risk 246–7
New classical model 140–5
New credit, sampling 111–12
Newton–Raphson 90
No-arbitrage 192–4
Non-bankrupt accounts 108–9
Non-life insurance
claim frequency 58
claim severity 58
claims reserving 61–3, *62*
predicting claim amounts 62
premium rating 58